Vertex Detectors

ETTORE MAJORANA INTERNATIONAL SCIENCE SERIES

Series Editor:

Antonino Zichichi

European Physical Society
Geneva, Switzerland

(PHYSICAL SCIENCES)

Recent volumes in the series:

Volume 24 **DATA ANALYSIS IN ASTRONOMY**
Edited by V. Di Gesu, L. Scarsi, P. Crane, J. H. Friedman , and S. Levialdi

Volume 25 **FRONTIERS IN NUCLEAR DYNAMICS**
Edited by R. A. Broglia and C. H. Dasso

Volume 26 **TOKAMAK START-UP: Problems and Scenarios Related to the Transient Phases of a Thermonuclear Fusion Reactor**
Edited by Heinz Knoepfel

Volume 27 **DATA ANALYSIS IN ASTRONOMY II**
Edited by V. Di Gesù, L. Scarsi, P. Crane, J. H. Friedman, and S. Levialdi

Volume 28 **THE RESPONSE OF NUCLEI UNDER EXTREME CONDITIONS**
Edited by R. A. Broglia and G. Bertsch

Volume 29 **NEW MECHANISMS FOR FUTURE ACCELERATORS**
Edited by M. Puglisi, S. Stipcich, and G. Torelli

Volume 30 **SPECTROSCOPY OF SOLID-STATE LASER-TYPE MATERIALS**
Edited by Baldassare Di Bartolo

Volume 31 **FUNDAMENTAL SYMMETRIES**
Edited by P. Bloch, P. Pavlopoulos, and R. Klapisch

Volume 32 **BIOELECTROCHEMISTRY II**
Edited by G. Milazzo and M. Blank

Volume 33 **MUON-CATALYZED FUSION AND FUSION WITH POLARIZED NUCLEI**
Edited by B. Brunelli and G. G. Leotta

Volume 34 **VERTEX DETECTORS**
Edited by Francesco Villa

A Continuation Order Plan is available for this series. A continuation order will bring delivery of each new volume immediately upon publication. Volumes are billed only upon actual shipment. For further information please contact the publisher.

Vertex Detectors

Edited by

Francesco Villa
Stanford University
Stanford, California

Plenum Press • New York and London

Library of Congress Cataloging in Publication Data

Vertex detectors / edited by Francesco Villa.
p. cm.—(Ettore Majorana international science series. Physical sciences; v. 34)
"Proceedings of a workshop for the INFN Eloisatran Project, held September 21–26, 1986, in Erice, Sicily, Italy"—T.p. verso.
Includes bibliographies.
ISBN 0-306-42798-2
1. Vertex detectors—Congresses. 2. Nuclear counters—Congresses. I. Villa, Francesco. II. Series.
QC787.V45V47 1988
539.7′7—dc19 87-34301
CIP

Proceedings of a workshop for the INFN Eloisatran Project, held
September 21–26, 1986, in Erice, Sicily, Italy

A Division of Plenum Publishing Corporation
233 Spring Street, New York, N.Y. 10013

Printed in the United States of America

PREFACE

It has been widely recognized that the "new physics" results from a high energy collider are related to the detector capabilities, and that future detectors must solve the problems presented by the new environment.

Vertex detectors, in particular, will have to sustain enormous rates, have a great resistance to radiation damage, while retaining good spatial accuracy. Promising technologies are emerging, and gaseous detectors are improving: this workshop was intended as a point of reference towards future detectors, with particular emphasis on experimental results achieved so far.

We wish to thank the Ettore Majorana Center for the splendid hospitality in Erice; and the secretaries of the conference, R. Nania and G. Anzivino for their hard work in collecting and organizing these proceedings.

F. Villa
Stanford Linear Accelerator Center
Stanford University

CONTENTS

Tracking at 1 TeV
A. Seiden 1

B Physics at PEP and SLC
A. Seiden 19

The MARK II Vertex Detectors: Status and Prospects
J. Jaros 37

The Mark-J Vertex Detector
H. Anderhub *et al.*, presented by M. Bourquin 71

A Modified Time Expansion Chamber as a Vertex Detector
G. Bari *et al.*, presented by C. Del Papa 95

Operation of Multidrift Tubes with Dimethyl Ether
R. Bouclier *et al.*, Presented by F. Sauli 101

Results From the MAC[1] Vertex Chamber
H. N. Nelson 115

Wire Chamber Aging and Wire Material
M. Atac 151

Charm Photoproduction Using a Silicon Vertex Detector
B. R. Kumar 167

Present and Future Semiconductor Tracking Detectors
G. Lutz *et al.*, presented by G. Lutz 195

Today and Tomorrow for Scintillating Fibre (SCIFI) Detectors
J. Kirkby 225

The INFN Eloisatron Project:
European LOng Intersecting Storage Accelerator
C. Aglietta *et al.*, presented by A. Zichichi 297

General Discussion 343

Index 349

TRACKING AT 1 TeV

A. Seiden

Santa Cruz Institute for Particle Physics
University of California, Santa Cruz, CA 95064

ABSTRACT

We discuss several aspects of designing a tracking detector for reconstruction of charged particles in 1 TeV jets in the central rapidity region. Typical angles between tracks in such jets are roughly quantified. Based on this characteristic, the parameters for an all silicon strip tracking system are defined. The approach taken can be used to design alternative tracking systems.

INTRODUCTION

In order to do physics with 1 TeV jets, characteristic of SSC physics, calorimetry is expected to become the dominant experimental technique. A question we can ask is whether tracking will be needed at all? Can we do everything with calorimetry? However, tracking is our only method to look inside a jet. If an interesting signal is seen at high $P_{\perp}$, beyond what's expected from QCD, we will probably want to study some details using tracking. For example, if a new object is related to the electroweak interactions, we might expect to find W's and Z's in the final state. If a new object is hadronic, for example, a supersymmetric hadronic partner, we can expect mostly quarks and gluons in the final state. Other possibilities are the existence of new longlived objects for which we would like to set the best possible limits on lifetimes and cross sections. This will require very good vertex detection. For heavy objects, such as a new quark decaying into t or b quarks, vertex detection of b quarks will be important.[1)] This requires looking at leading particles in the core of jets, which will be very challenging.

In addition to the above, measurement of charged particle multiplicity might be interesting. For example a 1 TeV W still decays into an average of 17 charged particles, like a W at rest. On the other hand, a 1 TeV energy quark gluon di-jet system with invariant mass near m_W will usually fragment into more than 75 charged particles. The W at very large energy might be like the τ at PEP, easy to isolate using multiplicity, provided it's produced at a large enough rate. The effectiveness of a multiplicity cut will depend on the fluctuations in the background multiplicity distribution and how large the production of signal to background is.

REQUIREMENTS FOR A TRACKING SYSTEM

To reconstruct leading particles in the core of jets will require a system which can resolve very closely spaced particle trajectories. In order to quantify this problem, without a sophisticated Monte Carlo simulation, we will try to define a track separation angle $\langle\theta\rangle$, at production, which characterizes the leading particles in a jet. In general these particles will have a momentum $\gg$ 1 GeV.

We look first at simulations of 1 TeV gluon and quark jets using QCD evolution models such as ISAJET. Figure 1 shows the expected mean multiplicity for gluon and quark jets. At 1 TeV, gluon jets are expected to have about 70 charged particles, quark jets about 40.[2)]

Figure 2 shows the energy dependence of $\theta_{1/2}$ for gluon and quark jets, where $\theta_{1/2}$ is the angle relative to the jet direction at which the energy flow drops by a factor of two.[2)]

At 1 TeV:

$\theta_{1/2} \simeq 20$ mrad for gluon jets.

$\theta_{1/2} \simeq 4$ mrad for quark jets.

Since a jet contains many particles, $\langle\theta\rangle$ is expected to be about an order of magnitude smaller than $\theta_{1/2}$. Thus a goal for a tracking system would be to be able to resolve particles produced at angles to each other of $\langle\theta\rangle \simeq$ .5 mrad. We will arrive at this number in another way below. Note, this implies that a two-hit segmentation of 500μm is required at a radius of 1m, which is very difficult for conventional drift chambers. Even a silicon strip tracking system could not fully resolve the particles until a radius $\gtrsim$ 20cm.

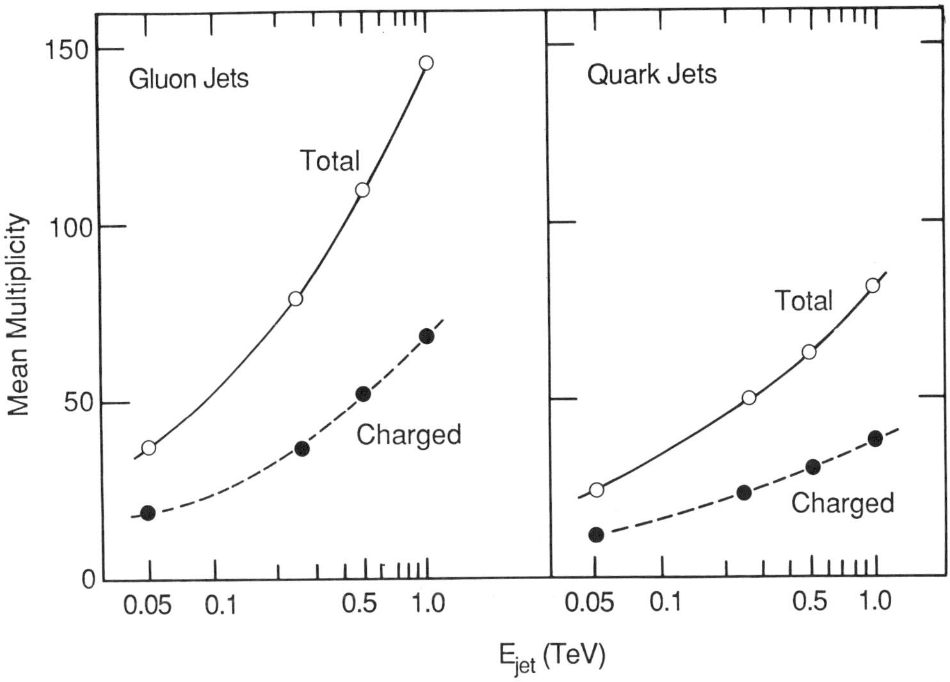

Fig. 1. Mean Multiplicity versus jet energy for jets initiated by gluons or light quarks.

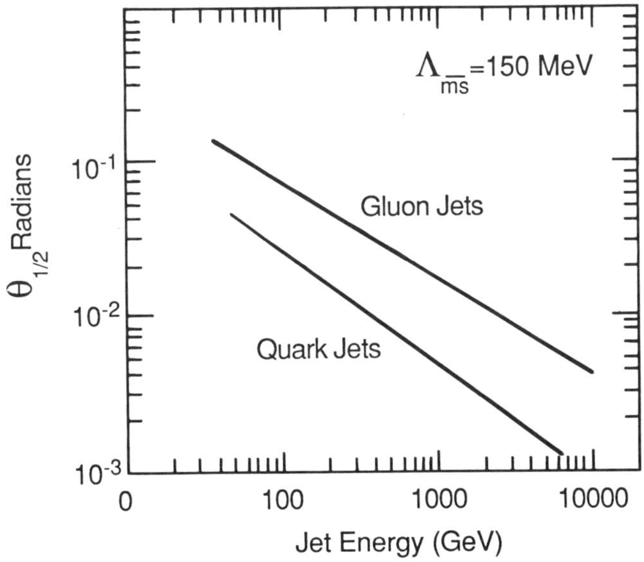

Fig. 2. Energy dependence of angle at which energy flow drops by a factor of two.

As an alternative method to arrive at $\langle\theta\rangle$, we can use the extensive Monte Carlo work on B tagging at the Z^0. Figure 3 shows the transverse momentum of particles from B decays. A typical value is about 500 MeV, rather similar to light quark jets at the same energy scale. Figure 4 shows the fraction of B's that can be reconstructed at the Z^0 using a tracking system with resolving angle $\Delta\phi$ between neighboring tracks and requiring that at least three charged secondaries are reconstructed. To achieve $> 90\%$ efficiency requires $\Delta\phi = 5$ mrad. Scaling each jet up to 1 TeV, a factor of 20, would give us a requirement of $\Delta\phi \simeq .25$ mrad for B reconstruction. Thus a choice of $\langle\theta\rangle = .5$ mrad to characterize neighboring particles in the jet core is a reasonable choice.

Using a typical P_T of 500 MeV we can estimate a minimum momentum, $P_{\rm min}$, which particles in the core of a jet have, as 500 MeV$/\theta_{1/2}$. Using $\theta_{1/2} \sim 10$ mrad gives $P_{\rm min} = 50$ GeV for a 1 TeV jet. Thus a tracking system will need a resolving power $\sim \langle\theta\rangle$ for tracks with $P \gtrsim P_{\rm min}$ but can reconstruct particles with $P \ll P_{\rm min}$ with a poorer segmentation since these are typically outside the core of the jet.

We have totally ignored the magnetic field in the above discussion. As long as tracks with $P > P_{\rm min}$ are bent by an angle $< \theta_{1/2}$, the presence of a magnetic field does not significantly help the track reconstruction problem. For $P_{\rm min} = 50$ GeV, $\theta_{1/2} = 10$ mrad, the magnetic field does not help substantially for reasonable choices for the field and tracker radius.

An additional important characteristic for a tracking system is the momentum resolution. Since essentially all jets have neutral particles, the tracking system will never see all the particles in a jet. There is, therefore, no point in having extremely good momentum resolution for the full reconstruction of a jet mass. For secondary vertices, to tell a D from a B, or for rough jet mass cuts, a $\Delta P/P$ of $\simeq 10\%$ for typical particle momenta in the jet should be adequate. To determine the sign of the charge of an e or μ will probably require a more restrictive $\Delta P/P$, since these particles can have the same momentum as a full jet. If we want to determine the sign at 2 TeV we need a $\Delta P/P \simeq .025\% P$ (GeV). This represents, for comparison, a factor of 8 improvement over the expected performance of the Mark II at the SLC.

DESIGN OF A 1 TeV TRACKING SYSTEM

Because of the very fine segmentation needed to resolve tracks, a tracker will probably have to use infrequent sampling to avoid ending up with an enormous number of channels.

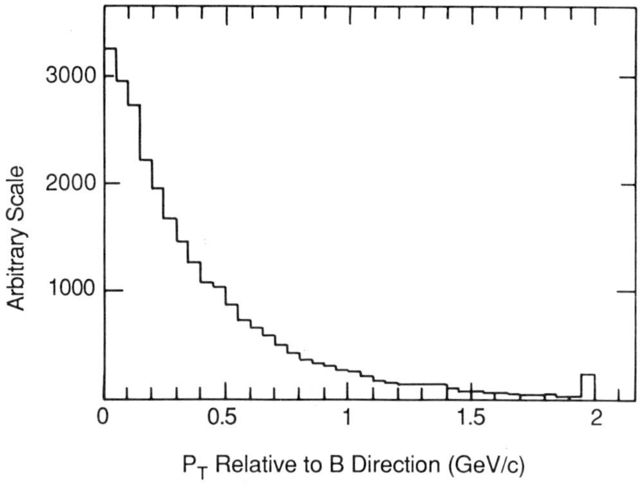

Fig. 3. P_T relative to B direction.

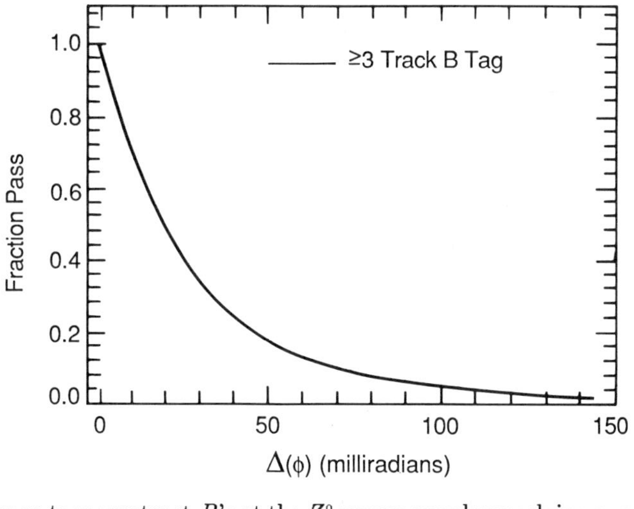

Fig. 4. Efficiency to reconstruct B's at the Z^0 versus angular resolving power of tracking system.

A model for the organization of such detectors is the vector tracking chamber, made of jet cells. The first of such chambers was used by the MARK III detector.[3)] Subsequent examples are the trackers for the MARK II, CDF, SLD and ZEUS. The vector tracking ideas work equally well for other detector elements, appropriately organized. We will look in some detail at the organization of a tracker made entirely of silicon strips. The numbers arrived at can be rescaled for other designs. The silicon strips are very attractive since they offer extremely good resolution in space and time. In addition, it is an area where much progress is being made and the detectors, for the scale discussed below, are still affordable in price.

Magnetic Field

The momentum resolution, ignoring multiple scattering, scales like

$$\frac{\sigma_m}{\sqrt{N_m}}\,\frac{1}{Br_m^2},$$

where σ_m = measuring error, N_m = number of measurements and r_m is the tracking length. We assume the errors for each measurement are the same and that the alignment of detectors can be controlled or measured to allow σ_m to be the intrinsic device resolution. To compare to a typical existing large detector, we will scale the various numbers from the MARK II at the SLC. We will assume that, compared to the MARK II, $\sqrt{N_m}$ is smaller by a factor of 1.5, r_m^2 is smaller by a factor of 16, and that σ_m is smaller by a factor of 40. r_m is chosen large enough to avoid making the B field too large. The choice for N_m will be justified by pattern recognition arguments in the next section. It is chosen to be as small as possible, consistent with good pattern recognition. If we left the B field unchanged at .5 tesla, we could then expect:

$$\frac{\Delta P}{P} \simeq .24\%P \times \frac{24}{40} = .15\%P$$

To recognize the sign of a lepton at 2 TeV we would have to go down another factor of about six. This gives, then, a B field of about 3.0 tesla. We will use below a value of 3.5 tesla, for which conveniently the radius of curvature of a track is:

$$\rho_{\text{track}} \simeq 1 \text{ meter per GeV}.$$

Tracker Parameters

We now specify the other parameters of the tracker:

(1) It is made of 16 layers of silicon strips organized into 8 pairs (superlayers) of detectors. The detectors are assumed to be double sided so the total number of measurements is 32. Half are assumed to be stereo and half axial measurements. Thus the detector can be thought of as 8 axial and 8 stereo superlayers. This can be compared to the MARK II chamber which has 6 superlayers of each type.

(2) The radius varies from 8 cm to 40 cm. The inner radius is determined by the amount of radiation damage that can be tolerated. For the SSC operating at a luminosity of $10^{33}\text{cm}^{-2}\text{s}^{-1}$, this inner radius corresponds to about 1/3 megarad per year.[4)] Clearly, the design of sufficiently radiation-resistant electronics is critically important. Note, that for high momentum tracks, the impact parameter error is still $\sim 5\mu$m, even starting 8 cm from the origin, since the long lever arm of the tracking device provides a very well measured angle.

(3) We assume each layer is 150μm thick and has double sided readout. This gives 2.5% of a X_0 of material for the silicon and probably about 4% X_0 in total if we include the other material in the detector. The multiple scattering contribution to the momentum resolution scales like: $\sqrt{\frac{\text{Material}}{X_0}\frac{1}{Brm}}$. For a 3.5 tesla field this number is comparable to the MARK II at the SLC, giving a $\frac{\Delta P}{P} \simeq 1.5\%$. Thus the momentum resolution would be multiple scattering dominated up to 60 GeV.

Pattern Recognition Requirements

We show in Fig. 5 two pairs of detector layers. The detector arrangement is specified by the distances δ and Δ. The conflicting demands on δ and Δ are as follows:

(1) Would like δ small enough to make matching of points on a track simple.

(2) Would like δ large enough that track angle (tangent vector) is well measured.

(3) Would like Δ big enough to fill a given space.

(4) Would like Δ small enough to allow frequent sampling

The basis for matching vector segments is shown in Fig. 6, below.[5)] The angle between the tangent vectors and a chord connecting the segments is the same for both vector segments belonging to the same circle.

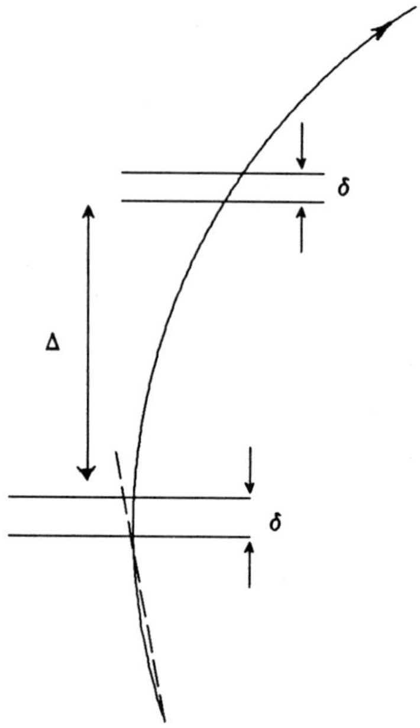

Fig. 5. A track passing through two adjacent layer pairs.

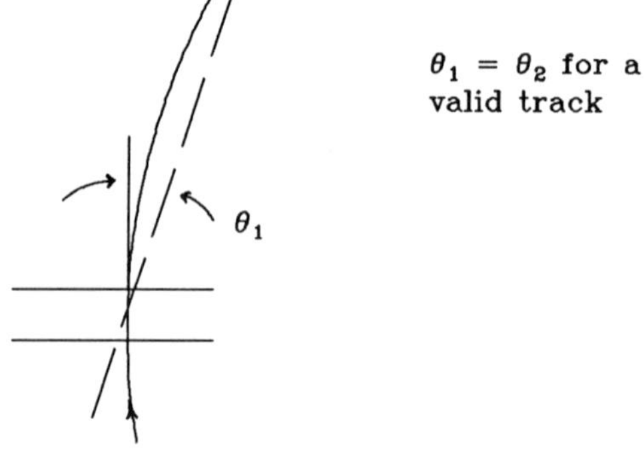

Fig. 6. Relation for vector segments lying on one track.

For the pattern recognition we require:

(1) Different tracks give distinct hits.

(2) Hits in the paired layers, separated by δ, can be locally associated into correct vector segments.

(3) Vector segments from different pairs can be correctly linked into tracks.

In terms of the angular track separation characterizing the physics, and the tracker parameters, we require that the tracker provides the resolution for the three steps above. We define:

ρ_{track}	=	Track's radius of curvature.
$\langle\theta\rangle$	=	Typical track separation angle.
σ_m	=	Position resolution of measuring element.
ε_m	=	Two track resolving power in space, locally, for a measuring element.
δ	=	Separation of elements used to measure track vectors.
r_m	=	Outer radius of device.
r_i	=	Mean radius of the ith pair.

From the physics considerations discussed earlier, $\langle\theta\rangle \simeq .5\text{mrad}$. For a silicon strip detector $\sigma_m \simeq 5\mu\text{m}$ and $\varepsilon_m \simeq 100\mu\text{m}$.

We can now check if our tracker can fulfill the three pattern recognition requirements.

(1) To see distinct hits from tracks we need:

$$\varepsilon_m < r_m \langle\theta\rangle \ .$$

This is satisfied since $100\mu\text{m}$ is $< 40\text{cm} \times .5\text{mrad}$. How many layers do we need? This should, of course, be determined with a real Monte Carlo program. We can expect that we need at least three at radii such that:

$$\frac{\varepsilon_m}{\langle\theta\rangle} < r_i \quad \text{or} \quad \frac{100\mu\text{m}}{.5\text{mrad}} = 20\text{cm} \ .$$

Thus taking 8 pairs with 8 cm $\leq r_i \leq$ 40 cm gives 4 pairs at radii $r_i >$ 24 cm, which is the mean radius. Thus the device should be able to find distinct hits from nearly all of the tracks.

(2) To match points into a segment we choose the simplest algorithm illustrated by the dashed line in Fig. 5. To each point in the inner layer of a pair we associate the point with the most similar ϕ angle in the outer member of the pair. For a curving track the deviation in space from a perfect match, using this algorithm, is given by:

$$\left[\frac{r_i}{\rho_{\text{track}}}\right]\delta \; .$$

However, in the jet core we can expect nearby hits from very stiff tracks with a mean spacing of $\langle\theta\rangle\, r_i$. These can cause incorrect associations in segment formation. Thus, the simple algorithm for matching points into vectored segments will work provided:

$$\frac{r_i}{\rho_{\text{track}}}\delta < \langle\theta\rangle\, r_i$$

$$\text{or} \quad \frac{\delta}{\langle\theta\rangle} \lesssim \rho_{\text{track}} \; .$$

This relation needs to be valid for all momenta within the jet core and provides the motivation for a small δ.

(3) The basis for matching vector segments has been illustrated in Fig. 6. Each angle is measured with a precision given by $\sqrt{2}\,\sigma_m/\delta$. Matching can be done if this accuracy is adequate. We show in Fig. 7 the trajectories of two very high momentum tracks separated by a production angle of $\langle\theta\rangle$. An incorrect match yields the dashed track whose angles don't match by an amount $\langle\theta\rangle$. Thus, a goal for the matching to work is:

$$\sqrt{2}\,\frac{\sigma_m}{\delta} \lesssim \langle\theta\rangle \; .$$

This provides the motivation for a large δ.

Note, a certain number of ambiguities can be tolerated since they can be solved using the information from several layers. However, in order to find tracks with a small number of layers and to keep the number of wrong associations small, $\sigma_m/\delta \leq \langle\theta\rangle$ is a reasonable criterion.

Before looking further at the case of tracking at 1 TeV, we can see how well the above criteria work for the upgraded MARK II at the SLC. At the Z^0 the B meson study gives $\langle\theta\rangle \simeq 5$mrad. For the large MARK II chamber, $r_m = 1.5$m. Thus, we need a two hit resolution $\varepsilon_m < 7.5$mm. It is, in fact,

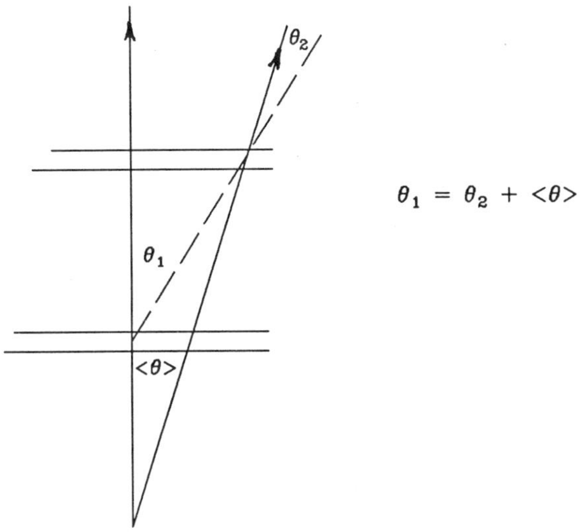

Fig. 7. Angular relations for a spurious track (dashed) created from hits from two real tracks.

about 4 mm. The Z^0 Monte Carlo actually indicates that we would have done even somewhat better in tracking efficiency for $\varepsilon_m = 2$mm, with very little improvement beyond this. For the cells in the chamber $\delta \simeq 5$cm. Thus easy segment association occurs for $\rho_{\text{track}} \gtrsim 5$ cm/5 mrad = 10 meters $\Rightarrow$ 1.5 GeV for the magnetic field used. This is adequate since softer tracks are generally spread from the jet core and not in a confused area. Finally $\sigma_m/\delta \simeq 150\mu\text{m}/5\text{cm} = 3$ mrad $\lesssim \langle\theta\rangle$, so segment association into tracks is usually unambiguous.

We now return to the silicon tracker for which, as discussed before, $\langle\theta\rangle \simeq .5$ mrad, $\sigma_m = 5\mu$m, $\varepsilon_m = 100\mu$m. Taking $\delta = 5$mm, we get easy association into vectors for:

$$\rho_{\text{track}} \gtrsim \frac{5\text{mm}}{.5\text{mrad}} = 10\text{m} \ .$$

This corresponds to 10 GeV at 3.5 tesla. It is substantially below the $P_{\min}$ of 50 GeV for tracks in the jet core estimated above. Note, that the deviation from a perfect straight line match of the points to be associated is only 100μm at a mean radius of 20 cm and ρ_{track}= 10m.

Finally, for matching vector segments into tracks we need $\sigma_m/\delta \lesssim \langle\theta\rangle$. For $\sigma_m = 5\mu$m and δ =5mm, we get $\sigma_m/\delta = 1$ mrad, which is $\sim \langle\theta\rangle$. This value might be a little large and perhaps a $\delta \simeq 1$cm would be better. Such choices would have to be looked at more carefully with a Monte Carlo simulation.

In the above we've assumed that tracks come from the origin when matching the two hits into a vector segment (but not in the segment matching). We show in Fig. 8 what happens for a track with an impact parameter b.

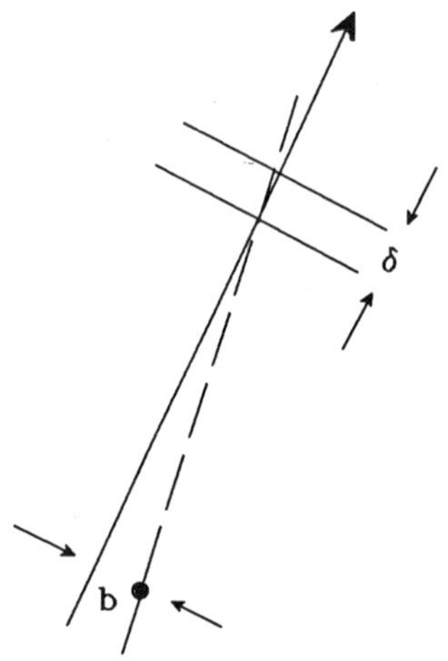

Fig. 8. Relation of points to be associated into a vector segment for a stiff track with impact parameter b.

The displacement from a straight line assuming a very stiff track is $d = \frac{b\delta}{r_i}$. For the decay of D or B mesons, $b \lesssim 300\mu$m. Therefore, $\delta = 5$mm, and $r_i > 8$cm implies $d < 20\mu$m; there is no problem for matching the hits into a vector. There is a problem for K_S^0, which might have to be picked up in a second pass if near the jet axis.

<u>z Measurement</u>

The optimum way to measure the z coordinate is a complicated question. We look at the use of stereo measurements. We assume each pair of strips is double sided and organized as shown in Fig. 9.

The goal for the pattern recognition is to measure z coordinates locally within each pair. Each of the 8 pairs making up the full detector would then have space coordinates and tangent vectors assigned locally to each track. These would then have to be matched from layer to layer.

The matching problem in z locally can come from two tracks which can be confused or from many tracks causing accidental matches. We consider the former first.

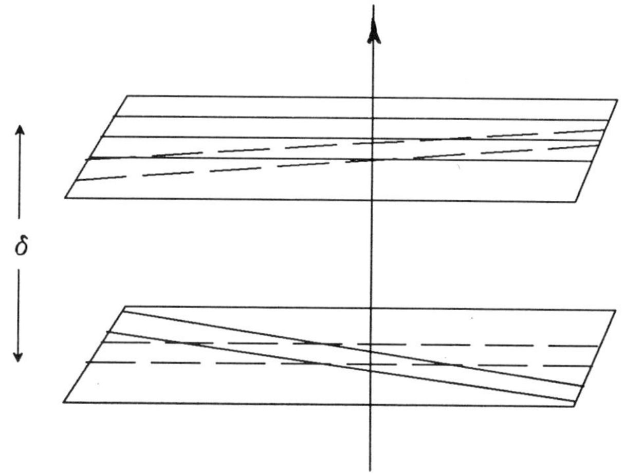

Fig. 9. Orientation of strips for a closely spaced pair of detectors.

We assume we have two tracks with mean axial coordinates x and x'. For simplicity we ignore the small displacement between layers (or correct for it approximately). Then the two stereo measurements in each layer would be, in terms of the stereo factor α:

$$x_u = x + \alpha z \qquad x'_u = x' + \alpha z'$$
$$x_v = x - \alpha z \qquad x'_v = x' - \alpha z'$$

The correct matches are then found to be the pairs that satisfy:

$$x = \frac{x_u + x_v}{2}, \quad x' = \frac{x'_u + x'_v}{2}.$$

The full 5μm precision of the device is available to do the matching. For exactly two correct hits per layer the pairing is therefore relatively easy.

The confusion from many tracks is more difficult to analyze. We can however minimize this by taking a very small stereo factor α. The idea is to trade off resolution in z for minimizing the track pairing problem. This does not significantly degrade the momentum or mass resolution for a solenoidal field. In addition, for z we only need 1mm resolution at the vertex to separate most multiple interactions. The very good 5μm resolution of the strips will yield a 1mm resolution in z for a very small stereo factor. Figure 10 shows the stereo arrangement needed for 1mm z resolution, assuming the tracking detector is made of 10cm long units.

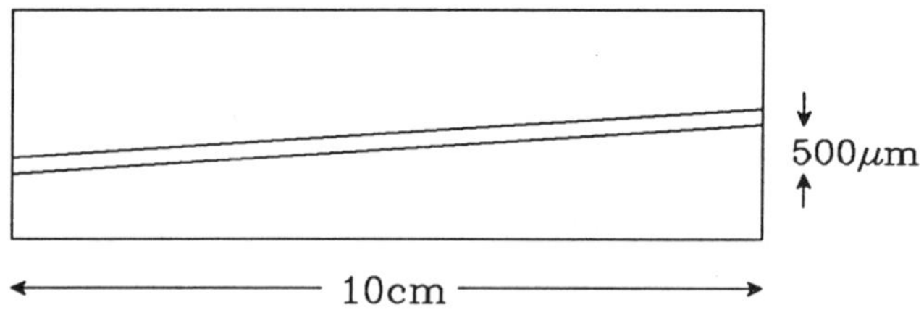

Fig. 10. Stereo displacement of 500μm over 10cm for stereo strips.

We get

$$\sigma_z = \sigma_m \left[\frac{10\text{cm}}{500\mu\text{m}} \right] = 1\text{mm} ,$$

for $\sigma_m = 5\mu$m. Thus tracks separated by $> 500\mu$m in the axial projection don't interfere with each other at all. This value is probably small enough to guarantee that nearly all stereo hits can be unambiguously associated or require pairing for the two track ambiguity only, which has been discussed above.

Lorentz Angle

A potential problem for the design presented comes from the Lorentz Angle associated with the drift of electrons or holes in the silicon. For a 150μm thickness of silicon, $B = 3.5$ tesla, we expect

25μm deflection for holes

100μm deflection for electrons.

The detectors can however be rotated in order to subtract 50μm from these deflections. Can the alignment be maintained for this case? What would the position resolution be in this case, since fluctuations in energy deposition along the track's path through the silicon can affect the position resolution? These questions need further study. They might provide the main reason for preferring a 2.5 tesla field rather than the 3.5 tesla value used above.

CONCLUSIONS

The use of a silicon tracker provides a nice match to pattern recognition and momentum measurement in 1 TeV jets. This is particularly true if tracks in the core of the jet are to be reconstructed with high efficiency. We list below some of the very difficult technical problems or questions that would have to be solved to make such a system:

(1) Can the alignment be maintained in order to realize the resolution $\sigma_m \simeq 5\mu$m?

(2) The power dissipation from the electronics, due to very many channels in a small space, provides a very difficult problem.

(3) Radiation damage, particularly to the readout electrons, is a difficult problem. This would presumably not be as serious for a large linear collider.

(4) For a *pp* collider, the storage of information at $\sim$ 50 Mhertz rate is a challenging problem. Data compression from the many channels is also needed.

REFERENCES

1. For some aspects of *B* meson detection, see B. Cox *et al.*, SLAC-PUB-4144 and F.J. Gilman and L.E. Price, SLAC-PUB-4138.
2. Proceedings of the ECFA-CERN Workshop on a Large Hadron Collider in the LEP Tunnel, p.200 and p.503, March 1984.
3. J. Roehrig *et al., Nucl. Instrum. & Methods* **226**, 319 (1984).
4. Report of the Task Force on Detector R and D for the Superconducting Super Collider, SSC-5R-1021, June 1986.
5. W. Atwood, SLAC SLD experiment, Internal Note #135.

Discussion A. SEIDEN (1st part)

D. RITSON

You described vertex chambers that are useful for improving central tracking capability and crude tagging only. Why not use modern technology to go for real vertex assignments like the FNAL fixed target experiments?

A. SEIDEN

I hope we will learn to do better on this issue. Unfortunately the lower momentum at the Z^0, as compared to FNAL fixed target experiments, significantly compromises the ability for separate vertex reconstruction because of multiple scattering. The hope at the SLC is ultimately to reduce the beam pipe from about 2,5 to 1 cm. This would be extremely helpful for full vertex reconstruction.

J. KIRKBY

Has a careful study been made of the possibility at SLC of detecting an absolutely clean sample of heavy flavour decays ($b\overline{b}$ or $c\overline{c}$) by means of a very close – in Si detector and requiring the decay to occur beyond this detector? This study would involve special masking for the synchrotron radiation and the possibility of an azimuthally asymmetric detector.

A. SEIDEN

Unfortunately there appears to be no chance of going to a few mm distance – which is what would be required. The hope is to go to 1 cm, where the effects of multiple scattering would be much reduced.

M. ATAC

How does $B\overline{B}$ identification relate to multitrack resolution?

A. SEIDEN

At the SLC you need an angular resolution for track separation of ~5 mrad to have good efficiency.

J. JAROS

Can't a cut on lifetime improve the signal to background ratio for the $B\overline{B}$ mixing signal of like sign leptons.

A. SEIDEN

The answer depends on $\Delta m/\Gamma$. In the limit of total $B^0s - \overline{B}^0s$ mixing, $\Delta m \gg \Gamma$ and then such a cut doesn't help. It should, however, certainly be looked at in the data since we don't know how large $\Delta m/\Gamma$ is.

P. WEILHAMMER

The method of using $B \Rightarrow \tau$ decay to measure $V^{b \Rightarrow u}/V^{b \Rightarrow c}$ at PEP is, according to your numbers, about as sensitive as the present upper limit of ~ .25 from CLEO and ARGUS. Will this method give much better results at LEP?

A. SEIDEN

This depends on particular event configurations and also instrumentation, better calorimetry; in particular recognition of missing energy (ν,τ) might help in this case significantly. Perhaps you could go down another factor of 5 in the limit with 10^6 $B-\overline{B}$ events. This should be studied with a Monte Carlo for each detector.

P. WEILHAMMER

How good is the experimental evidence for exclusive B – decays into charged particles to be as small as $5x10^{-3}$ for those decays which can be fully reconstructed? This number seems to be extremely small and reminds me of a similar situation as in D decays branching ratios 3 years ago.Now D branching ratios are up by a factor 2.

A. SEIDEN

I expect that these branching ratios are good to 50% from errors on ARGUS and CLEO measurements. In general, these experiments are only able to reconstruct about 30 B decays per 10^5 $B-\overline{B}$ events in several very clean channels. Adding channels with significant background the number is about 100 B decays. That is why I conclude that you need at least 10^6 $B-\overline{B}$ events to do significantly more physics with exclusive B decays.

M. ATAC

Would not you combine your proposed silicon strips tracking system with some tilted cells of drift chamber layers to improve calorimetry considering high multiplicity with neutrals at one TEV? Drift chamber wires with tilted layers would give you prompt signals (within 10 nsec) to associate the tracks with silicon layers.

A. SEIDEN

It would be a reasonable possibility. One would then bring the calorimetry as close in as possible – to perhaps about 1 m, and fill the region from 40 cm to 1 m with a tracking device of resolution ~50 μm. This would allow a smaller B field (perhaps 2 Tesla), which would have some advantages.

B PHYSICS AT PEP AND SLC

A. Seiden

Santa Cruz Institute for Particle Physics
University of California, Santa Cruz, CA 95064

ABSTRACT

In the coming year or two initial data from the SLC and the luminosity upgraded PEP machine should provide reasonably large numbers of B decays. We look below at the prospects of addressing some of the physics issues of the B system with the expected data.

PHYSICS ISSUES

For the B mesons the primary questions one would like to answer are:

(1) Lifetime, averaged over B^-, B_d^0, B_s^0.

(2) If possible, measure the individual lifetimes.

(3) Mixing for $B_d^0 - \overline{B}_d^0, B_s^0 - \overline{B}_s^0$, and CP violation. This would be interesting for $D^0 - \overline{D}^0$ as well.

(4) Ratio of matrix elements for: $b \to u / b \to c$.

(5) Branching ratios for exclusive decay modes.

Other interesting questions which will arise for production in e^+e^- annihilation:

(1) b fragmentation function.

(2) In three jet events, where the b jets are identified, the remaining gluon jet could be studied in order to understand gluon fragmentation.

(3) At the Z^0, a measurement of the vector and axial vector couplings of the b are an interesting test of the Standard Model.

(4) If new particles are produced, do they decay to a b? This is expected in the case of a top quark, $t \to b$, or $H^0 \to b\bar{b}$, if H^0 is a standard Higgs in the mass range 11 GeV $\leq m_{H^0} \leq 2m_t$.

Table 1 presents the number of $B\overline{B}$ per running year expected at various accelerators, as well as the event purity and overall event rates. Because of the signal to noise ratio, it is likely that most information in the next few years will come from e^+e^- machines, although initial evidence for $B^0_s - \overline{B}^0_s$ mixing has in fact come from the UA1 experiment at the SPS $p\bar{p}$ collider.[1)] The CESR machine will probably provide the best place to measure exclusive B decays given the rates and event purities. The SLC and PEP machines will do best in cases where the motion of the B is important. The SLC, and eventually LEP, should provide unique information on Z^0 couplings and heavy new particle sources of B mesons, if they exist below the Z^0 mass.

TABLE 1: B PRODUCTION FROM DIFFERENT SOURCES

Source	$\sqrt{S}$ (GeV)	σ_{tot}	Total Rate (Hz)	Fraction B's	$B\overline{B}$ 10^7 sec
TEV II (Fixed Target)	45	50 mb	10^7	10^{-6}	10^8
TEV I ($\bar{p}p$)	2,000	100 mb	10^5 ($\mathcal{L} \simeq 10^{30}$)	10^{-4}	10^8
SSC (pp collider)	40,000	200 mb	10^7 ($\mathcal{L} \simeq 10^{32}$)	10^{-3}	10^{11}
CESR (e^+e^-)	10.6	5 nb	.3 ($\mathcal{L} \simeq 10^{32}$)	.25	7×10^5
PEP (e^+e^-)	29	.4 nb	.04 ($\mathcal{L} \simeq 10^{32}$)	.09	4×10^4
SLC (e^+e^-)	93	32 nb	.03 ($\mathcal{L} \simeq 10^{30}$)	.14	4×10^4
LEP (e^+e^-)	93	32 nb	.03 ($\mathcal{L} \simeq 10^{31}$)	.14	4×10^5

ISOLATION OF B SIGNALS

For PEP and the SLC, the selection of B events can be made through the detection of displaced decay vertices or by selecting events with leptons with large transverse momentum relative to the direction of the jet they are in. These strategies have been looked at in Refs 2 and 3. We will be using the results of these references, although a detailed analysis of the optimum way to fully use the information of the very good resolution vertex devices of the future does not exist.

Of crucial importance for secondary vertex detection is the impact parameter resolution. We quote the expected values for this quantity, as well as the expected momentum resolution, for the upgraded MARK II at the SLC and the upgraded TPC at PEP.

	Impact Parameter Resolution σ_b	Momentum Resolution σ_P/P
MARK II (Si vertex detector + micro-jet vertex chamber)	$\sqrt{\left(\frac{34\mu m}{P(\mathrm{GeV})}\right)^2 + (5\mu m)^2}$	$\sqrt{(1.4\%)^2 + (.2\% P(\mathrm{GeV}))^2}$
TPC (Straw tube vertex chamber)	$\sqrt{\left(\frac{50\mu m}{P(\mathrm{GeV})}\right)^2 + (40\mu m)^2}$	$\sqrt{(1.1\%)^2 + (.28\% P(\mathrm{GeV}))^2}$

The first term in each expression comes from multiple scattering and presents a serious limitation to the ability to find secondary vertices. To limit this effect one needs to get as close as possible to the colliding beam vertex. The MARK II numbers assume a minimum radius of the vertex detector of about 3cm. The SLD experiment will attempt to reduce this number to a value between 1 and 2cm, depending on a detailed understanding of backgrounds at the new SLC machine.

The average impact parameter expected for secondaries from various decaying particles is given below:

Particle	Mean Flight Distance, $c\tau$
τ	$90\mu m$
$D^{\pm}$	$300\mu m$
D^{0}	$120\mu m$
D_S	$100\mu m$
B	$330\mu m$

For reasonable detection efficiency we want: $\sigma_b \leq \frac{c\tau}{4}$ or $80\mu m$ for B mesons. For the MARK II this will be satisfied for all secondary particles with momentum $\lesssim$ 450 MeV. We will look later at the expected tagging efficiencies, given the resolutions quoted above.

Figure 1 shows two B mesons emerging from an annihilation event. The secondary vertices are shown as V_1 and V_2, the primary vertex is V_3. The tracks originating from V_3 are not shown, those from the B mesons are shown, including errors on the measured trajectories. The luminous region is shown as an ellipse. For the SLC this region is negligible in size; for PEP it is about $500\mu m \times 65\mu m$. Strategies for B tagging that have been investigated are the measurement of the displacement of V_1 from V_2 for PEP or the presence of several large impact parameter tracks from the well defined SLC interaction point.

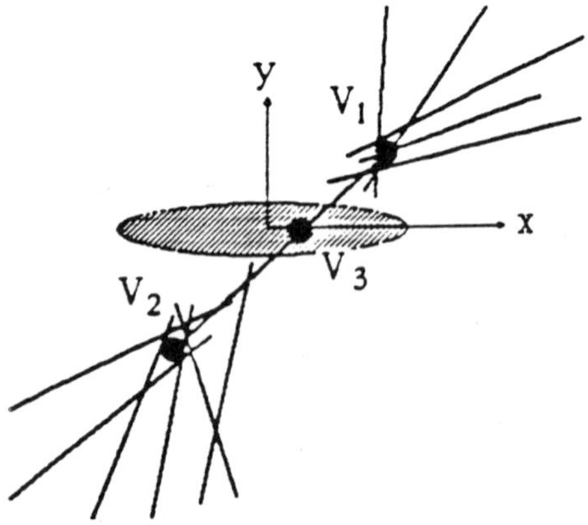

Fig. 1. Vertex configuration for B^+B^- produced in e^+e^- annihilation.

Detailed Numbers for PEP

We assume we start with $4 \times 10^4 B\overline{B}$ events, which is 9% of the total. The vertex related tagging procedure then is as follows:

(1) Divide event into two jets.

(2) Calculate the position of the best vertex for each jet. Although this includes some tracks from the primary vertex, a cut on the distance between the two vertices, $D_{vtx_1-vtx_2}$, serves to eliminate most of the $u\bar{u}, d\bar{d}, s\bar{s}$ events. Figure 2 shows the distribution of $D_{vtx_1-vtx_2}$ for $b\bar{b}$ events and $u\bar{u}, d\bar{d}, s\bar{s}$ events based on a jet Monte Carlo.

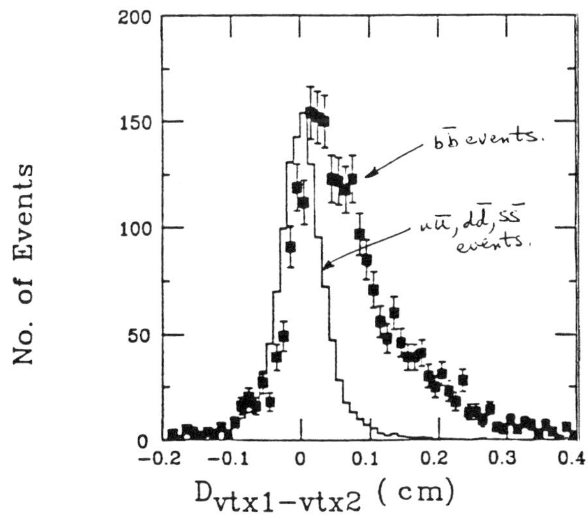

Fig. 2. Vertex displacement between two jets for $b\bar{b}$ and light quark jets.

(3) Find the most likely primary vertex position by intersecting the line joining V_1 and V_2 and the ellipse in which the interaction must occur, as shown in Fig. 1. Each track is then reassigned to the vertex it comes closest to. We require the invariant mass of the tracks with the largest decay distance to be > 2 GeV and large sphericity in the rest frame of each potential B jet. These cuts serve to reduce the initially large $c\bar{c}$ background.

The results are given in the table below where numbers for the presently used lepton tagging cuts are also presented. We assume both e and μ can be used.

TABLE 2. B TAGGING AT PEP

Tag	Factor Enhancement	Tag Efficiency	$\#B\overline{B}$ Events
Lepton (e and μ) P and P_T cuts	25	8%	2,500
Secondary Vertex, mass and sphericity cuts	25	15%	6,000

DETAILED B TAGGING NUMBERS FOR MARK II AT SLC

We outline below the tagging strategy employed for studying B detection with the MARK II at the SLC:

(1) Divide the event into two jets.

(2) For each jet select tracks with impact parameter b in the range:

$$3\sigma_b \leq b \leq 2\text{mm} .$$

This is measured for each track relative to the precisely known beam–beam collision point.

(3) Require at least three tracks satisfying the large impact parameter criterion (2) in one of the jets.

(4) Invariant mass of the three or more large impact parameter tracks $\geq$ 1.95 GeV.

These criteria result in a tag efficiency about three times larger than at PEP, due to the higher track momenta and the known event origin. There is about 10% background without any cuts on event shapes or finding of secondary vertices.

Based on Monte Carlo studies, and the above criteria, we can display the B tag efficiency in terms of the B kinematic parameters. Figure 3 shows the B tag efficiency as a function of the B decay length, while in Fig. 4 the tag efficiency is shown as a function of the B momentum. Figure 5 shows the loss in tag efficiency as a function of the detector's ability to separate tracks with closely spaced azimuthal angle, ϕ. The detector at the Z^0 needs to be able to resolve tracks separated by $\gtrsim$ few milliradians.

Table 3 gives the results for B tagging at the Z^0. We assume that we start with $4 \times 10^4 B\overline{B}$ events, 14% of the total number of events.

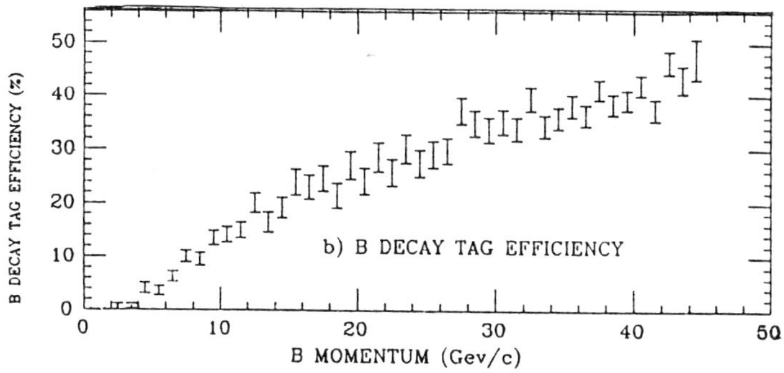

Fig. 3. B tag efficiency versus decay length.

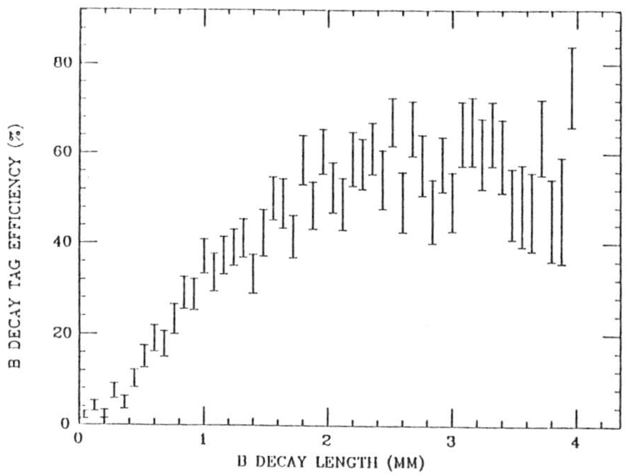

Fig. 4. B tag efficiency versus momentum.

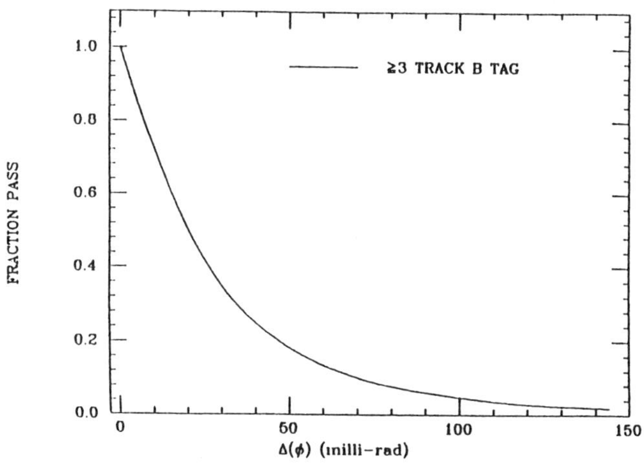

Fig. 5. B tag efficiency versus angular multihit capability of tracking detector.

TABLE 3. B TAGGING AT SLC

Final Number of $B\overline{B}$ Events	Background from $c\bar{c}$	$u\bar{u}, d\bar{d}, s\bar{s}$
17,000	1,500	400

For completeness we quote also the results for charm tagging, which should also be of interest at the Z^0. The strategy that has been looked at is as follows:

(1) Divide event into two jets.

(2) For each jet select tracks with impact parameter b in range:

$$3\sigma_b \leq b \leq 2\text{mm} .$$

(3) Require each jet to contain one or two high impact parameter tracks. Invariant mass of the high impact parameter pair should be < 2 GeV. Tracks are not counted if within 40 MeV of K^0 mass.

Starting with 2.8×10^4 $c\bar{c}$ events, 10% of the total, yields 6,000 $c\bar{c}$ events with a background of 3,500 events from $b\bar{b}$ and 3,000 events from $u\bar{u}, d\bar{d}$ and $s\bar{s}$. There is substantially more background for $c\bar{c}$ than $b\bar{b}$. The $c\bar{c}$ events should be of interest in measuring the Z^0 couplings to the c quark.

Exclusive charm decays can also be reconstructed by adding further invariant mass cuts. As an example, to select $D^0 \to K^-\pi^+$ one uses the $c\bar{c}$ data sample discussed above and additionally:

(1) Calculate invariant mass of the high impact parameter pairs in each jet, cutting on $|m_D - m_{K\pi}| \leq 37.5$ MeV (1.5σ).

(2) Jet with $D \to K\pi$ required to have exactly 2 high impact parameter tracks.

(3) Opposite hemisphere jet should have no more than 2 high impact parameter tracks, with mass ≤ 2 GeV.

Analysis looking for $D^\pm \to K^\mp\pi^\pm\pi^\pm$ or $D_S^\pm \to \phi\pi^\pm$ is done in an analagous fashion. The results are given in Table 4 below:

TABLE 4. TAGGING OF EXCLUSIVE D DECAYS

Process	# Signal Events	Background $b\bar{b}$	$c\bar{c}$	$u\bar{u}, d\bar{d}, s\bar{s}$
$D^{\circ} \rightarrow K\pi$	330	150	230	50
$D^{\pm} \rightarrow K^{\mp}\pi^{\pm}\pi^{\pm}$	230	400	230	90
$D_S^{\pm} \rightarrow \phi\pi^{\pm}$	20	20	20	0

The exclusive decay events can be used to tell c from $\bar{c}$ jets. Not yet included in any of the analyses above is the formation of distinct vertices from the high impact parameter tracks and the requirement that the reconstructed D momentum correctly point from the primary to the secondary vertex. These criteria should further reduce the background and will probably be needed for better background rejection.

Analogous reconstruction of exclusive B decays should be very difficult because of the small value of the B branching ratios into convenient, simple, final states. As an example, we consider using the channels $B^{\circ} \rightarrow D^{\pm}\pi^{\mp}$ or $B^{\pm} \rightarrow D^{\circ}\pi^{\pm}$ for B reconstruction. The branching ratios for these are $\sim 5 \times 10^{-3}$. Choosing a simple D final state to reconstruct adds another branching ratio $\sim 4 \times 10^{-2}$. The product is therefore $\sim 2 \times 10^{-4}$, before considering losses due to reconstruction efficiency. These numbers imply that one needs $\gtrsim 10^{6} B\overline{B}$ events to end up with a reasonable number of exclusive B decays. This physics will therefore probably be the domain of CESR and DORIS for the next few years. The same conclusion is even more true for measuring $\frac{b \rightarrow u}{b \rightarrow c}$ via exclusive channels.

PHYSICS USING TAGGED B SAMPLES

Lifetimes

The B lifetime, averaged over B species, has typically been determined using about 200 events. With thousands of events, we can expect to reduce the statistical errors by a large factor $\sim$ 5 to 10. The measurement will then be systematics limited. This error can probably be brought down to about 5% (similar to the error on τ, D°, D^{+} lifetimes) as compared to the present error on the B lifetime of about 15–20%.

How about individual types of B's: B^+, B^0, and B^0_S? It is unlikely that enough exclusive B decays can be reconstructed at PEP or the SLC to make individual lifetime measurements. It is likely that eventually 1000 B's can be identified (tagged) by adding together enough decay modes at CESR. The untagged B can then be looked at for a semileptonic decay. One expects 1000 tagged B^+ or B^0 to be associated with about 100 opposite side electrons each. Since the final states at CESR are $B^0\overline{B}^0$ or B^+B^- at the lowest energy open bottom resonance, the semileptonic branching ratios for B^0 and B^+ can separately be measured. Taking ratios then gives a measurement of the lifetime ratio accurate to about 15%.

An alternative method to identify the B species would be if a strong correlation exists between the initial B and the final D produced. For example, there may be a preference for $B^0_d \to D^+ + X$, $B^- \to D^0 + X$, $B^0_S \to D^+_S + X$. This would allow the B type to be tagged without reconstructing the state X, giving a much larger data sample. Considering that both D and D^* states are produced, it is likely that the best correlation exists in the case $B^0_S \to D^+_S + X$. We can expect at the SLC that the 17,000 $B\overline{B}$ tags will yield about 5,000 B^0_S. Reconstructing the $D^\pm_S$ via its $\phi\pi^\pm$ decay, which should have an efficiency of about 2%, would yield ~100 events. Using the other tracks in the B^0_S decay we can measure its lifetime with an error which will probably be about 30–50%.

Note that this B^0_S sample selected by the D^+_S decay can be used to look for mixing. The recoil B should yield about 10 semileptonic decays whose charge is a measure of mixing. Unfortunately the number of events is marginal. LEP should provide a bigger data sample for a mixing search, although their B tagging efficiency will not be as good.

Mixing and $b \to u$ Transition

Probably more important than an extremely accurate B lifetime measurement will be a measurement of the K–M matrix element $\frac{V_{b\to u}}{V_{b\to c}}$ and a mixing measurement. Note that CP violation for the K^0 in the Standard Model involves a product of all K–M angles. We would therefore have a problem if $V_{b\to u}$ gets too small. In particular, $\left|\frac{V_{b\to u}}{V_{b\to c}}\right| \ll .1$ would be a problem for the Standard Model if the top mass is not enormous.

Mixing is sensitive to high mass objects which can participate in virtual transitions between neutral meson pairs. What is expected is very strong mixing for $B^0_S - \overline{B}^0_S$, moderate to little mixing for $B^0_d - \overline{B}^0_d$ and mixing at the 10^{-4} level for $D^0 - \overline{D}^0$. The D^0 system, since such a small result is expected, might well be the most sensitive system to look for new physics.

The search for mixing at PEP has been analyzed in some detail by Fridman and Schwarz.[4)] The signal is same sign leptons with the ratio:

$$\frac{N^{++} + N^{--}}{N^{+-}} \neq 0 \; .$$

Based on a full Monte Carlo, what can be expected for complete $B_S^0 - \overline{B}_S^0$ mixing is 150 like-sign dileptons from 40,000 $B\overline{B}$ events. The signal to background is equal to .4, with most of the background from $B\overline{B}$ events. Thus, establishing a signal will require a lot of confidence in the Monte Carlo, as is the case for the mixing signal from $\bar{p}p$ collisions.

A measurement of $V_{b\to u}/V_{b\to c}$ using exclusive decays suffers from too few events. An alternative approach that has been looked at for PEP is a search for $B^{\pm} \to \tau^{\pm}\nu$. The branching ratio expected is

$$Br(B \to \tau\nu) = 1.6\% \left(\frac{f_b}{200 \text{ MeV}}\right)^2 \left|\frac{V_{b\to u}}{V_{b\to c}}\right|^2 \; .$$

If this is measurable, it would establish $|V_{b\to u}/V_{b\to c}|$ within the uncertainty coming from f_b (perhaps 50%). Depending on background, it is expected that a limit can be set near 10^{-3}, which corresponds to $|V_{b\to u}/V_{b\to c}| < 1/4$, near its present upper limit. Thus, it is unlikely that this measurement will be competitive with CESR.

B Production Characteristics at SLC

The 17,000 tagged $B\overline{B}$ events expected at the SLC should provide an excellent sample for studying the strong and weak interactions. In the area of the strong interactions the data should allow a large statistics measurement of b fragmentation using, for example, the lepton spectrum. Even more interesting are three jet events where the identification of the b jets allows the unambiguous isolation of the accompanying gluon jet. We would expect about 1,000 events with identified gluon jets. These should allow the study of the angle and energy of such jets as well as gluon fragmentation for jets with reasonably large energy. Little is known experimentally about gluon fragmentation, so this should prove very interesting.

We give in Table 5 the expected vector and axial vector couplings for the Z^0 to the various fermions.

TABLE 5: FERMION COUPLINGS OF Z^0

Particle	a	v
ν	-1	1
e, μ, τ	1	$-1 + 4\sin^2\theta_W \simeq -.08$
u, c, t	-1	$1 - \frac{8}{3}\sin^2\theta_W \simeq .39$
d, s, b	1	$-1 + \frac{4}{3}\sin^2\theta_W \simeq -.69$

The combination $|a|^2 + |v|^2$ for the b can be measured using $\Gamma_{b\bar{b}}$. The accuracy will be limited by the knowledge of the tagging efficiency for B detection, yielding a 10% measurement. As a further test of the Standard Model couplings, av can be measured using the charge asymmetry. For a given flavor quark f, this can be defined as

$$A_{ch} = \frac{\frac{d\sigma}{d\Omega}(\theta)|_f - \frac{d\sigma}{d\Omega}(\theta)|_{\bar{f}}}{\frac{d\sigma}{d\Omega}(\theta)|_f + \frac{d\sigma}{d\Omega}(\theta)|_{\bar{f}}}$$

$$= \frac{2\cos\theta}{1+\cos^2\theta}\left(\frac{2a_f v_f}{a_f^2 + v_f^2}\right)\frac{\left[\frac{2a_e v_e}{a_e^2+v_e^2} + P_e\right]}{\left[1 + P_e\frac{2a_e v_e}{a_e^2+v_e^2}\right]}$$

on the Z^0 pole. In this expression P_e is the beam polarization.

For leptons:

$$\frac{2a_e v_e}{a_e^2 + v_e^2} \simeq -.16$$

so the asymmetry is very small for the $\mu^+\mu^-$ final state.

For the b system we get the largest effect, since:

$$\frac{2a_b v_b}{a_b^2 + v_b^2} \simeq -1 \ .$$

Thus we expect $\sim$.16 asymmetry without polarization and $\sim$50% with polarization. This measurement requires that we measure the sign of the b via the sign of a final lepton or by reconstructing the D from a B decay.

In addition to direct production of B, new particles may decay to B. In fact this characteristic will be crucial for our interpretation of a new

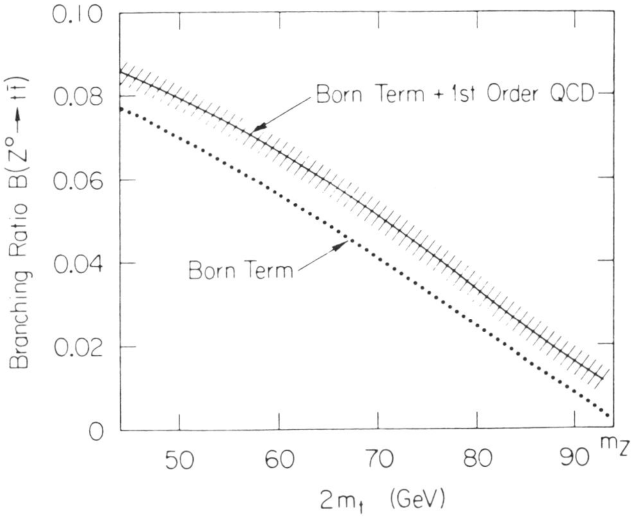

Fig. 6. $Z^0 \to t\bar{t}$ as a function of m_t.

particle. The new particle which may likely be produced is the top quark. If $m_t \simeq 35$ to 40 GeV, $t\bar{t}$ will be $\sim$4% of the events,[5)] as shown in Fig. 6. This would give about 10,000 $t\bar{t}$ events at the SLC.

We shall assume that a reasonable fraction of the events can be isolated using the spherical shape of the event or by selecting high transverse momentum leptons. For a t, the lifetime is expected to be so short that the leptons come from the primary vertex, unlike the case of potential background from a B or D decay. For a b', in the case of a fourth generation with $m_b' < m_t$, the lifetime might be reasonably long, which is a characteristic to look for. In this case the lepton might not come from the primary vertex.

For a new quark, we can in principle establish a t versus b' just using the lepton spectrum, which is different for the two choices. Assuming a decay into light quarks, then for x = fractional lepton momentum:

$$\frac{d\Gamma^t}{dx} = \frac{G^2 m^5}{16\pi^3} x^2 (1 - x)$$

$$\frac{d\Gamma^{b'}}{dx} = \frac{G^2 m^5}{96\pi^3} x^2 (3 - 2x)$$

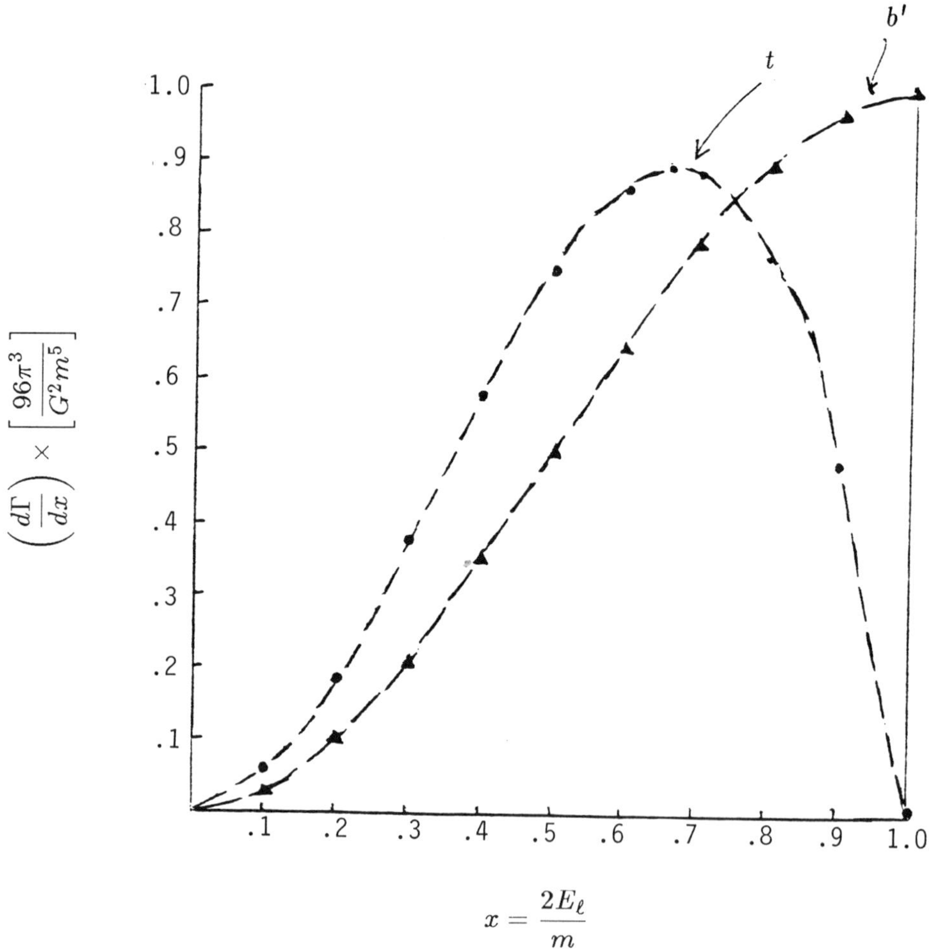

Fig. 7. Lepton spectrum for a t or b' parent quark.

ignoring QCD corrections.[6] These spectra are shown in Fig. 7.

Assuming a 40 GeV t quark mass, Fig. 8 shows the expected B momentum for a B from t decay, which peaks at about 13 GeV. Using Fig. 4, we see that this is still a large enough momentum to be in the region of reasonable B tag efficiency, which is expected to be about 15%. This value is large enough that it should be relatively easy to establish the $t \to b$ pattern.

Finally, we note that a standard Higgs should decay mostly into $b\bar{b}$ if its mass is between 11 and $2m_t$. It would be produced via $Z^\circ \rightarrow H^\circ e^+e^-$ yielding at the SLC about 40 events for $m_H = 11$ GeV, falling to 10 events at $m_H = 30$ GeV. If such a Higgs exists a detailed investigation will require the full luminosity of LEP.

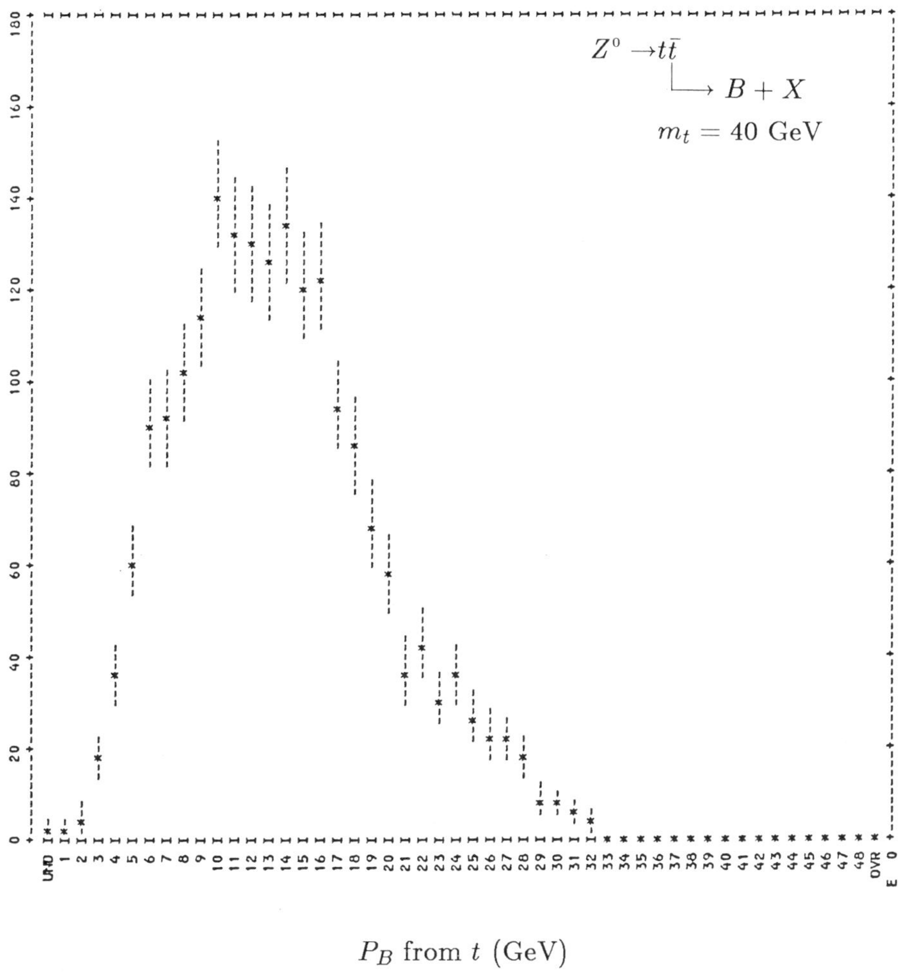

Fig. 8. B momentum resulting from decay of a 40 GeV t produced at the Z°.

REFERENCES

1. C. Albajar *et al.*, CERN-EP/86-209.
2. *Proceedings of the Workshop on e^+e^- Physics at High Luminosities*, SLAC-283, 1985.
3. K. Hayes, *B* Tagging at the SLC. MARK II/SLC Note #73, 1984.
4. A. Fridman and A. Schwarz, *Phys. Rev.* **D32**, 1650 (1985).
5. J.H. Kühn, A. Reiter and P.M. Zerwas, *Z* decays to top quarks, *Nucl. Phys.* **B272**, 560 (1986).
6. G. Altarelli *et al.*, *Nucl. Phys.* **B208**, 365 (1982).

Discussion A. SEIDEN (2 nd part)

A.H. WALENTA

There is a detailed study of the ARGUS group which indicates that the reconstruction efficiency for secondaries vertices is considerably improved if the second coordinate has the same high precision. You did not consider the second coordinate: could you comment on this subject?

A. SEIDEN

If the precision in both coordinates is as good as in the one, you have to have an improvement. If the measurements in the two coordinates, are poorer than in the one coordinate then it would require a detailed study as to where the best results would occur. One possibility would be to include good z measurement on the inside of the detector only, and optimize the rest for pattern recognition.

F. SAULI

You did not mention rate in your discussion of tracking detectors at SSC. Most of the drift chamber used today just do not seem to have good enough time resolution to permit efficient tracking (typical memories of 200 nsec or more, compared with the expected 20 ns or so collision rate).

A. SEIDEN

I agree, that's why the silicon is so attractive. If one had a mixed system, with drift chambers on the outside, then the silicon could be used to match onto the in time tracks and partially alleviate this problem.

P. JARRON

You propose the use of CCD as buffer before trigger. At 100 MHz rate of the frequency clock you will get about 100 mW per storage line channel due to capacitive load of CCD pixels. This power dissipation seems to be very difficult to reduce.

A. SEIDEN

I agree, it is a very serious problem.

G. LUTZ

1) How do you propose storing the information until trigger decision?
2) How many channels has your proposed device?
3) Comment: we have readout electronics at MPI working at 2 mW/channel
4) Comment: I would like to stress the importance of precise measurement in two coordinates for being able to do vertex topology reconstruction on an individual event basis.

The device would have ~2$10^6$ channels if 25 μm pitch is maintained throughout. The readout is a very serious problem. If one were at an e^+e^- machine, solutions close to what already exist would work. For a hadron collider one needs a readout which functions like a CCD but without requiring a lot of power.

THE MARK II VERTEX DETECTORS: STATUS AND PROSPECTS*

John A. Jaros

Stanford Linear Accelerator Center
Stanford University
Stanford, California 94305

I. INTRODUCTION

The art of detecting heavy quark and lepton decay vertices among the collision products of e^+e^- interactions has developed rapidly in the past ten years. To date, there have been three generations of detectors employed or under construction for measuring and identifying heavy quark and lepton decays. Figure 1 plots the history of the field since the beginning of the PEP/PETRA era in terms of the impact parameter resolution of the detectors. The large central tracking chambers common to the PEP/PETRA detectors were optimized for momentum resolution and reliable jet reconstruction rather than impact parameter resolution. Even so, the 500 μ resolution typical of these devices was adequate for the first tau[1] and B[2] lifetime measurements. In both cases, the resolution was much larger than the average impact parameter of the decay products ($\sim$ 100 μ), so the success of the measurements depended on isolating the decays of interest on the basis of event topology, kinematics, and lepton identification. This was relatively easy in the e^+e^- environment. Because of the poor resolution, the first measurements were just at the limit of statistical significance. Imperfect knowledge of the resolution function added considerable uncertainty to the systematics of the measurements.

* Work supported by the Department of Energy, contract DE – AC03 – 76SF00515.

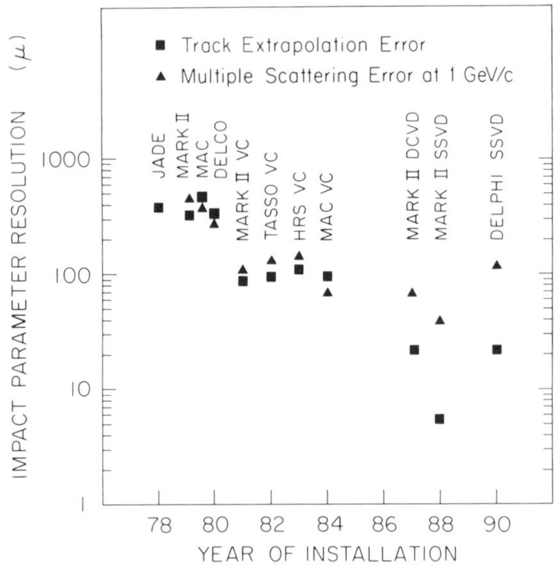

Fig. 1. The impact parameter resolution for existing and proposed collider vertex detectors plotted vs. the year of installation.

The Mark II vertex chamber was the first of the collider "vertex detectors" built expressly for detecting and measuring decay vertices. It boasted sufficient resolution to "detect" a heavy quark decay. The resolution was comparable to the typical impact parameters being measured, resulting in about a $\sqrt{2}$ loss in the statistical precision for lifetime measurements. The Mark II vertex chamber and other second generation detectors have reliably measured the charm, bottom and tau lifetimes, and searched for long-lived anomalies, but they still depend on the cleanliness of the e^+e^- environment to achieve good signal/noise. Decays two or more lifetimes out on the exponential tail can be isolated, so these devices can tag heavy quark jets at low efficiencies. Resolution effects can easily masquerade as physics effects, so a detailed knowledge of the resolution function is required. Despite this limitation several groups[3] have reported τ, B, and D lifetimes with errors below 15%.

Third generation detectors are presently under construction or being planned at TRISTAN, SLC and LEP. They will have impact parameter resolutions significantly smaller than the expected average impact parameters from charm and bottom, and so will measure lifetimes without the loss of statistical precision. The devices will cleanly tag decays that have lived at least one mean lifetime. They will measure lifetimes and KM mixing angles, search for new heavy quarks and neutrinos, search for the Higgs, and study $B\overline{B}$ oscillations as a function of proper decay time. They will *not* provide an emulsion-like view of the interactions, with its unique association of track to vertex. That will be left to succeeding generations.

This paper will concentrate on the second and third generation Mark II vertex detectors. It will first discuss the Mark II PEP vertex chamber, focusing on our operating experience and our present level of understanding the chamber's performance. Physics results from the chamber are reviewed, and a progress report on a refined measurement of the B lifetime is presented. The rest of the paper is devoted to a technical discussion of the new Mark II Drift Chamber Vertex Detector for the SLC, which is under construction. Chamber design, construction techniques, and prototype performance are discussed. A generation 3+ vertex detector using silicon microstrip detectors is also being built for the Mark II detector, to be used in conjunction with the drift chamber detector. A thorough discussion of this device can be found in Ref. 4.

II. THE MARK II PEP VERTEX CHAMBER

A. Design

The PEP vertex chamber[5] was designed to optimize impact parameter resolution for measurements of heavy quark and lepton lifetimes. The detector consisted of two bands of axial drift cells, the first (four layers deep) at an average radius 11.4 cm from the beams, and the second (three layers deep) at 31.2 cm radius. By measuring points at both small and moderate radii, the chamber permitted track extrapolation to be essentially independent of the less precise central-tracking chamber behind it. To minimize multiple coulomb scattering, the chamber was built directly around a 7.5 cm radius beryllium beam pipe, 0.6% of a radiation length thick including its inner liner of 50 μm titanium which absorbed fluorescence X-rays from nearby synchrotron masks. The chamber was roughly 1.2 m long. The wire load was supported by the outer aluminum shell, which was securely fixed to 2″ thick aluminum endplates.

The drift cell was a roughly hexagonal wire structure using 150 μ diameter aluminum field wires. (See Fig. 2.) The exact cell geometry varied cell-to-cell in order to hold the anode-to-field wire distance constant (5.3 mm) throughout the chamber. This had two nice features: (1) tracks (which are essentially radial in the chamber) were always approximately perpendicular to a sense-field layer, so corrections for track angle were unnecessary; (2) a single time-distance relation characterized the entire chamber. The wires were positioned with 15 μm accuracy, so survey and software corrections were unnecessary.

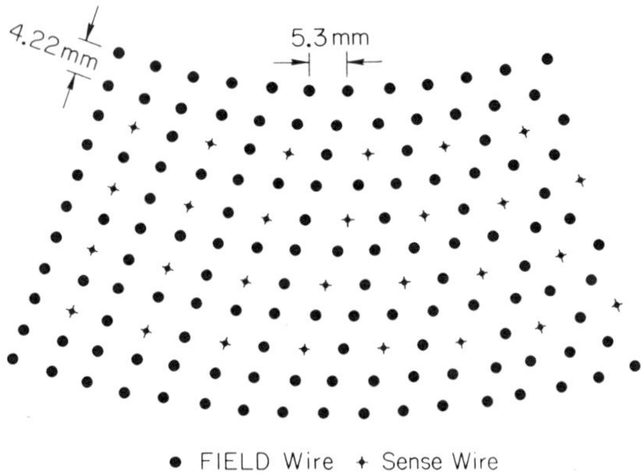

Fig. 2. Arrangement of anode and cathode wires in the Mark II vertex chamber. One-tenth of the four-layer inner band is shown.

The chamber achieved 100 μm resolution per measurement in 1 atm argon/ethane (50/50). This resulted in an impact parameter resolution slightly under 100 μm at high momentum. The multiple coulomb scattering contribution to the impact parameter resolution was 95 μ/P (GeV/c). Roughly 75% of tracks in jet-like multi-hadronic events could be reliably reconstructed and pass tight quality cuts. Losses in efficiency came about when two or more tracks traversed the same drift cell.

B. Operating Experience

The chamber was installed at the PEP storage ring at SLAC in September 1981. Although the pressure vessel was designed for 2 atm operation, leaks associated with the HV feedthroughs, which doubled as pressure feedthroughs, prevented operation at pressures above a few pounds per square inch. Drift velocities in the argon/ethane are sufficiently well saturated that temperature and pressure fluctuations did not affect the time-distance relationship. Prototype tests performed at 2 atm pressure gave improved spatial resolutions ($\sigma \sim 70\ \mu$). However, since multiple scattering errors dominated the impact parameter resolution for most tracks, the fact that we were forced to operate the chamber at 1 atm, with consequently lower resolutions, had very little impact on the physics.

The chamber began routine data taking in November 1981. Anode wires were run at ground potential, the cathode at –2.25 kV, which corresponds to rather high gain on the sense wires. Typical currents were 200 nA/wire in the innermost wires coming primarily from electromagnetic shower spray and beam gas events. During the Spring of 1982, these currents increased to nearly 500 nA/wire as background levels in PEP increased. After a week or two, one section of the inner band of wires developed "glow discharge" problems, which we assumed were due to Mälter discharges. The chamber would hold high voltage if the radiation load were low, but would draw large currents and subsequently trip the high voltage supplies in actual beam operations. We lowered the chamber voltage to 1.95 kV in the damaged section (and 2.1 kV elsewhere in the chamber), and added preamplifiers that lowered the threshold sensitivity to about 100 μV, but were forced to run without the damaged section of the chamber in the late spring. The problem wasn't cured until, at the suggestion of M. Atǎc, we added about 1.6% ethanol to the gas. With the alcohol, it was possible to run the damaged section of the chamber; though we chose to do so at the lower voltages. The chamber operated very reliably during the next two years of data taking. In all, the device logged more than 200 pb^{-1} of data, all at the e^+e^- center of mass energy of 29 GeV.

C. Performance

The Mark II PEP Vertex Chamber resolution is roughly equal to the average impact parameter expected in heavy quark and heavy lepton decays. Thus an observed decay length distribution is the convolution of the primordial exponential distribution with a near-Gaussian resolution function, where the mean decay length is approximately equal to the decay length resolution. A maximum likelihood fit to this distribution has the unfortunate property that the fitted decay length is sensitive to the estimated resolution. So we judged it important to understand the resolution and its possible biases.

The study of the chamber's performance has taken several years and involved many people.[6] Much of the work reported here was done by R. Ong. Three indicators have been used to gauge the performance. The resolution per layer is determined by looking at the cell-by-cell residuals to track fits. Distributions of track χ^2 in the vertex chamber have allowed the global chamber performance to be studied. And the average distance of closest approach to the beam center for Bhabha scattered e^+e^- pairs allows a good check for biases. This is shown in Fig. 3a for an early data set. The figure shows systematic offsets in the one-fifth of the chamber which was run at lowered voltages, and a few isolated offsets with biases in the 50 to 80 μ range. Careful fitting of the time distance relation in the low voltage section, with particular attention to edge effects, eliminated the systematic offsets. A survey of the distribution of residuals (i.e., fitted minus measured drift distance) for all the cells in the chamber, led us to single out a few off-centered and otherwise anomalous distributions. These were corrected by changing time offsets or wire positions, or they were subsequently eliminated from the fits. Figure 3b shows the distance of closest approach distribution after

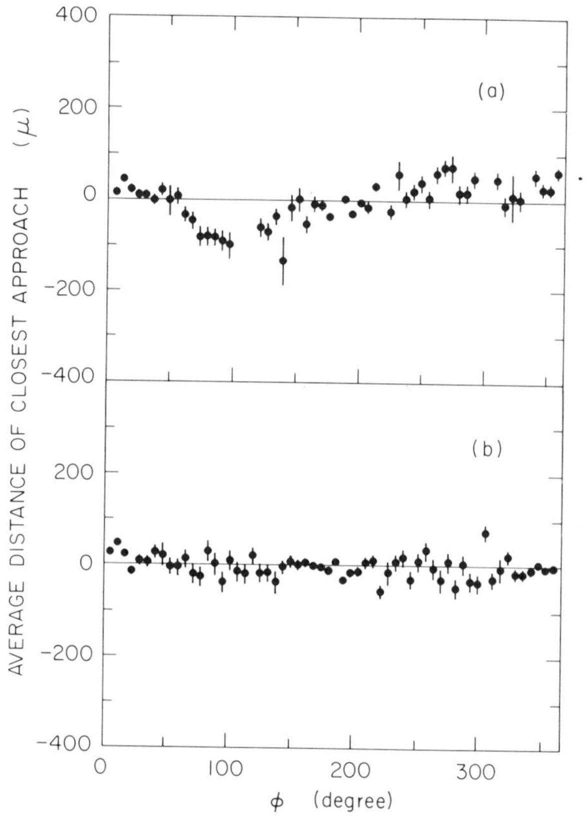

Fig. 3. Average distance of closest approach to the beam center for Bhabha scattered electrons vs. azimuthal angle ϕ. The distribution is shown before (a) and after (b) corrections mentioned in the text.

these fixes, and demonstrates that the remaining systematic biases are at the 40 μ level or below.

Figure 4a shows the vertex chamber track χ^2 distribution for Bhabha tracks, clean, back-to-back events which are easily reconstructed. The curve is the canonical χ^2 distribution for 5 degrees of freedom. Except for an excess of tracks at very large χ^2, the distribution is reasonably well fit, demonstrating that the tracking resolution in the chamber is understood. The resolution is considerably degraded in hadronic events, however, as can be seen from the poor fit in Fig. 4b. This problem is due to cross-talk on the preamplifiers, which slews a given drift time when a neighboring cell has recorded a slightly earlier drift time. After correcting for these nearby hits, the tracking χ^2 distribution is given by Fig. 4c, which shows that even in the hadronic environment the resolution is reasonably well understood.

A final diagnostic is shown in Fig. 5, which shows the distance of separation between two Bhabha tracks after they have been extrapolated to the production

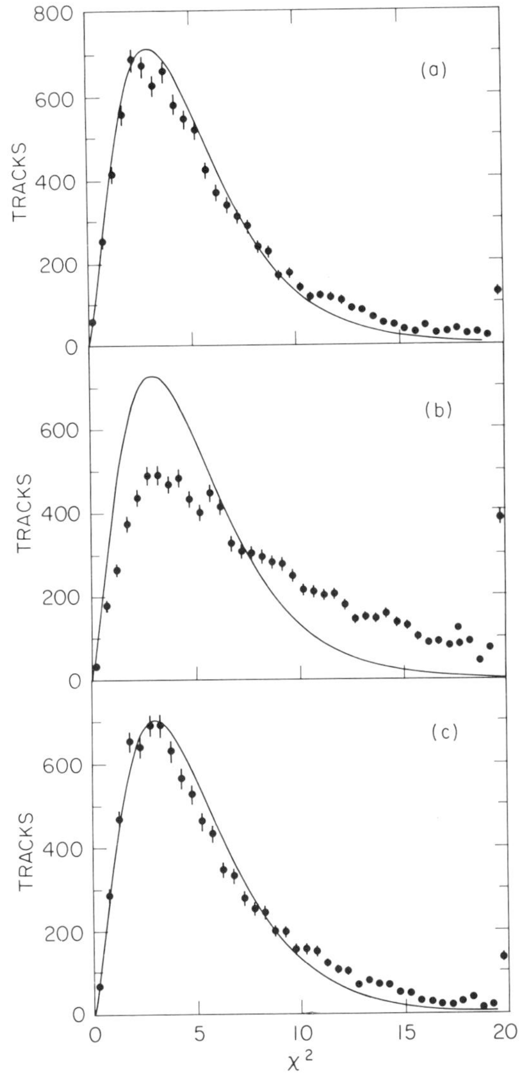

Fig. 4. χ^2 distributions for track fits in the vertex detector for (a) Bhabha scattered electrons; (b) hadronic tracks before cross-talk corrections; (c) hadronic tracks after cross-talk corrections. The solid curve is the canonical χ^2 distribution for five degrees of freedom normalized to the total number of tracks.

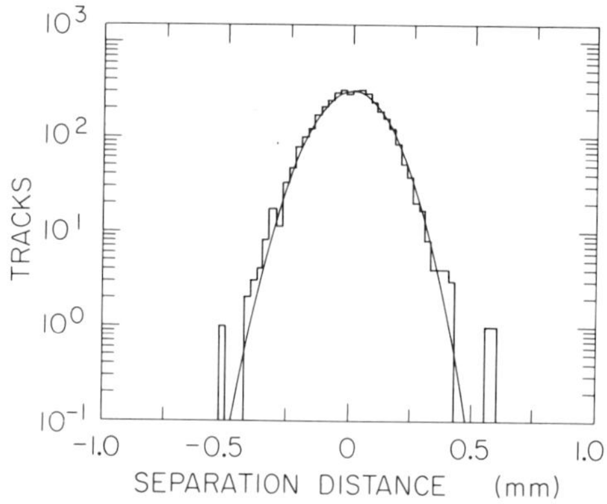

Fig. 5. Projected distance between back-to-back tracks measured in the vicinity of the primary interaction point in Bhabha scattering events. The width of the distribution is 122 μ, implying the impact parameter resolution is 85 μ.

point. The Gaussian curve, which is predicted on the basis of the chamber resolution and the known geometry, correctly describes the observed distribution over several decades, with very little tail. The resolution at the origin is nicely described by a Gaussian of width 85 μ. Similar studies with lower momentum tracks in events of the form $e^+e^- \to e^+e^-\mu^+\mu^-$ and $e^+e^- \to e^+e^-e^+e^-$ have been used to confirm that the multiple scattering contribution to the resolution is 95 μ/P (GeV/c). Impact parameter errors arising from systematic errors in wire placement or momentum errors are negligible in this chamber.

D. Physics Results

The Mark II vertex chamber, and the other detectors of comparable resolution at e^+e^- machines, have produced the first measurements of the tau and bottom hadron lifetimes, confirmed earlier measurements of charmed particle lifetimes, and searched for anomalous decay vertices. Resolution of the second generation devices has not permitted reliable charm tagging, although B tagging at efficiencies of a few percent looks feasible.

The tau lifetime has been measured[7] by isolating a clean sample of tau decays into three charged particles, which is accomplished with simple topological cuts at PEP/PETRA energies. The projected decay length is determined from the known decay vertex and the beam collision point. The tau energy is known (1/2 E_{cm}) and the direction is accurately approximated by the direction of the three-charged-pion system, so the proper lifetime can be deduced. At PEP, the tau decay length averages about 650 μ; the typical measurement

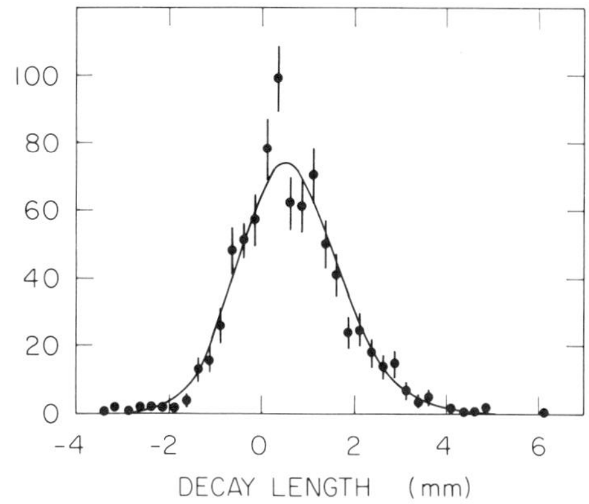

Fig. 6. Decay lengths of tau leptons produced in the reaction $e^+e^- \to \tau^+\tau^-$ at $\sqrt{s} = 29$ GeV.

resolution was 1000 μ. The integrated luminosity of 200 pb^{-1} resulted in about 800 $\tau \to \nu 3\pi(n\pi^0)$ decays, whose decay lengths are plotted in Fig. 6. The measured lifetime was $\tau_\tau = 2.86 \pm 0.16 \pm 0.25 \times 10^{-13}$ sec, in good accord with the theoretical expectation, confirming $e - \mu - \tau$ universality, or in some people's minds, the reliability of a technique with rather modest resolution.

The lifetimes of the D^0 and D^+ mesons have been measured,[8] using much the same technique. The essential step in the measurement was the isolation of a relatively clean sample of mesons. This was accomplished by first identifying D^{*+} mesons (and their charge conjugates). Decays into $D^0\pi^+$ with subsequent D^0 decays into $K^-\pi^+$ and $K^-\pi^+\pi^0$ gave 67 fully reconstructed D^0 mesons. D^{*+} decays into $D^+\pi^0$, with the D^+ decaying into $K^-\pi^+\pi^+$ gave 23 D^+ mesons. Figure 7 shows the measured lifetimes. The observed distributions, which are the convolutions of the exponential decay distributions with our Gaussian-like resolution, are decidedly offset from zero decay length, and just as decidedly nonexponential. The resulting lifetimes, $\tau_{D^0} = 4.7^{+0.9}_{-0.8} \pm 0.5 \times 10^{-13}$ s and $\tau_{D^+} = 8.9^{+3.8}_{-2.7} \pm 1.3 \times 10^{-13}$ s, are in good agreement with other experiments, and overwhelmed by recent advances in statistics and resolution available in fixed target runs.[9]

The discovery[2,10] that the B lifetime is of order 1 picosecond, and not one or two orders of magnitude shorter as had been theoretically expected, underscored the importance of second generation vertex detectors. At PEP and PETRA energies, where the B has sufficient boost to travel a measurable distance, it has so far proved impossible to fully reconstruct B mesons. Instead, experiments have measured the impact parameter of leptons from B semi-leptonic decay, which are

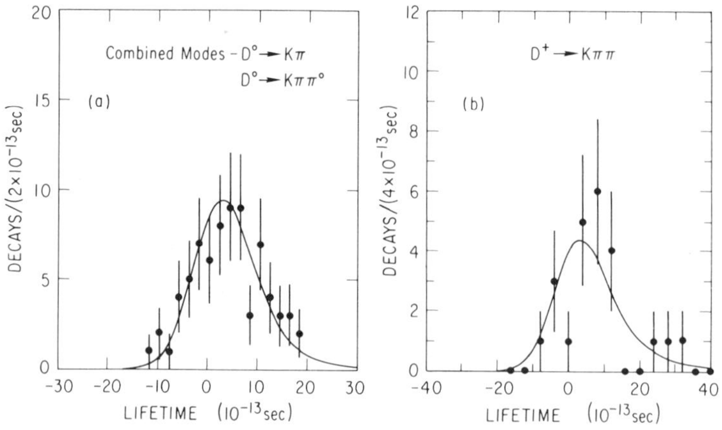

Fig. 7. Proper lifetimes of D^0 and D^+ mesons produced in e^+e^- annihilation at $\sqrt{s} = 29$ GeV.

identified by having large momenta transverse to the hadronic jets in the event. Sample purities are better than 50% B. The impact parameter is measured with respect to the *average* beam position, as shown in Fig. 8. The impact parameter resolution thus depends on both the track error and the components of the vertical and horizontal beam size perpendicular to the track direction. The large ($\sim$ 400 μ) horizontal beam size dominates the resolution for tracks within about 45° of vertical. Figure 9 shows the distribution of signed impact parameters measured in the Mark II experiment.[11] The B lifetime deduced from the distribution was $0.85 \pm 0.17 \pm 0.21 \times 10^{-12}$ sec. A second technique, with more statistical precision but larger systematic uncertainties was also used.

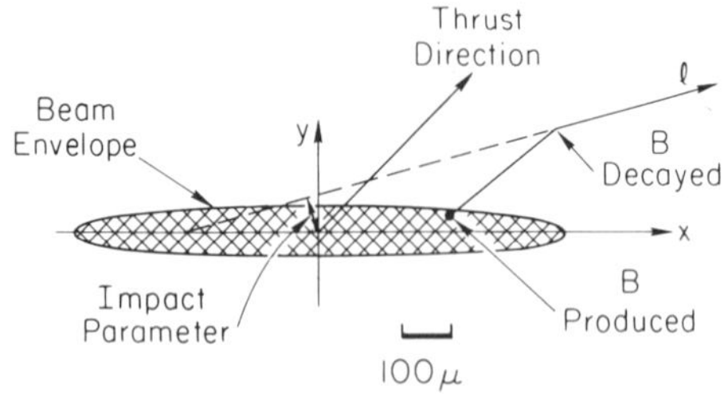

Fig. 8. The projected lepton impact parameter is defined with respect to the average beam position as shown. The thrust direction is used to approximate the B meson direction. The impact parameter is signed positive when the intersection of the lepton direction and the thrust vector corresponds to a positive decay length.

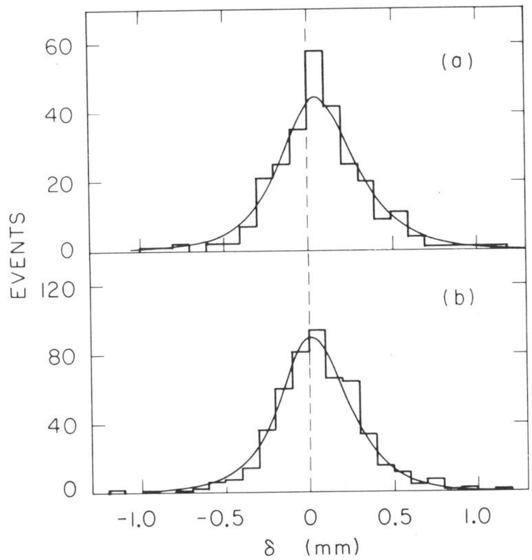

Fig. 9. Lepton impact parameters for (a) high transverse momentum leptons, which come principally from B meson decays; and (b) low transverse momentum leptons, which come primarily from charm decays. The distributions are slightly asymmetric. The mean of the distribution in (a) is $80 \pm 17\ \mu$.

We tagged $e^+e^- \rightarrow b\bar{b}$ events by selecting events with high transverse momentum leptons, and then vertexed all the well-measured tracks in each jet to form a "jet vertex." Although such a "vertex" contains an admixture of tracks from the primary interaction point, the B decay, and subsequent D decay tracks, the resultant decay length does provide a measure of the B lifetime. Since the B and D decays account for about two-thirds of the charged particles per jet, and for most of the high momentum, and consequently low multiply-scattered and high resolution tracks, B lifetime effects dominate the distribution. Figure 10 shows the jet vertex decay lengths; the resulting B lifetime is $1.25^{+0.26}_{-0.19} \pm 0.50 \times 10^{-12}$ s. Both lifetime determinations are in reasonable agreement with other measurements.[12]

Searches for events with anomalous vertex topologies are in their infancy. We have searched,[13] unsuccessfully, for one class of events with a particularly striking signature. The integrated luminosity at PEP, about 200 pb^{-1}, is sufficient to produce about 70 events of the form $e^+e^- \rightarrow \nu\bar{\nu}$ per neutrino generation. We searched for heavy neutrino production, assuming the heavy neutrino would decay sequentially by mixing with the known leptons. The fact that the weak couplings of all leptons are of universal strength argues that the mixing between

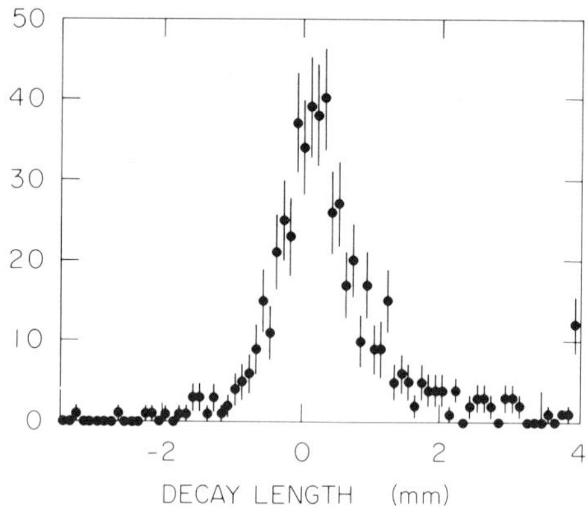

Fig. 10. Jet vertex decay lengths in events with high transverse momentum leptons.

them and any new heavy neutrino are most likely small. Hence the decays might plausibly be inhibited and the resultant decay lengths observable. So we searched for pair-produced objects, one of which decayed between 2 mm and 10 cm of the interaction point in events with no evidence for particle production at the primary interaction point. No events consistent with heavy lepton production were found. Assuming that the heavy neutrino decays like a sequential lepton, and characterizing its coupling to each lepton family in turn by a small Cabibbo-like angle ϵ, the search excludes the existence of heavy neutrinos in certain regions of the mass-mixing angle plane. Figures 11 and 12 show the regions excluded at the 90% Confidence level.

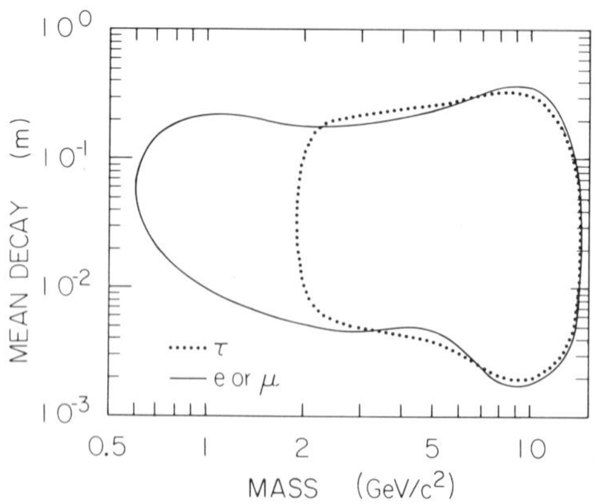

Fig. 11. Heavy neutrinos which mix with e or μ (solid line) or τ (dotted line) are excluded at the 90% confidence level within the contours shown.

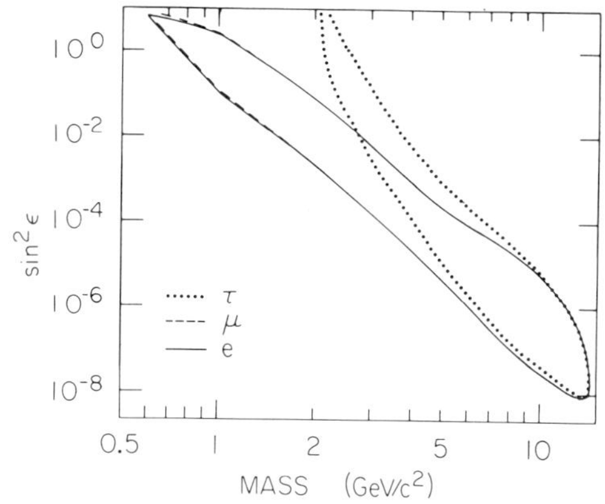

Fig. 12. Excluded regions for heavy neutrinos coupling to e (solid line), μ (dashed line), or τ (dotted line) in terms of the heavy neutrino mass and mixing angle.

A particularly striking event observed in the PEP running is shown in Fig. 13, which shows tracks extrapolated to the vicinity of the primary interaction point, which is indicated by the 1 sigma contour of the beam ellipse. Track number 8 is identified as a muon with 1.1 GeV/c of transverse momentum, tagging the event as a likely example of $e^+e^- \rightarrow b\bar{b}$. Tracks 6, 7, 8, and 14 form a vertex roughly 2 mm from the interaction point. The other tracks in the event group at the opposite end of the beam ellipse. Track 13, having low momentum, has large measurement errors and is compatible with having come from either vertex. The other tracks are measured with roughly 100 μ accuracy. The invariant mass of the tracks which group together in the separated vertex is 4.25 GeV/c^2, well above charm mass, substantiating the hypothesis that this is a long-lived B decay.

E. Update on the B Lifetime Measurement

The measurement of the B lifetime discussed above is being refined by R. Ong of our collaboration. The principal shortcoming of our previous result is that the lepton impact parameter was measured with respect to the average beam position, not the actual production point. As discussed above, this makes the resolution unusually large for vertical tracks. Ong has developed a technique for determining the production point using the other tracks in the event which improves the resolution significantly.

The method is as follows. Each tagged event is divided by a plane perpendicular to the thrust axis to separate particles in one jet from those in the other. A jet

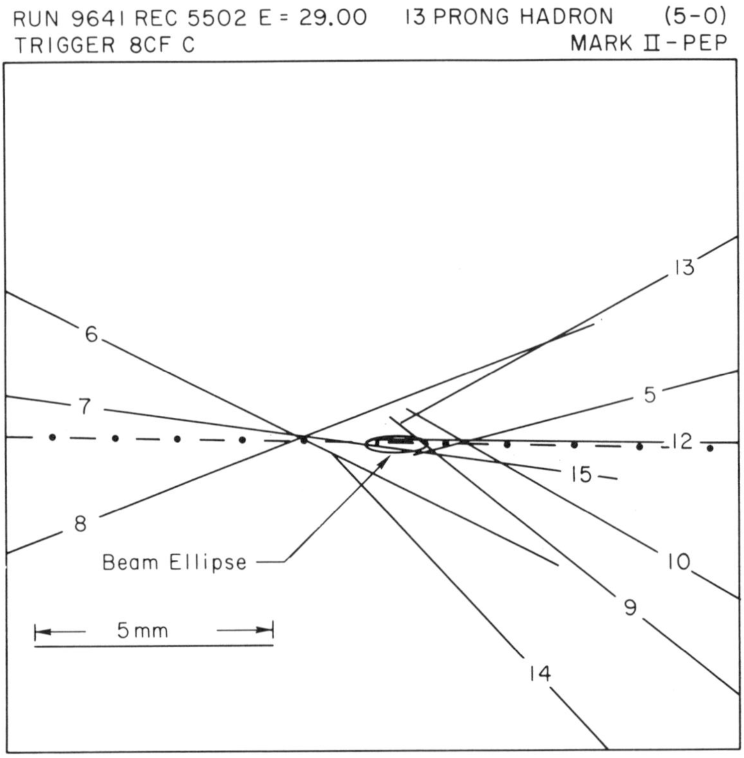

Fig. 13. B decay candidate as seen in the Mark II vertex chamber. The event is described in detail in the text.

vertex is formed from all the well-measured tracks in the jet with momenta above 600 MeV/c, with at least two tracks required for the vertex. The high transverse momentum lepton is included in the vertex. The mean number of tracks per vertex is 3.6. Of course some of the particles come from the primary interaction point, and some from the B and D decays. Generally speaking, the presence of particles from secondary (or tertiary) decays "pulls" the location of the jet vertex along the B direction, (which is well-approximated by the thrust direction) to a point intermediate between the production and decay points. Pulls transverse to the B direction are small, and average to zero. So to find the production point, we take the measured jet vertex, extrapolate back to the beam ellipse along the measured thrust direction, and compute the most likely production point, taking into account the vertex and beam error matrices and uncertainties in the thrust direction. A very thorough simulation with Monte Carlo techniques has shown

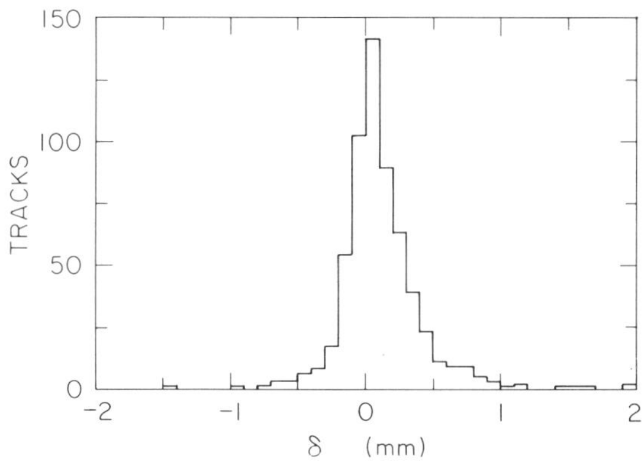

Fig. 14. Lepton impact parameters measured with respect to the primary interaction point.

this estimate of the production point is unbiased for decay lengths below 3 mm, and that the efficiency for finding a jet vertex is constant out to ≤ 4 mm decay lengths. The method significantly improves the resolution and roughly doubles the available statistic. Figure 14 shows the resulting impact parameter distribution for leptons. The mean of the distribution is $110 \pm 11.9\ \mu$, 9.3 standard deviations from zero. The full analysis of this result is still in progress,[14] and will be reported soon. The resulting B lifetime is close to 1 picosecond.

III. THE MARK II DRIFT CHAMBER VERTEX DETECTOR FOR THE SLC

The Mark II Collaboration is planning to install a drift chamber vertex detector for its 1987 running at the SLC, and to add a microstrip detector to the apparatus the following year. The silicon device is intended to have three layers of detectors, with 12 detectors per layer, and will use 4 microplex readout chips per layer. It is discussed in detail in Ref. 4. The following describes the new drift chamber in some detail.

Figure 15 shows a cut-away drawing of the chamber. It consists of ten pie-shaped jet cells which extend radially between 5 and 17 cm. There are 38 active sense wires in each jet-cell, spaced 2.90 mm apart. The chamber is 55 cm long, and supported by its 1.3 mm thick Beryllium core of radius 4.5 cm, and an outer aluminum shell. It has dual endplates, the first for accurate wire positioning, the second a pressure head. The chamber will operate between 2 and 3 atm pressure with a "slow" gas, i.e., one with unsaturated drift velocity. The drift field will be

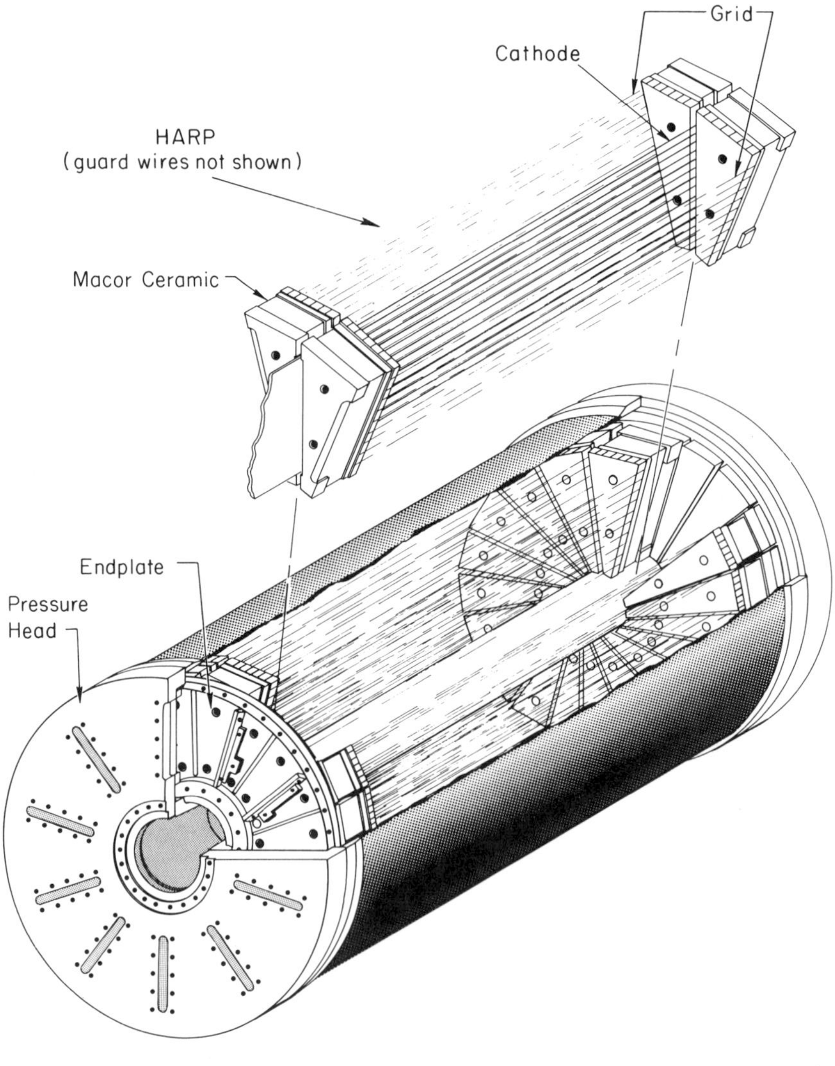

Fig. 15. Cut-away drawing of the Mark II drift chamber vertex detector.

approximately 1 kV/cm atm, and the drift velocity between 5 and 10 μ/ns. The use of a slow gas has several advantages. First, since diffusion remains at the thermal limit in slow gases, the resolution is very good. The accuracy per measurement will be 30 – 40 μ, averaged over 2 cm of drift. Secondly, since the time between pulses from closely spaced tracks will be more than five times greater than in a fast gas, the electronic problems of resolving such tracks and timing them to the required accuracy become tractable. The principal disadvantage of using slow gases is that the electron drift velocity, and thus the chamber's final calibration, depends on gas composition and density and the magnitude of the electric field. Special care is required to ensure stable chamber operation. The chamber is being constructed by Mark II collaborators from LBL, SLAC, and the University of Colorado, and visitors from KEK.[15]

The SLC environment is ideal for precision vertex detection. The very small beam size will confine the primary interaction to a circle with a radius of a few microns, orders of magnitude smaller than the beam spot in a storage ring. The small beam phase space allows the beam pipe radius to be pushed within 2.5 cm of the beams, thus minimizing the inevitable multiple coulomb scattering. And the repetition rate is sufficiently low that ions can be cleared before the next pulse, eliminating the possibility of space charge effects in the chamber.

A. Cell Design

Our design is a variant of the traditional jet cell,[16] incorporating planes of field shaping wires, called grid planes, on either side of a central plane of alternating anode and potential wires. The design is shown in Fig. 16 along with the field lines. The anode and potential wires are spaced 1.45 mm apart; the grid wires lie directly above the potential wires and are spaced 1.80 mm from them. The anode/potential wire spacing is basically dictated by the desired track-pair resolution. The operating voltages in 3 atm of CO_2/isobutane (92/8) are rather high. The sense wire is held at +3.191 kV, the potential wires at ground, and the grid wires at −0.504 kV. The grid wires serve two functions. They improve the track-pair resolution by "focusing" the electrons onto the anode wire. This minimizes differences in the length of the electron drift trajectories. It also minimizes the variation in field strength, and consequently drift velocity, along the different drift paths. The goal, of course, is isochronous collection. Figure 17 shows the equal-time contours for the cell, plotted on a distance scale appropriate to saturated and unsaturated drift velocities. Ninety-five percent of the charge is collected within about 10 ns in a saturated gas, translating into a track-pair resolution of 500 μ. In an unsaturated gas, the fact that the longest trajectories have the lowest electric fields, worsens this to about 800 μ.

The presence of four independent electrodes (anode, potential wire, grid, and cathode) allows the gain and drift field to be specified, and the cell "tune" to be optimized for isochrony or sense wire electrostatic stability. In practice our

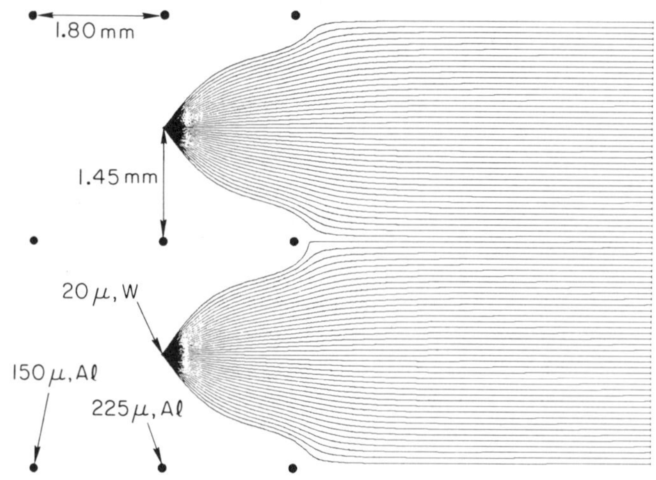

Fig. 16. Detail of the drift cell showing the "focusing" of field lines from the cathode. Tracks which enter the cell are approximately perpendicular to the field lines.

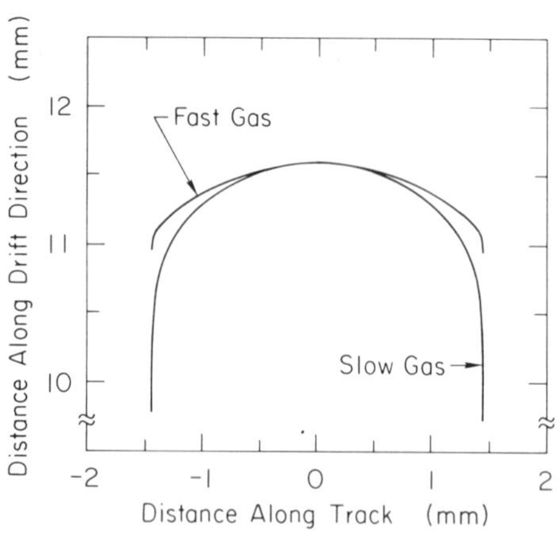

Fig. 17. Electrons deposited along the contours shown are collected isochronously. The offset in the drift distance is arbitrary. The width of the contour is fixed by the sense wire spacing of 2.9 mm.

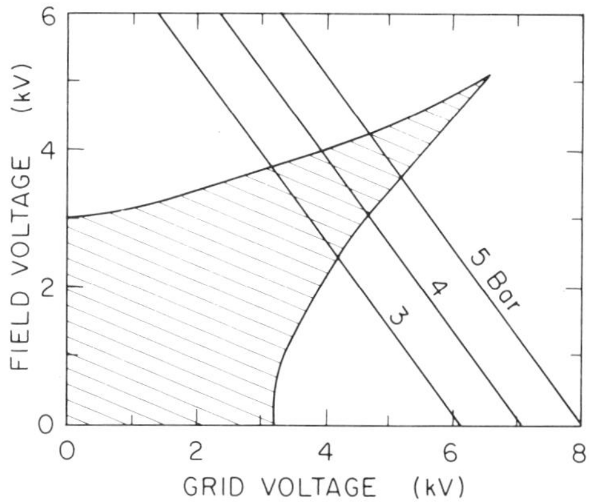

Fig. 18. The sense wire is electrostatically stable for the range of potential wire and grid voltages shown. The chamber operating points in 3, 4, or 5 bar CO_2/C_4H_{10} (92/8) are shown as lines of constant gain.

operating voltages are nearly optimal for both stability and isochrony, so the presence of the grid wires markedly improves the electrostatic stability of the sense wire. Consequently, the chamber could be run at higher gas pressures (with their higher operating voltages), or the wire length could be increased. The range of grid and potential wire voltages (relative to the anode voltage) for which the chamber is stable is shown in Fig. 18.

B. Prototype Results

The performance of this cell has been measured[17] in a prototype chamber with 22 sense wires, the central 8 of which were instrumented. The anode/potential wire spacing was 2.0 mm (not 1.45 mm as in our final chamber), and the cathodes were planar, and 3 cm distant from the grid plane. Resolutions were measured by studying the residuals to straight-line fits through the eight drift cells. All the measurements were made with cosmic rays. Figure 19 shows the observed resolution as a function of drift distance in 3 atm of CO_2/isobutane (92/8) and in argon/ethane (50/50). Drift times were measured by threshold discriminating the outputs of fast amplifiers. The resolutions in the slow gas range from below 20 μ, just outside the grid, to over 40 μ at 2.5 cm drift distance and are proportional to the square root of the drift distance as one expects when diffusion dominates the

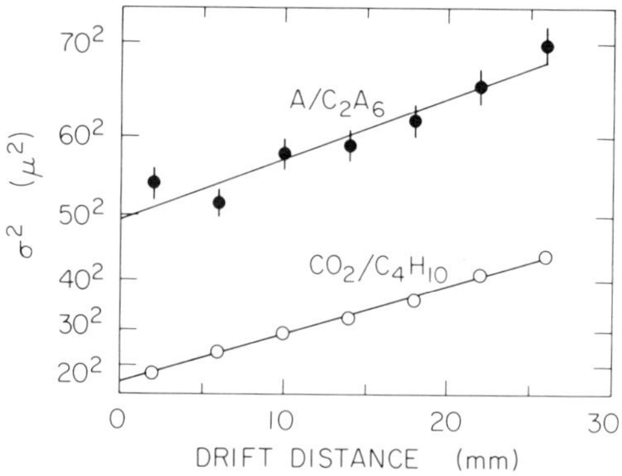

Fig. 19. Resolution squared vs. drift distance at 3 bar pressure in A/C_2H_6 (50/50) and CO_2/C_4H_{10} (92/8) measured in the prototype chamber.

resolution. Surprisingly, the diffusion term in argon/ethane is nearly identical, but the resolution is considerably worse, ranging from 50 to 70 μ.

The resolution, averaged over 2 cm of drift length, is plotted as a function of pressure in Fig. 20. Approximately $\sigma(p) \propto p^{-.7}$. The bulk of this dependence presumably comes from the dependence of the diffusion term on pressure ($\propto p^{-.5}$), the remainder from the improved electron statistics. Figure 21 shows there is little dependence of the average resolution on E/p. In fact, the diffusion contribution decreases as the drift field is increased, but this decrease is compensated by worsening resolution in the grid-anode gap. This result may well be dependent on the type of gas used.

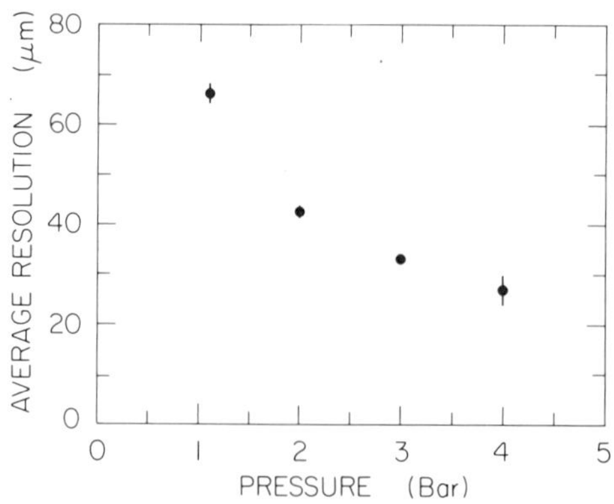

Fig. 20. Dependence of the average resolution on operating pressure.

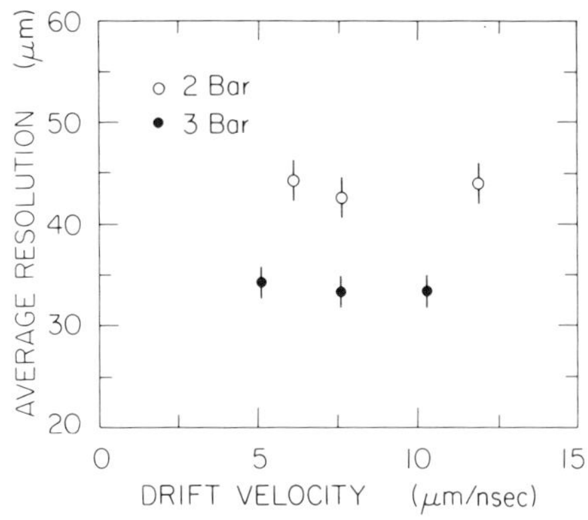

Fig. 21. Dependence of the average resolution on drift velocity at 2 and 3 bar pressures.

Two studies have confirmed that the track-pair resolution is as expected. In the first, we used 100 MHz waveform digitizers to study the average pulse shape for a minimum ionizing particle. It is shown in Fig. 22, where the drift velocity was 7.5 μ/ns. The "error bars" represent the rms of the average pulse height in a single time bin, and show the presence of large fluctuations in the tail of the pulse. Little signal is seen beyond 150 ns, which corresponds to about 1200 μ. The second study used the familiar technique of superposing two independent waveforms at a known separation, and using a simple pattern recognition algorithm to detect the presence of the second peak. Any such algorithm will mistake late arriving ionization in the tail of the first pulse for a second hit;

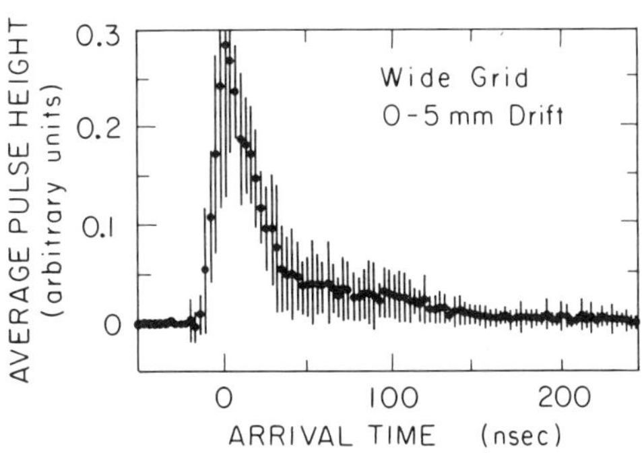

Fig. 22. Average pulse waveform in the prototype chamber at 3 bars pressure. The drift velocity was 7.5 μ/ns. The error bars indicate the root-mean-square of the distribution of pulse heights in any bin.

thus an algorithm has associated with it some "fake rate." Figure 23 shows that a legitimate second pulse is identified with nearly full efficiency beyond 1200 μ in the prototype chamber. The track-pair capability in the actual chamber should be better by the ratio of sense-field spacings, i.e., (1.45/2) (1200) = 900 μ. Of course a reliable evaluation of the track-pair resolution awaits studies in dense hadronic jets.

C. Field Cage Design

The arrangement of wire planes in a jet cell is shown in Fig. 24. The jet cell consists of the grid-anode-grid sandwich, cathode wire planes on either side of it, field shaping wires at the inner and outer radii, which help minimize the end effects, and electrode surfaces attached to the inner Beryllium core and the outer aluminum skins. The cell is tilted with respect to the radial direction by slightly more than 15° as measured at the inner wire radius. This feature solves the left-right ambiguity, provides self-calibrating constraints from tracks which pass through cell-boundaries, and provides relatively uniform response for tracks at all azimuth. The cathode wires are graded in potential between about –5 kV and –15 kV in order to create a very uniform electric field throughout much of the cell volume. Potentials on the inner and outer guard wires and on the other electrode surfaces are shown to minimize the end effects. The resulting field is uniform within ± 0.1% over more than two-thirds of the cell. Potentials on the inner and outer HV "skins" are near –2 kV and –10 kV, respectively.

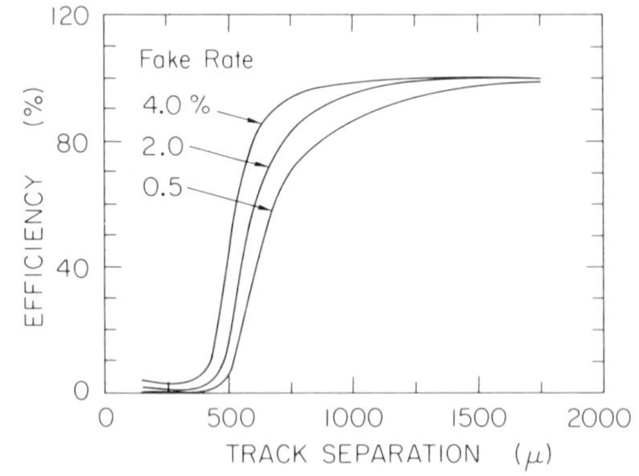

Fig. 23. Efficiency for detecting a second track as a function of its separation from the first.

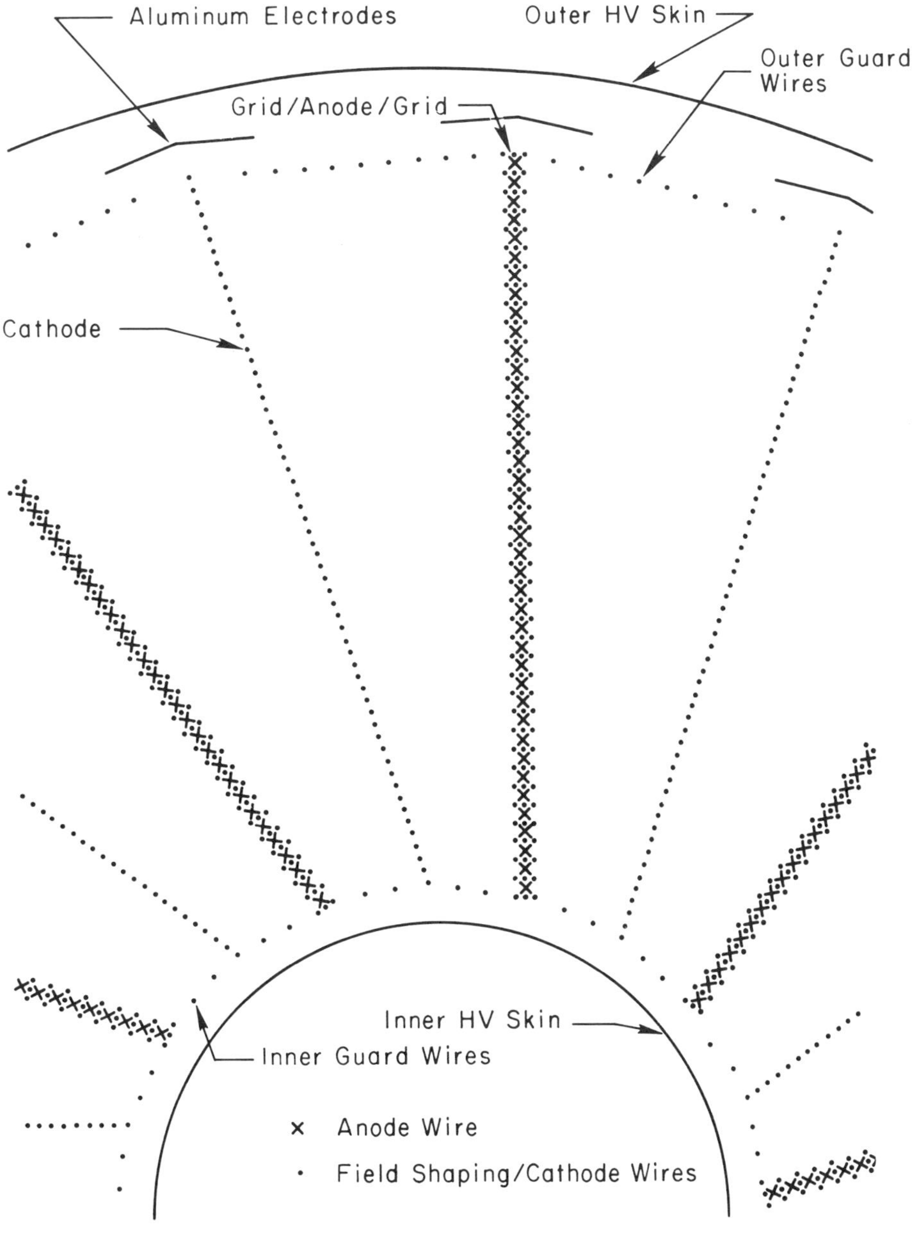

Fig. 24. Wire and electrode placement in the jet cells.

End effects are of course worst at the outer radius; to help minimize them aluminum electrodes are positioned over cathode and anode planes, and held near −15 kV and −2 kV, respectively.

D. Mechanical Construction

To achieve good spatial accuracy in a drift chamber which utilizes a non-saturated drift velocity requires first a device which is mechanically and electrostatically stable, and second, a device which can be calibrated. Since the full calibration of such a device requires accurate knowledge of the positions of each anode wire and to a lesser extent the cathode and grid wires too, we have adopted a technique for building the chamber which seeks to minimize the number of variables needed to characterize all the wire locations.

Wires are strung, a plane at a time, on very high precision grooved bars, which locate the wires with an accuracy better than 3 μ. The entire plane of wires is attached to a ceramic foundation with epoxy, the wires "floating" $\sim$ 100 μ above the ceramic, and captured in glue. Thus instead of dealing with the azimuthal and radial coordinates of each wire at each end of the chamber, we need to know only the two coordinates which fix the positions of the innermost and outermost wires in each plane.

The accuracy of the wire placement on the precision bars is shown in Fig. 25, where both optical and capacitive measurements have been used for the wire position measurements. The figure shows deviations from the expected wire positions in the wire plane. The wire positions are in general accurate to the 2 μ level, considerably better than is needed.

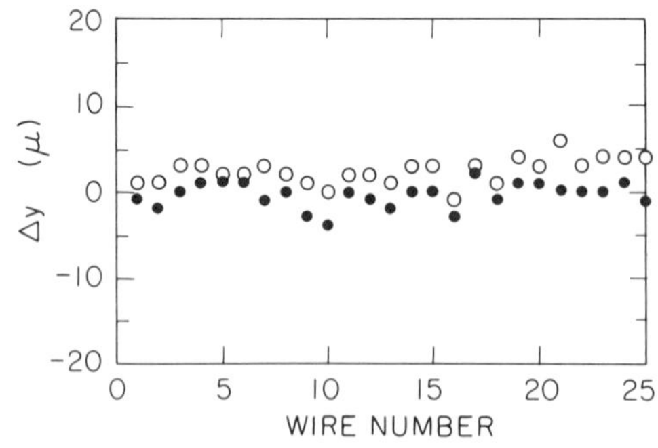

Fig. 25. Deviations of wires from their nominal positions in the plane of the wires. The open (closed) circles correspond to measurements made with a capacitive probe (optical microscope).

Fig. 26. Cathode plane assembly showing ceramic founcations, kapton circuitry, and pressure connector.

The chamber construction is illustrated in Fig. 26. The jet cells are modular, made up of the ceramic foundation and wire planes mentioned above. They include flexible kapton printed circuits to bus all the signal and cathode wires past the positioning endplate and through a pressure feedthrough to amplifiers or high voltage resistor chains. These are mounted outside the chamber gas volume in order to isolate temperature generating components from the gas. A small mechanical mockup of one sector of the chamber is shown in Fig. 27. The anode wires are held on a small stainless steel foundation which butts against the endplate behind the ceramic foundations.

Kapton printed circuitry with factory installed kapton overlay has proved an effective way to bus large numbers of high voltage traces. The circuits typically

Fig. 27. Mechanical mockup of the DCVD showing two grid planes sandwiching the anode plane, guard wires, positioning endplates, and pressure connectors.

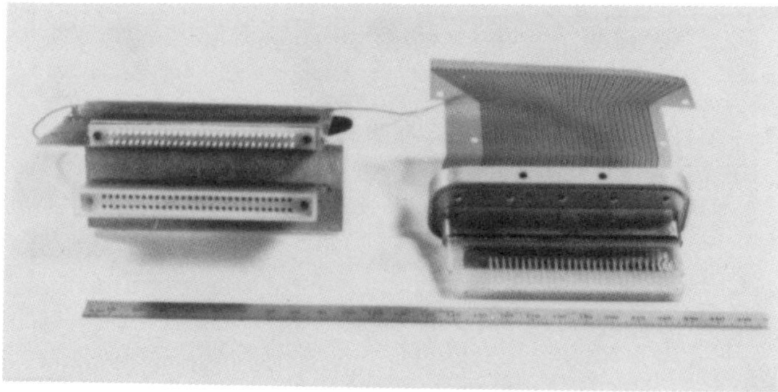

Fig. 28. At left, multipin high voltage connector; at right, high voltage feedthrough consisting of kapton circuitry, potted into an aluminum pressure flange and connected to a high voltage connector.

have trace-to-trace resistances above $10^{13}\ \Omega$ and can handle potential differences of several kilovolts across spacings of ~ 1 mm. Figure 28 shows the kapton circuitry and the pressure and high voltage connections. The high voltage connector holds off voltages of 10 kV between adjacent pins with leakage currents well below 1 nA. It can be run at voltages up to 16 kV with discharge rates well below 0.1 Hz.

Establishing the wire positions in the chamber is a two-part operation based on a system for reproducibly locating the ceramic foundations on smooth surfaces. Our system uses two "sockets" in each of the mating parts. The sockets are simply cylindrical pieces of steel into which a conical surface has been machined. (See Fig. 29.) When a precision ball bearing is captured between the mating sockets, and the sockets of one piece have been properly matched to those in the other, the parts will mate reproducibly to the micron level. A master template, also shown in the figure, defines the relationship between the pairs of sockets which fix the anode, cathode, and grid plane locations. Wire location proceeds as follows. First, the wire planes are accurately positioned with respect to the socket locations on the ceramic foundations and then secured in place with epoxy. This is accomplished using a large, high precision measuring microscope. Second, the ceramic foundations are transferred to the chamber endplates, where a pattern of mating sockets has been established by replicating the master template pattern ten times, once for each jet cell. This is done using a dividing head to fix the angle of rotation between one cell and the next. We hope to place wires in the chamber to within 10 μ of their nominal positions.

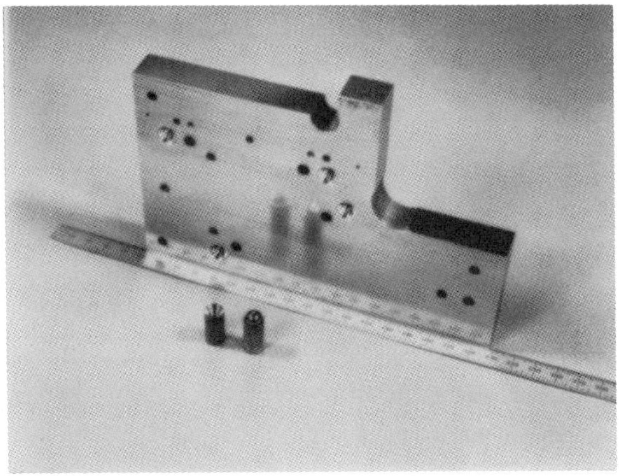

Fig. 29. Master locating template, showing conical sockets and ball bearings.

One needs to understand wire positions in the chamber to the level of 5 to 10 μ to exploit fully the resolution inherent in the gas. This follows from the fact that the average resolutions per wire is about 30 μ, and roughly 12 wires will measure a common feature of the track, so the potential spatial precision will be about $30\,\mu/\sqrt{12} \sim 10\mu$.

E. Electronics

The readout electronics consists of a preamplifier mounted on the chamber, a postamplifier and shaper, and a wave form digitizing system. The sense wires, which are run at positive high voltage, couple capacitively through a 100 pF capacitor to a hybrid preamp designed by Radeka and Stefani.[18] It has a 3 ns rise time, 40 ns fall time, and a noise level of 150 μV, and is mounted directly outside the chamber gas volume. The preamp output is carried by about 15 feet of coax to a postamp mounted on the outside of the Mark II detector. It is similar in design to the postamp[19] developed for the Mark II Central Tracking Chamber, and includes shaping circuits to cancel the preamp fall time and the 1/t tail in the pulse. Its output runs via twisted pair to a modified version of the Mark II dE/dx flash-ADC module.[20] The new module incorporates a new fast memory chip, the Hitachi 10474-10, which extends the memory at 100 MHz to 10 μsec. The vertex detector FADC module is packaged 16 channels per single width Fastbus module, and uses a 6-bit flash ADC, the TRW–1029. The modules are read out with the SLAC Scanner Processor.[21]

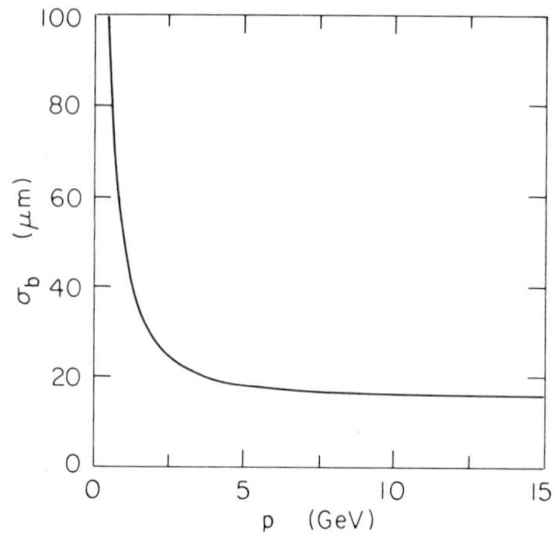

Fig. 30. Predicted impact parameter resolution as a function of track momentum.

F. Expected Performance

Figure 30 shows the impact parameter resolution of the new vertex detector used in conjunction with the Mark II central tracking chamber. The multiple scattering contribution is something worse than 60 μ/P (GeV/c), and the ultimate resolution at high momentum is about 15 μ. Studies of B-tagging at the SLC[22] show that roughly 33% of B decays can be cleanly tagged.

Hadronic jets from Z^0 decays are dense with charged particles, so adequate track-pair resolution is essential for successful vertex detection. One loses 10% of tracks when the angular segmentation is ± 10 mr which corresponds to a track-pair separation of ± 500 μ at the innermost radius of our chamber, r = 5 cm. As discussed above, the track-pair resolution achievable with our chamber is about 1000 μ. The fact that the jet cells are tilted effectively halves this number, however. In a chamber where the anode plane was radial, a given track could be obscured by tracks on either side of the anode plane which were collected slightly before the track in question. In a tilted cell (see Fig. 31), a track can be obscured by another track on the same side of the anode plane which has a slightly smaller drift time, but tracks which passed on the opposite side of the anode will appear to intersect the original track at sizable angles and obscure only a small part of it. Figure 31b shows the tracks which can be reconstructed after track-pair resolution effects are taken into account. A Monte Carlo simulation has shown that the B tagging efficiency in our chamber is about 10% lower than in a device with perfect track-pair resolution.

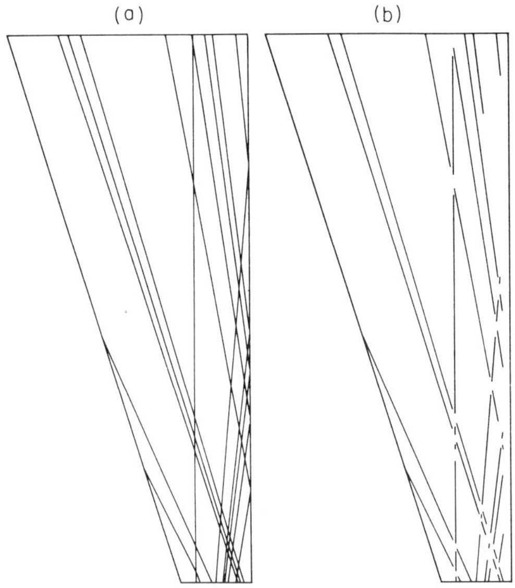

Fig. 31. Monte Carlo simulation of tracks passing through a jet cell with (a) perfect track-pair resolution; (b) 1500 μ track-pair resolution. See text for details.

IV. CONCLUSIONS

The art of detecting the decay vertices from heavy quarks and leptons is comparatively new at electron-positron storage rings. So far, drift chambers positioned just outside the vacuum pipes which surround the interaction region have provided the first accurate determinations of the tau and bottom lifetimes, and confirmed earlier measurements of charmed particle lifetimes. "Second generation" vertex detectors have demonstrated the feasibility of tagging heavy flavors by observing decay vertices, and are being used to search for anomalous decay topologies. These chambers have modest resolution on the scale of the effects they seek to measure, but are now well-understood and reliable tools.

A new generation of vertex detectors, considerably more ambitious, is under construction for experiments at SLC and LEP. They boast impact parameter resolution improved by a factor of four or more over previous detectors, and sub-millimeter track-pair resolution. The Mark II collaboration hopes to reach these goals with a high pressure precision drift chamber, and eventually surpass them with the addition of a silicon microstrip detector.

REFERENCES

1. G. J. Feldman *et al.*, Phys. Rev. Lett. **48**, 66 (1982).
2. E. Fernandez *et al.*, Phys. Rev. Lett. **51**, 1022 (1983).
3. See, for example, M. G. D. Gilchriese and D. M. Ritson in Proceedings of the 23rd International Conference on High Energy Physics, Berkeley, July 16 – 23, 1986.
4. A. Litke, in Proceedings of the International Conference on Advances in Experimental Methods for Colliding Beam Physics, Stanford, California, W. Kirk, editor, to be published in Nucl. Instr. and Meth.
5. J. A. Jaros, in Proceedings of the International Conference on Instrumentation for Colliding Beam Physics, Stanford, California, 1982, edited by W. Ash (SLAC Report No. 250).
6. D. Amidei, D. Baden, L. Gladney, L. Golding, J. Jaros, R. Ong, G. Trilling.
7. D. Amidei *et al.*, to be published in Phys. Rev. D. Preliminary results are given in J. A. Jaros, in Proceedings of the International Conference on Physics in Collision IV, Santa Cruz, California, Aug. 22 – 24, 1984, edited by A. Seiden, Editions Frontieres. First results were reported in J. A. Jaros *et al.*, Phys. Rev. Lett. **51**, 955 (1983).
8. L. Gladney *et al.*, Phys. Rev. D **34**, 2601 (1986).
9. J. C. Anjos *et al.*, Phys. Rev. Lett. **58**, 311 (1987).
10. N. S. Lockyer *et al.*, Phys. Rev. Lett. **51**, 1316 (1983).
11. J. A. Jaros, in Proceedings of the International Conference on Physics in Collision IV, Santa Cruz, California, August 22 – 24, 1984, edited by A. Seiden, Editions Frontieres.
12. See, for example, D. M. Ritson in Proceedings of the 23rd International Conference on High Energy Physics, Berkeley, July 16 – 23, 1986.
13. C. Wendt *et al.*, Phys. Rev. Lett. **58**, 1810 (1987).
14. R. Ong *et al.*, to be published in Phys. Rev. D.
15. J. Alexander, R. Dennis, P. Drell, W. Ford, R. Fuzesy, R. Harr, K. Hayes, D. Hinshaw, C. Hoard, D. Hutchinson, J. Jaros, D. Koetke, S. Odaka, T. Ohama, R. Ong, M. Perl, J. Smith, G. Trilling, S. Wagner, P. Weber, S. White.
16. W. Farr *et al.*, Nucl. Instr. and Meth. **156**, 283 (1978).
17. J. Alexander *et al.*, Nucl. Instr. and Meth. **A252**, 350 (1986).

18. BNL Hybrid IO-354.
19. D. Briggs *et al.*, IEEE Trans. **NS-32**, 653 (1985).
20. D. Bernstein *et al.*, IEEE **NS-33**, 86 (1986).
21. H. Brafman *et al.*, IEEE Trans. **NS-32**, 336 (1985).
22. K. Hayes, Mark II/SLC Note #73 (unpublished internal memorandum).

Discussion J. JAROS

J. KIRKBY

A trivial question: why it is planned to delay the Si vertex detector relative to the drift chamber vertex detector until a year after startup of MARK II at SLC? It is clear that the Si has better performance than the drift chamber: a measurement improvement on the impact parameter of between 2 and 4 and also a factor of 10 improvement in 2 track resolution. Moreover we now have experiments which have successfully used Si microstrip trackers and a VLSI (microplex) which works.

J. JAROS

The improvement coming from addition of silicon is much less than you assert. For most tracks, which have their resolution limited by multiple scattering, the resolution improves by about 50%, as do the B tagging efficiencies. The two – track resolution of the silicon device is much better, but the drift chamber will measure between 90 and 95% of tracks with its poorer segmentation; i.e. the drift chamber two – track resolution is adequate for the physics. We have chosen to install both devices for several reasons. First, it was not possible to have a silicon detector ready at the time of SLC turn on. Second, the silicon device, in order to achieve higher impact parameter resolution than the drift chamber, must reside at much smaller radius than the drift chamber where backgrounds may be dangerously high. Third, although some of the silicon components had been tested at the time of our decision, up to now no one has successfully built a large solid angle, compact, and mechanically precise silicon detector. It is a challenging project with real physics potential, but not a conservative choice for an experiment that hopes to beat the world to Z^0 discoveries. Finally, a stand – alone silicon system should have more layers to ensure eough redundancy for unambignous track matching with the central detector. Three layers is already an ambitious project. One may also ask, why then include a silicon device in MARK II at all. It provides some insurance against the prospect that the drift chamber fails to perform up to specifications, it may improve resolution and it offers some useful experience in a technique that becomes more powerful as the beam pipe radius shrinks.

G. CHARPAK

Could you say a few words on the reliability problems you have had with your drift chambers?

J. JAROS

Our problems with the PEP vertex detector were two – fold. First, we have intented to operate the chamber at 2 bars pressure. Leaky feedthrus – a bad design for a pressure vessel – prevented this resulting in an average chamber resolution of 100 μm instead of about 70 μm. Given, however, the importance of multiple scattering in PEP physics, this degradation in chamber performance was hardly noticed in physics performance. Our real reliability problems stemmed from running at PEP with high backgrounds. One of the five sections of the inner wire layers, which drew 1/2 μA/wire during beam operations, developed "glow" problems: i.e. the chamber would hold voltage if the current draw were small, but any appreciable ionization load would cause a rapid excalation in chamber currents, and a consequent high voltage trip. We ran with lowered voltages and lowered thresholds for some time, and eventually turned off the bad chamber section. It was fully recovered only a year later after we tried – at Atac's suggestion – bubbling our gas (argon/ethane 50/50) through ethanol. This worked an instant cure on the "sick" section of the chamber. We ran the next one and 1/2 years without major difficulties, although we kept gain in the "injured" chamber section lower than the rest of the chamber. We had no problems with wire breakage, but had some recurrent problems with remnants of sense wire left in the chamber at the time of construction. We finally succeeded in removing the bad wire pieces. A poor design for cable relief resulted in losing about 3% of our signal cables.

F. PIUZ

At what threshold expressed in mean number of primary electrons will you operate the chamber?

J. JAROS

Let me answer as best I can. Our prototype studies, which used threshold discriminations and TDC's, were typically run with thresholds of about 1/20 the average peak pulse height. Assuming 90 primaries/cm (at 3 bar), and a collection region of about 1/5 cm contributing to the peak pulse height would put the threshold at about the 1 primary (i.e.cluster) level. We will run our SLC vertex detector with FADC readout and run the chamber at a gain of $\sim 5 \times 10^4$.

THE MARK-J VERTEX DETECTOR

H. Anderhub[6], H. Anders[8], S. Ansari[7], A. Boehm[1], M. Bourquin[5], J. Burger[4], M. Chen[4], V. Commichau[1], M. Deutschmann[1], M. Dhina[4], K.J. Draheim[1], J. Fehlmann[6], D. Fong[4], W. Friebel[10], U. Gessner[1], K. Hangarter[1], R. Hausammann[5], U. Herten[1], H. Hofer[6], K. Hofmann[1], M.M. Ilyas[4], B. Krause[1], R. Leiste[10], Q.Z. Li[6], D. Linnhoefer[1], X. Lue[6], S. Masson[1], M. Nusbaumer[5], H. Nierobisch[1], W.D. Nowak[10], M. Pohl[6], R.R. Rau[2], D. Ren[6], E. Roderburg[9], M. Rohde[3], H.J. Schreiber[10], J. Schug[1], B. Stoehr[6], S.C.C. Ting[4], M. Tonutti[1], G. Viertel[6], S.X. Wu[4], M.F. Wyne[7], and B. Wyslouch[4]

Presented by M. Bourquin

1) R.W.T.H., III Physik Inst., 5100 Aachen, Fed. Rep. of Germany
2) Brookhaven Nat. Lab., Upton, NY 11973 USA
3) DESY, 2000 Hamburg, Fed. Rep. of Germany
4) Massachusetts Inst. of Technology, Cambridge, MA 02139 USA
5) Univ. of Geneva, 1211 Geneva, Switzerland
6) Eidg. Techn. Hochschule, 8093 Zurich, Switzerland
7) Pakistan Atomic Energy Commission, Islamabad, Pakistan
8) CERN, 1211 Geneva, Switzerland
9) Univ. Gesamthochschule Siegen, 5900 Siegen, Fed. Rep. Germany
10) AdW der DDR, 1615 Zeuthen, German Democratic Republic

Abstract

A high-resolution drift chamber based on the time expansion principle has been built as a vertex detector for the Mark-J experiment at DESY. Results on the chamber performance obtained from test beam measurements and first results from running at PETRA are reported.

INTRODUCTION

The Mark-J vertex detector consists of a new type of cylindrical drift chamber based on the time expansion principle (TEC) [1]. It has been built in order to gain experience running at an e^+e^- collider in view of the L3 experiment at LEP [2], where a TEC has been proposed as the central tracking chamber. The following criteria have to be fulfilled by the L3 chamber :

- a length in the beam direction of 1000 mm
- a diameter of 1000 mm
- a spatial resolution of 30 microns for 1 cm drift length
- a double-track resolution of 600 microns.

The TEC aims at reaching these high resolutions by means of the following features :

- low-field drift regions (drift velocity around 5 microns per nsec) separated from high-field detection regions
- analog shaping of the anode signals to cancel the ion tails
- 100 MHz flash-ADC to digitize the anode pulse shapes in order to determine the drift times by a center of gravity method
- drift chamber gas with lowest possible diffusion (cool gas) and small Lorentz angle.

First prototype tests with planar chambers in 1983 [3] showed promising results with single and double-track resolutions of around 30 and 300 microns, respectively.

The MARK-J TEC is a cylindrical chamber with an outer diameter of 242 mm (Fig. 1). It has already been described in detail in ref. [4] and [5]. Its main parameters are given in table 1. The chamber is built on the fully supporting beam pipe (diameter 100 mm) which consists of 1.4 mm carbon fiber-epoxy with a 0.2 mm thick inner coating of stainless steel. Two circular end-plates are glued on the pipe at a distance of 600 mm. The sensitive volume of the chamber is divided into 12 drift segments with a 4 mm detection gap in the center, inclined by 4° in order to resolve the left-right ambiguities.

The chamber is operated with a gas mixture of 80% CO_2 and 20% i-C_4H_{10}, at a pressure of 1.9 bar, which has the required low longitudinal diffusion. The gas system is schematically shown on Fig. 2. The gas in the closed-loop circuit is regularly pumped through "OXYSORB" in bypass to allow a maximum flow of 60 liters per minute. This keeps the oxygen impurities below 0.1 ppm and the water content to less than 1 ppm.

Table 1: Parameters of the MARK J vertex chamber

inner diameter	100 mm
outer diameter	242 mm
sensitive wire length	576 mm
number of segments	12
number of sense wires per segment	14
total number of sense wires	168
spacing between anodes	2.4 mm
spacing between potential wires	2.54 mm
spacing between grid wires	0.6 mm
gas mixture	CO_2-i-C_4H_{10} (80 %-20 %)
pressure	1.9 bar
drift velocity	7 micron/ns

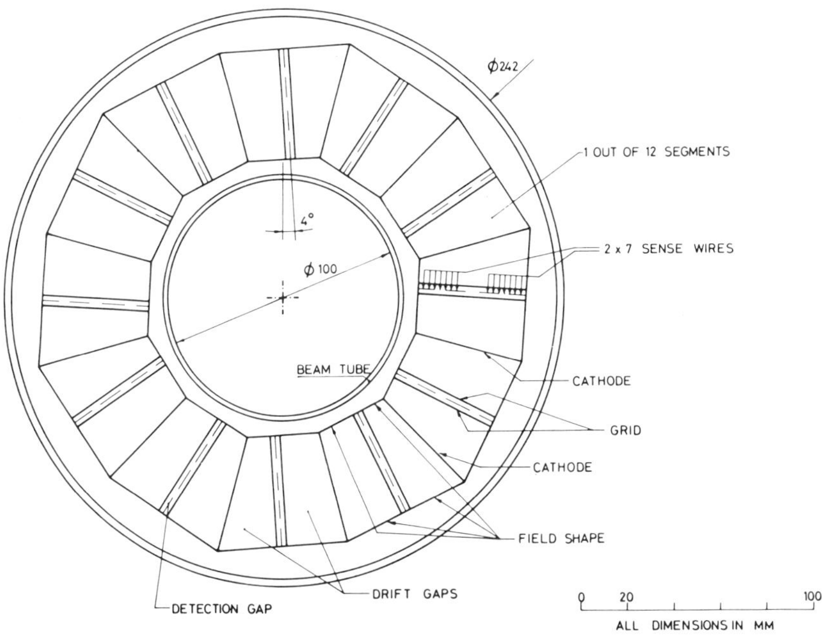

Fig. 1 MARK-J Vertex Chamber (Cross-section)

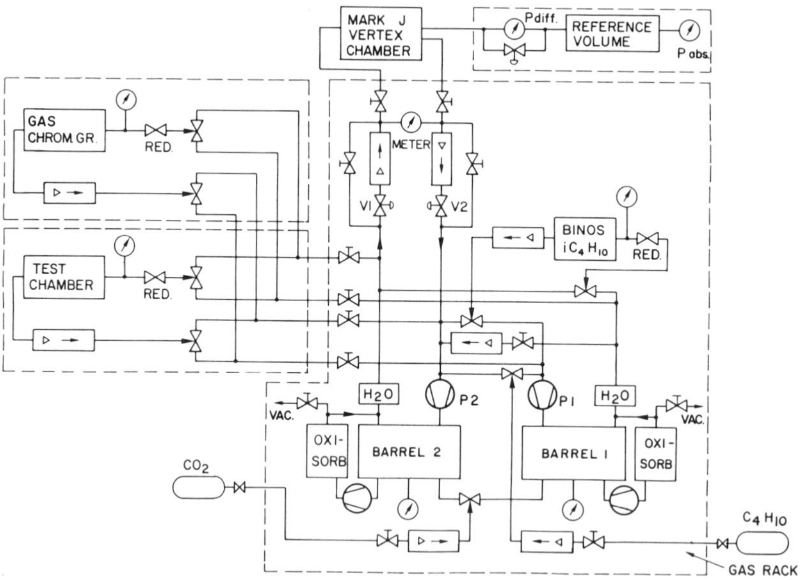

Fig. 2 Closed-loop gas system

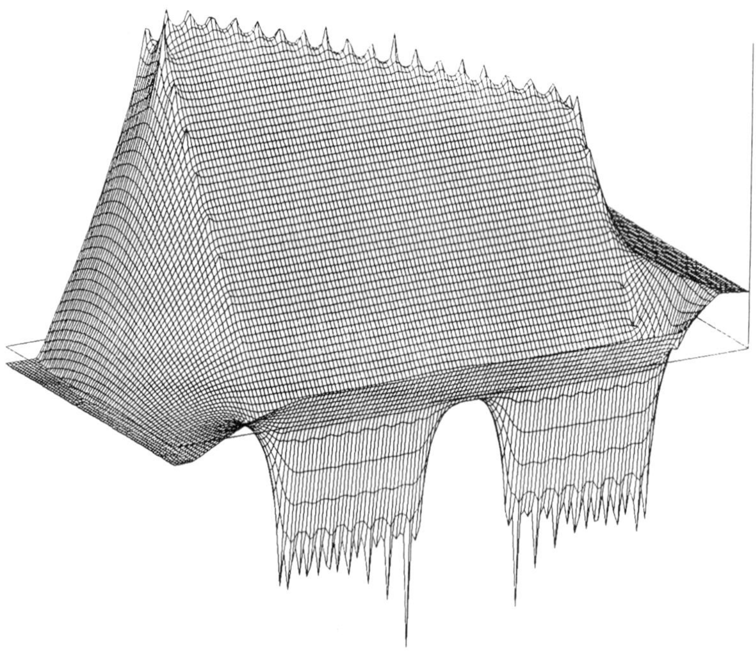

Fig. 3 Computed potential and field lines in a half segment. The detection gap is in the foreground

The high voltage on cathode, field shaping, grid and anode wires applied to produce a drift field as homogeneous as possible, has been optimized by a computer program [6]. Fig. 3 represents the computed potential and field lines.

The anode wires are capacitively coupled to fast, current-sensitive preamplifiers of the MSD-2 type [7], mounted on one end-flange of the chamber. The signals are sent to cable equalizer amplifiers [8], which reshape, amplify and drive them to amplifier shapers located in the counting room, where the ion-tails are canceled [3, 9]. Each output channel of the shapers is fed into a 100 MHz 6-bit FADC [10], which samples the incoming signal at 10 ns intervals and stores the digitized amplitudes into a fast 1 kbyte-deep ECL RAM. To each FADC a Data Reduction Processor (DRP) is associated, whose function is to reduce the large amount of raw data. For this experiment, a mode was selected where only data above a certain threshold and additional channels before and after the signal are retained. An overall data reduction factor of typically 17 is achieved.

2 * 6 anodes are equipped with digital readout electronics for each of the 12 chamber segments. The readout of the total of 144 DRP modules distributed over 16 CAMAC crates is accomplished with a VAX-11/750 computer configuration assisted by 16 intelligent CAMAC auxiliary crate controllers (ACC) [11] based on Fairchild's 9445 microprocessor as event-builders for each crate. The TEC readout is independent from the MARK-J system except for the trigger to start digitization.

MEASUREMENTS IN TEST BEAMS

A single-segment-prototype chamber and two opposite segments of the MARK-J TEC have been exposed to a beam of 5 GeV/c electrons prior to installation in the detector (see ref. [4] and [5]). On the average, over all inner anodes measured, the resolution is 35.5 microns at 1 cm drift length and remains below 50 microns over the homogeneous drift region (Fig. 4). The measured values of drift velocities form the basis for the TEC calibration in the MARK-J detector after correction for different temperatures and pressures.

The dependence of the resolution on the azimuthal angle ϕ of the track with respect to the detection gap is shown in Fig. 5. In MARK-J, the angle for tracks originating from the interaction point is limited to 19 degrees.

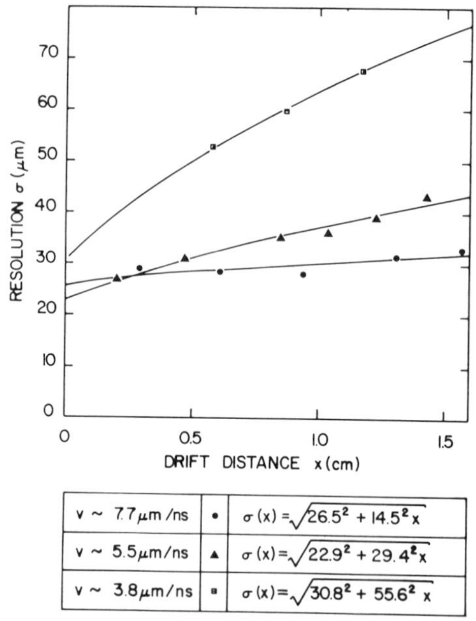

Fig. 4 Resolution vs drift distance

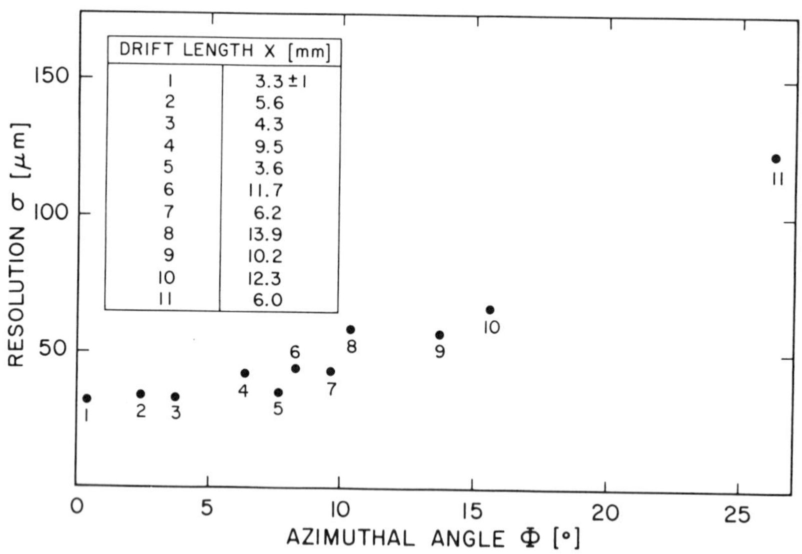

Fig. 5 Resolution vs azimuthal angle

The double-track resolution has been studied with multiple tracks extrapolated between inner and outer groups of anodes [12]. A value of (300 ± 50) microns has been obtained for a track separation at which 50% of the tracks are correctly resolved (Fig. 6).

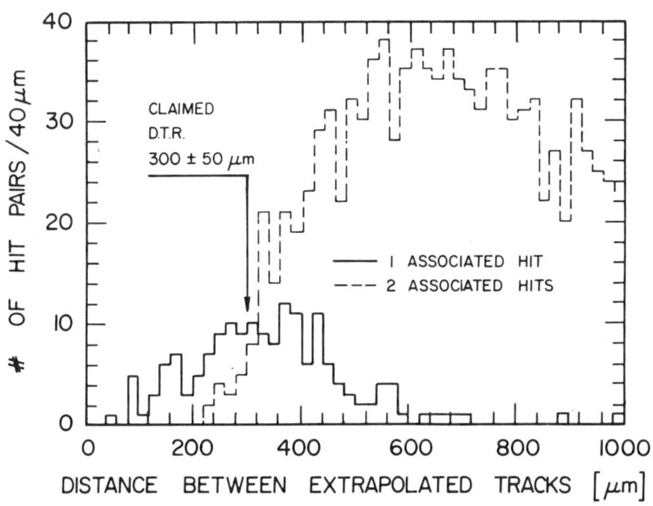

Fig. 6 Double-track resolution (DTR)

Due to a default in the fabrication process of the beam pipe by industry, the 0.2 mm steel inliner collapsed when exposed to both outer pressure of 2 bars and vacuum inside. It could be removed and replaced by a 2 mm thick aluminium tube just before installation into the MARK-J detector (December 1985). By means of an electro-mechanical resonance technique [13], it was verified that the proper tension (about 80 g) of a selection of the anode wires was not affected by this incident (Fig. 7).

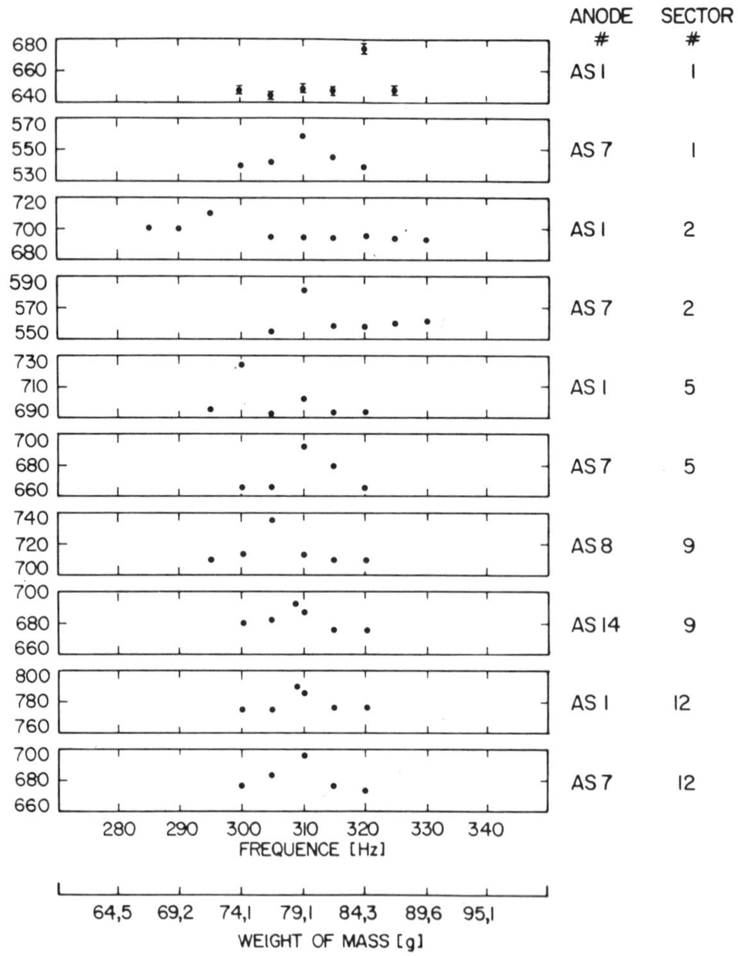

Fig. 7 Wire tension measurement

OPERATION EXPERIENCE WITH THE MARK-J CHAMBER

In the MARK-J TEC, the drift velocity is very sensitive to the operating conditions. Fig. 8 shows the temperature dependence which amounts to 2.5%. for 1° K temperature change. The sensitivity to isobutane content is 1% for 1% change (Fig. 9). The change in drift velocity due to various changes in high voltage settings is represented in Fig. 10.

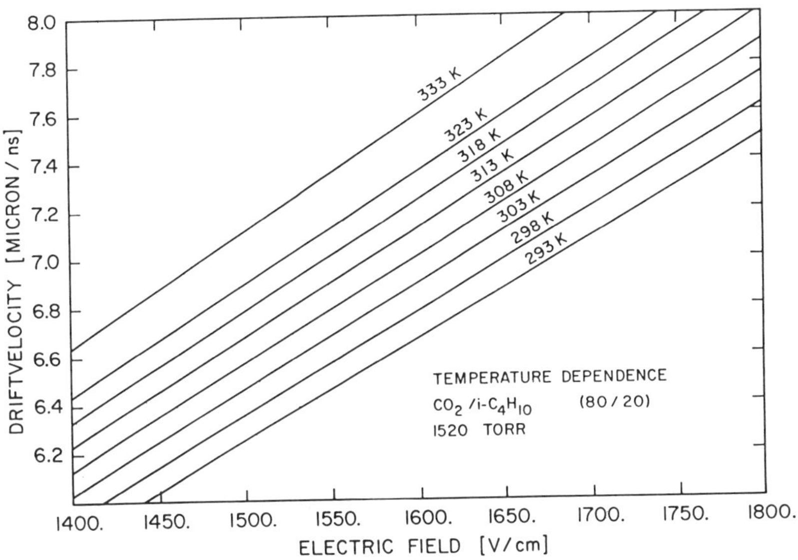

Fig. 8 Drift velocity variation as function of electric field

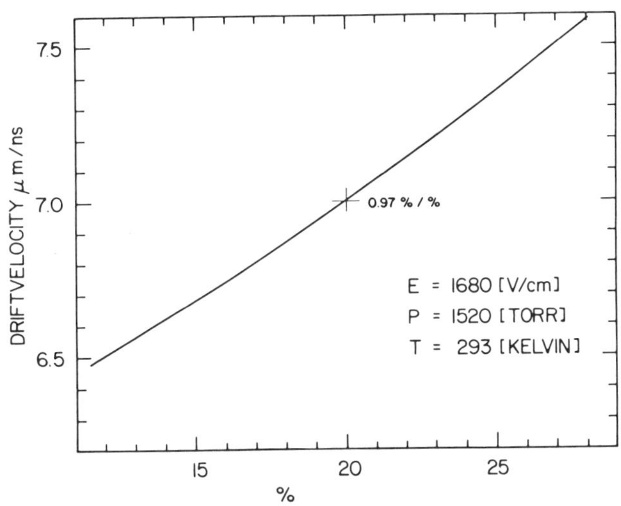

Fig. 9 Drift velocity as function of isobutane content

HV STANDARD: ANODE (2968 V) I-FOCUS (1778 V) O-FOCUS (2115 V) GRID (0 V) CATHODE (4225 V)

CHANGE OF VOLTAGE COMPARED TO STANDARD [%]					
OUTER FOCUS	0.0	0.0	0.0	+ 3.8	- 5.6
INNER FOCUS	0.0	0.0	0.0	+ 3.8	- 5.6
ANODE	0.0	+ 1.1	+ 3.1	0.0	0.0
GRID	0.0	0.0	0.0	0.0	0.0
CATHODE	0.0	0.0	0.0	0.0	0.0

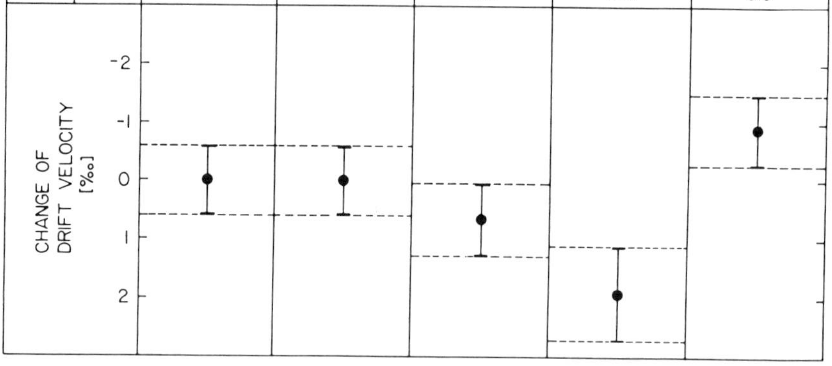

Fig. 10 Change of drift velocity with high voltage settings

During the operation of the chamber, the temperature is monitored at gas inlet and outlet, on the beam pipe, and the chamber flanges. The pressure is controlled within ± 0.3 mbar. Various instruments monitor the gas composition : "BINOS" measures the isobutane, "PANAMETRICS" the water content, and "TELEDYNE" the oxygen impurities. The high voltage settings are checked every 10 s.; if out of range (current and voltage), data taking is disabled and all voltages of the corresponding group are turned off. The drift velocity can be monitored by means of a laser beam in a test chamber supplied in parallel by the same gas system (Fig. 2).

After the installation into the MARK-J detector, the chamber was first run with cosmic ray triggers. The online display for such an event is shown on Fig. 11a. (data provided by the DRP's of one segment) and on Fig. 11b. (calculated hits in the R ϕ plane). In this exemple, the inner part of the track passes through the detection gap, where the drift velocity is larger. The signals appearing on the right-hand side are reference signals obtained from time-marker signals injected capacitively on the focus wires at a well-defined time after the digitization starts, derived from the beam crossing trigger.

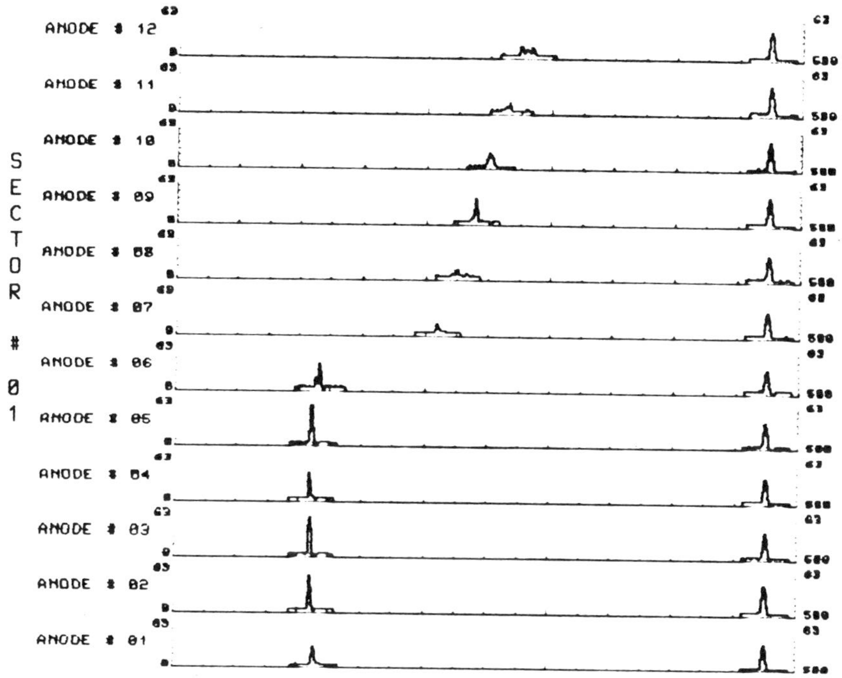

(a)

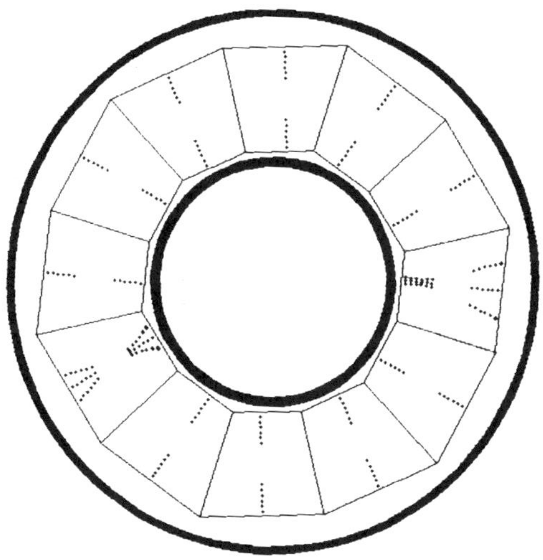

(b)

Fig. 11 Example of cosmic ray trigger

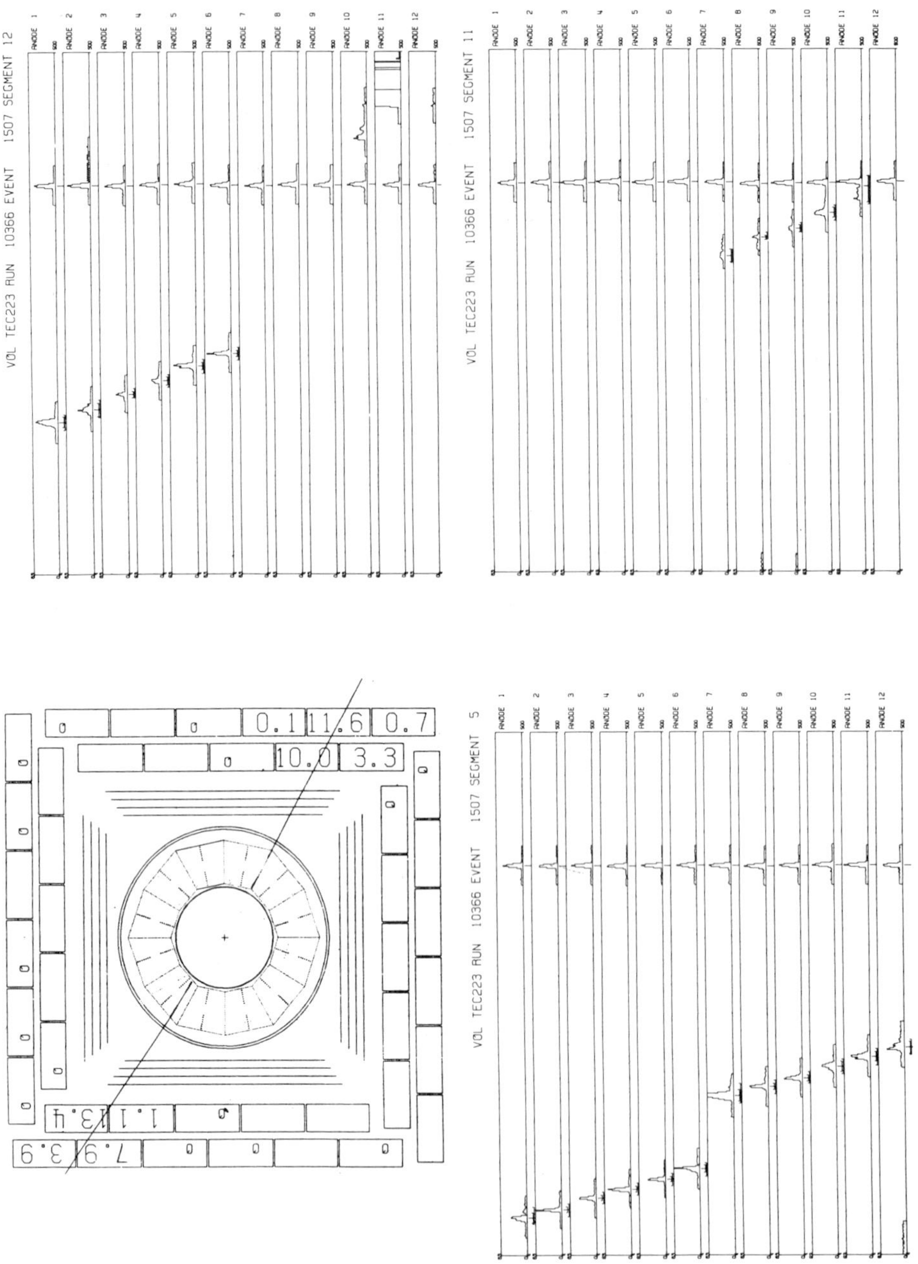

Fig. 12 Example of reconstructed Bhabha event

The analysis of the first data taken at PETRA is in progress. An example of a reconstructed Bhabha event, identified by the MARK-J electromagnetic calorimeter is shown on Fig. 12. Two tracks are visible in the TEC, one of which crossing the cathode plane. Fig. 13 displays the wire efficiency, with values around 90%, as a function of the polar angle θ measured with respect to the beam direction. Fig. 14 shows the peak area of the signals as a function of the drift length with respect to the anode plane. The area is observed to be constant, except in the detection gap.

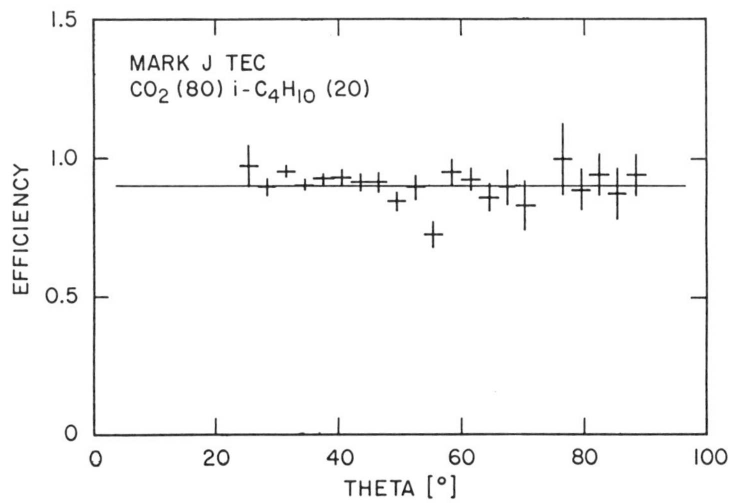

Fig. 13 Efficiency versus polar angle

From the reconstructed tracks, the inner chamber resolution was estimated. Fig. 15 shows the distribution of residuals from Bhabha events summed over drift distances and angles. An average of 40 microns is obtained, as expected from test beam measurements. Fig. 16 displays the resolution as a function of drift distance. It is found to be, as expected, a slowly increasing function of the drift distance. Fig. 17 is an example of a reconstructed hadronic event.

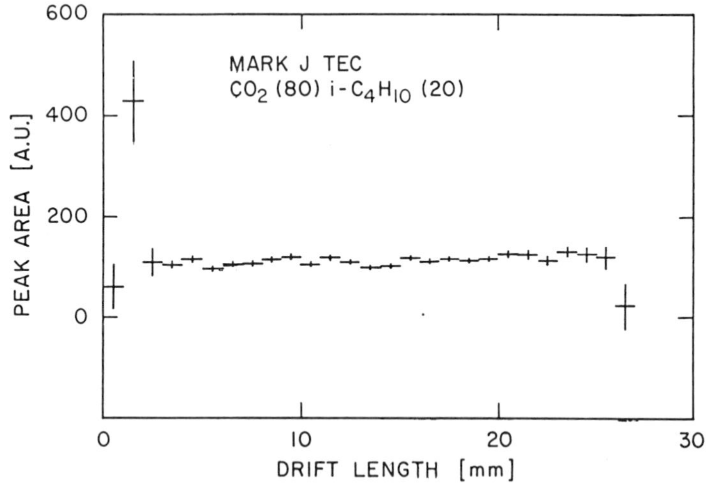

Fig. 14 Peak area versus drift length

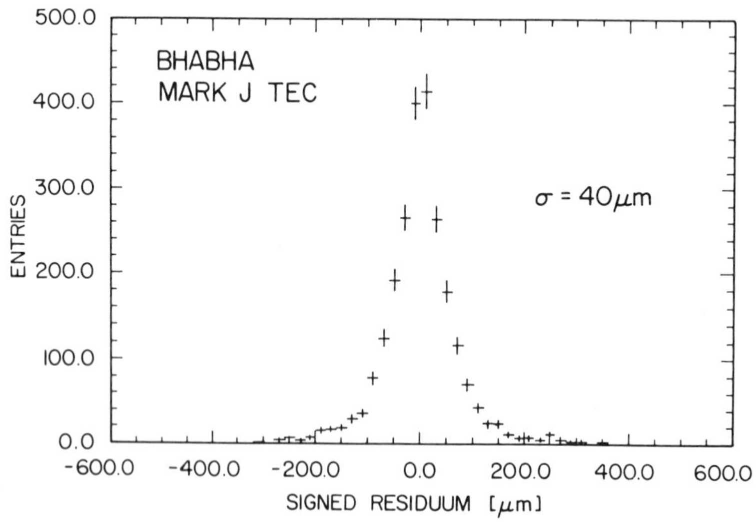

Fig. 15 Signed residuum

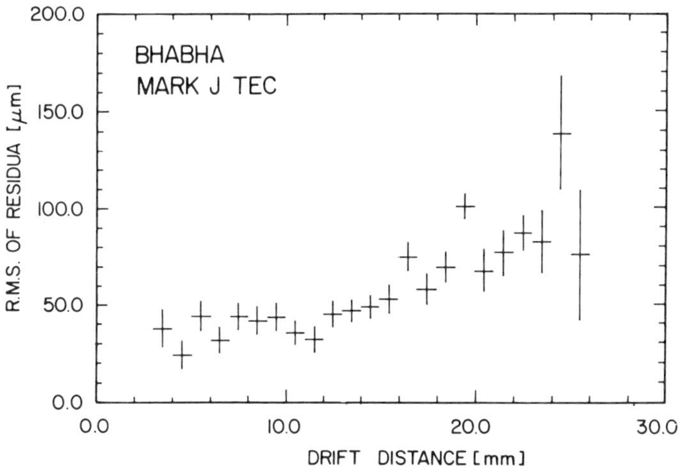

Fig. 16

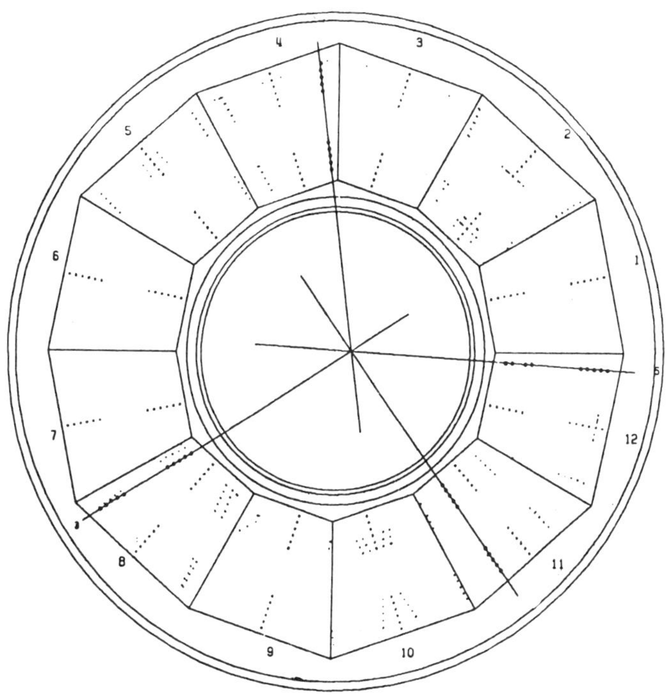

Fig. 17 Example of reconstructed beam trigger event

After the first months of operation, it was observed that the values of the mean pulse integrals were gradually decreasing, although the mean pulse widths remained stable (Fig. 18). The efficiency dropped substantially, particularly on anode wires located in the horizontal plane (the plane of the machine). An increase of the high voltage settings and the addition of water (~ 140 ppm) in the gas mixture did not cure the problem.

It was then decided to replace the isobutane by argon and to add alcohol instead of water. As alcohol can affect the Oxysorb, a new open gas system was implemented (Fig. 19) with the following features : gas flushed through (no loop), new type of Oxysorb, micro filters downstream of the Oxysorb, injection pump for the addition of ethanol, reference chamber in the same system.

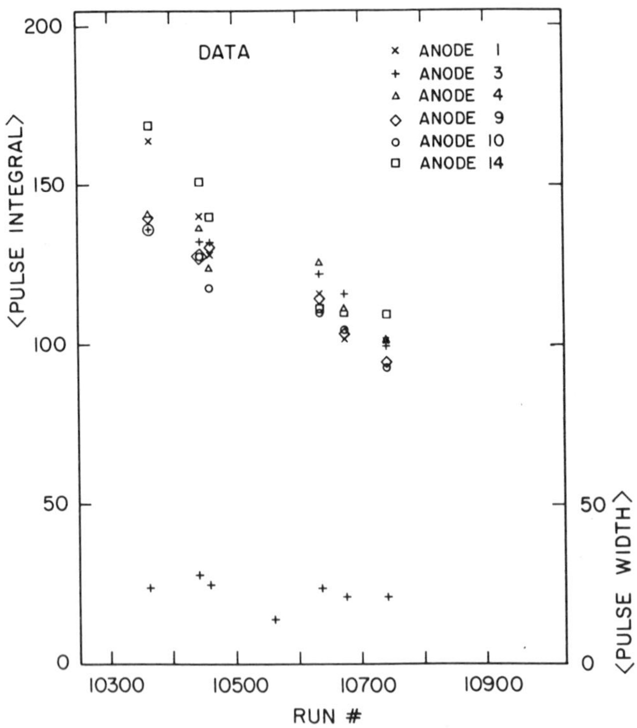

Fig. 18

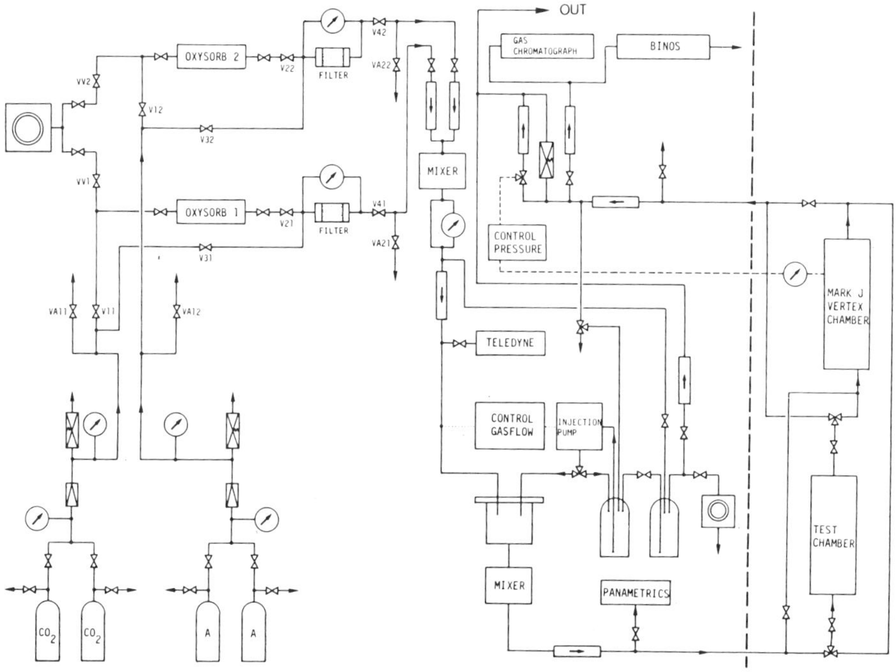

Fig. 19 Open gas system

After a general retuning of the analog electronics, the chamber was operated again with cosmic rays. With a gas composition CO_2 (70%) -Ar(30%) at 1.65 bar an efficiency of 90% was again obtained for all wires, except in three half-segments when the high voltage settings could not be brought to nominal values (Fig.20). The resolution averaged over all tracks in one segment was found to be 65 microns (Fig. 21). The difference with the results quoted earlier (e.g. Fig. 15) is due to the lower gas pressure and to the smaller charge deposited by cosmic rays with respect to Bhabha electrons. The dependence of the peak integral with the polar angle θ is shown on Fig. 22. One observes the expected peak integral increase with the larger number of electrons collected, proportional to the sampled track length, when θ decreases. In the same figure, one notes that the efficiency is constant with θ.

At the present time (September 1986), the chamber is running with beam at PETRA. A gas mixture of CO_2 (85%) -Ar(15%) with ~ 3% of injected ethanol is used.

A 1% change in ethanol content has been measured to change the drift velocity by 3%.

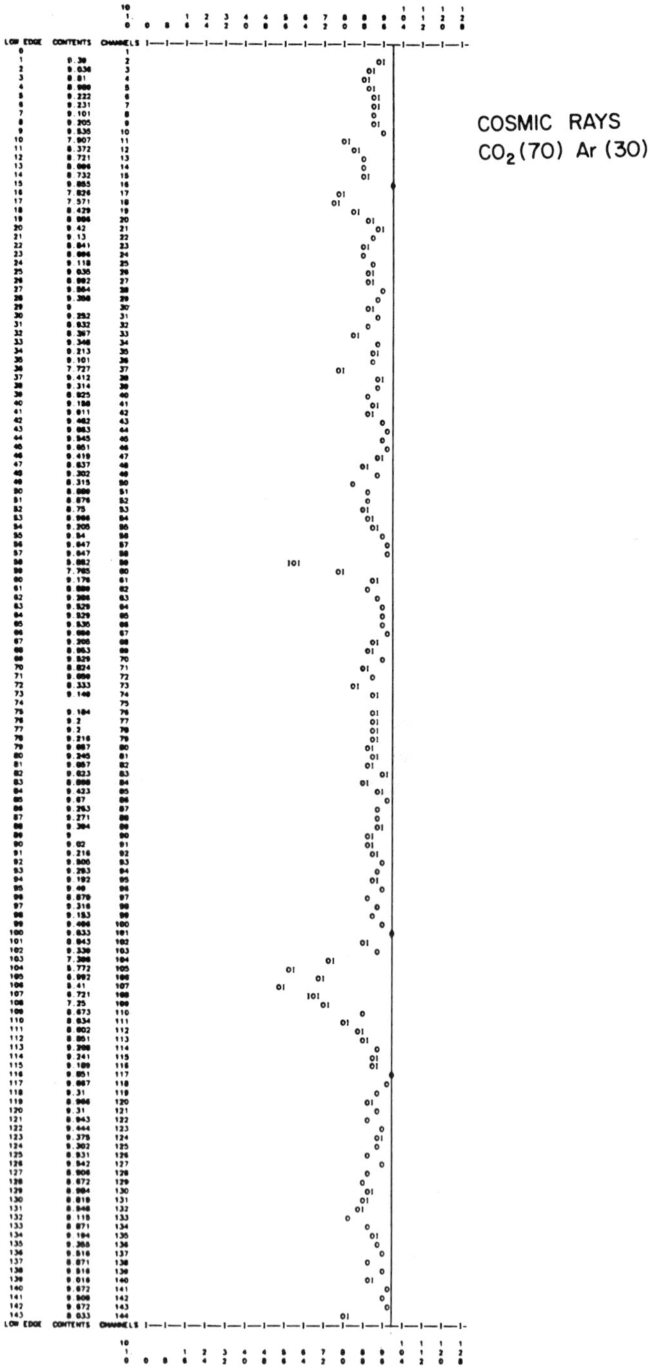

Fig. 20 Wire number versus efficiency.
The straight line represents 100%

CONCLUSIONS

The primary goal of the MARK-J Vertex Chamber is the first testing of a TEC prototype in a realistic e^+e^- collider environment. At the early stage of running with CO_2 and isobutane, the MARK-J TEC confirmed the test beam results and reached its design

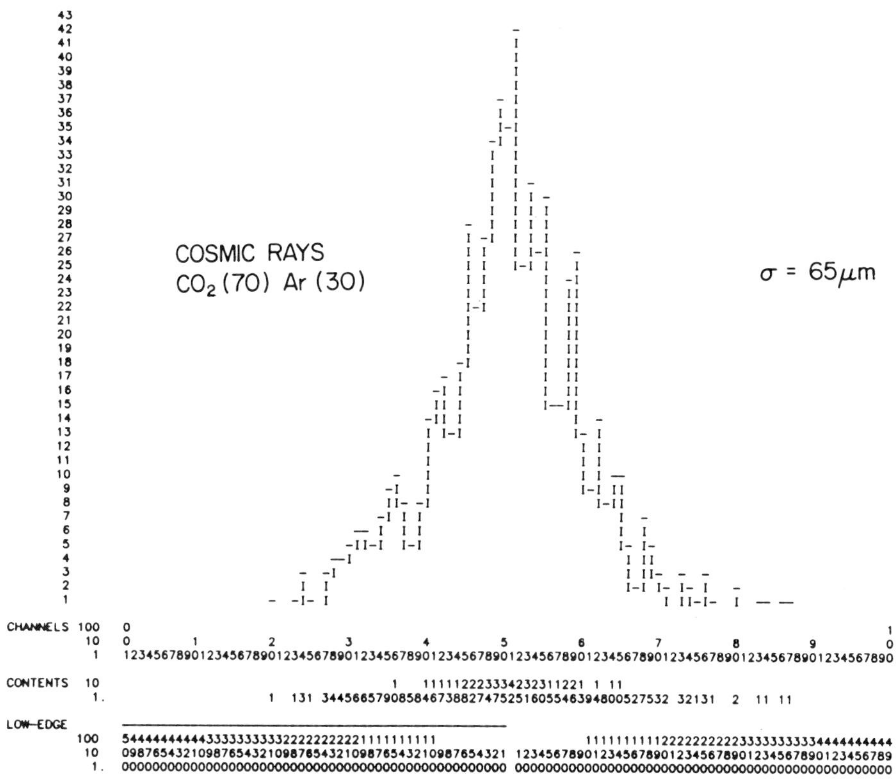

Fig. 21 Signed residuum

goal. Due to lack of space, it was not possible to built in calibration devices into MARK-J, which makes the determination of the calibration parameters (e.g. drift velocities) very difficult. Analyses are in progress, using Bhabha events, hadrons and cosmic rays.

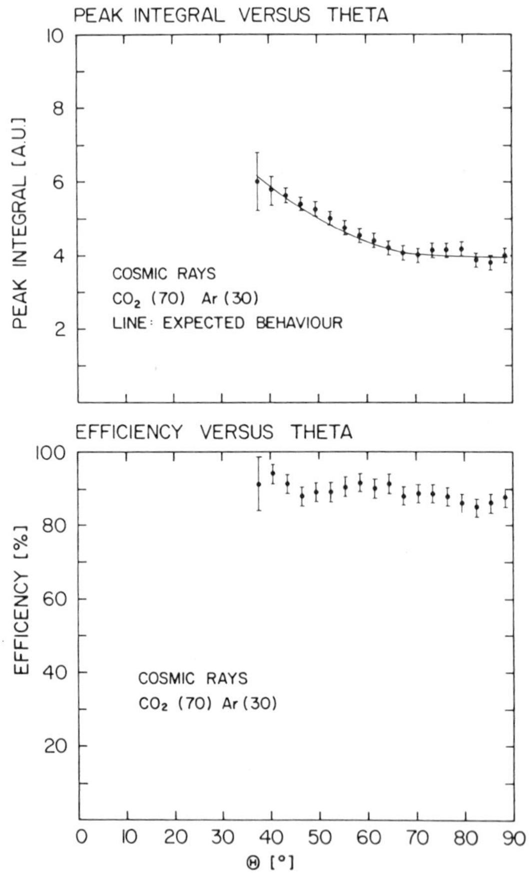

Fig. 22

The deterioration of the chamber performances which followed is still being investigated. It could be due to one of the following causes :

- too small a beam pipe (MARK-J has the smallest diameter pipe at PETRA)
- mechanical problems
- faulty Oxysorb (chromium trioxyde has been found in the gas system)
- aging effects in the chamber.

An answer is expected when the chamber is opened after the final PETRA shutdown.

As regards the readout and data acquisition of the fairly large volume of data generated by the TEC, it is found that the system of 144 FADC's and DRP's works without any difficulties. The DRP's and the 16 ACC's provide a sufficient reduction factor. The VAX data acquisition load is minimal (< 15%) for a typical rate of 8 Kbyte at 2Hz.

The experience obtained with this prototype chamber now forms the basis for the design of the L3 TEC.

REFERENCES

[1] A.H. Walenta, IEEE Nucl. Sci. NS-26, Nr. 1 (1979) 73

[2] L3 Technical Proposal, LEPC/83-5/p4, 1983, CERN, Geneva, Switzerland

[3] V. Commichau et al., Nucl. Instr. Meth. A235 (1985) 267

[4] H. Anderhub et al., Nucl. Instr. Meth. A252 (1986) 357

[5] H. Anderhub et al., A Time Expansion Chamber as a Vertex Detector, 3d Pisa Meeting on Advanced Detectors, Castiglione della Pescaia, Univ. Geneva preprint UGVA-DPNC (1986)/ 10-115

[6] J. Fehlmann, "WIRCHA" : A Program Package to Simulate Drift Chambers, ETH-Zürich (1985)

[7] P. Jarron, M. Goyot, Nucl. Instr. Meth. 226 (1984) 156

[8] H. Anders et al., CERN-DD 85-20 (1985)

[9] J.A. Paradiso, Compilation of Data on the Current TEC Analog Electronics, ETH-Zürich (1983)

[10] V. Commichau et al., Internal Report, III. Phys. Inst. RWTH Aachen (1984)

[11] M. Bourquin et al., Nucl. Instr. Meth. A239 (1985) 562

[12] B. Wyslouch, Mark-J Internal Note, 31 July 1985

[13] D. Ren et al., Internal Report, "Measurement of the Mechanical Tension for the TEC Anode Wire", ETH-Zürich IHP/86/05 (1986).

Discussion M. BOURQUIN

F. SAULI

You mentioned that you are switching to $A-CO_2$. Because of ageing problems in the original TEC gas CO_2 – i C_4H_{10}. What is known about drift properties in this gas? In particular, as far as I know, this mixture does not satisfy the TEC condition, of a gas having small drift velocity at high fields (to reduce diffusion). One can have small drift velocity in any gas at small fields, but only a few (e.g. DME) allow to have small diffusion also.

M.BOURQUIN

The drift velocity in $A-CO_2$ has been measured as about 7 μm/ns for the same voltage as for $CO_2-i-C_4H_{10}$. Further studies are under way.

M. ATAC

As we heard from you, ethanol cured the ageing of your wires. This is nice. Have you done tests using gas with isopropyl alcohol? I highly recommend it. You will see less electron attachement with isopropyl alcohol. It will require less amount of isopropanol for same quenching effect of UV photons.

M. BOURQUIN

Thank you for your comment.

F.G. HARTJES

Why did you change iC_4H_{10} to A, because of the impossibility of laser calibration?

M. BOURQUIN

Mixture was not changed because of calibration problems but because of ageing.

A.M. WALENTA

How do you measure the efficiency and how large is the electronics noise?

M. BOURQUIN

The quoted efficencies include all effects from hardware to the presently not yet optimised software. The noise as measured by the fluctuation of the late line is about 1 channel out of 64.

F. PIUZ

Could you comment about "double hit" fraction from single track?

M. BOURQUIN

The peaks from single tracks do not show " double hits".

H. NELSON

How much luminosity will you collect with the vertex detector, and can we look forward to physics results?

M. BOURQUIN

Unfortunatily PETRA is due to be turned off on November 3rd. This does not leave much time for original physics measurements.

A MODIFIED TIME EXPANSION CHAMBER AS A VERTEX DETECTOR

G. Bari, M. Basile, G. Cara Romeo, R. Casaccia, L. Cifarelli, F. Cindolo, A. Contin, G. D'Ali, C. Del Papa, S. De Pasquale, G. Iacobucci, G. Laurenti, G. Maccarrone, T. Massam, F. Motta, R. Nania, F. Palmonari, E. Perotto, G. Prisco, G. Sartorelli, G. Susinno, L. Votano, M. Willutzky and A. Zichichi

Dipartimento di Fisica dell'Universita, Bologna, Italy

Istituto Nazionale di Fisica Nucleare, Bologna, Italy

CERN, Geneva, Switzerland

Istituto Nazionale di Fisica Nucleare, LNF, Frascati, Italy

Istituto di Fisica, Universita di Palermo, Italy

1. Introduction

The design presented is one of two proposals for a Vertex Detector for the experiment ZEUS at HERA.

It is essentially a standard jet cell using a "cool" gas: dimethyl ether (DME) [1] and incorporating some of the concepts of a Time Expansion Chamber. The technique is entirely original, making use of hybrid circuits to implement both the electronics and the mechanics.

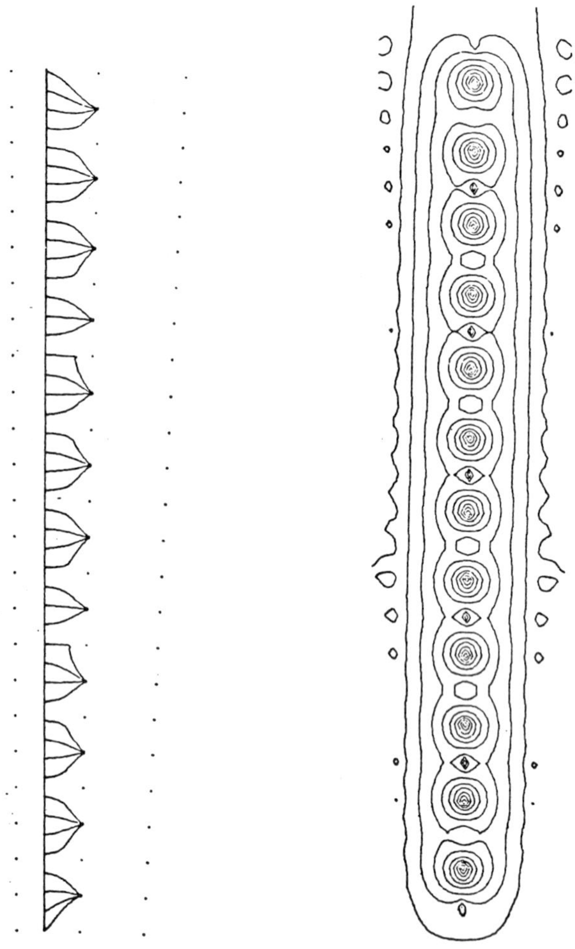

Figure 1. Shape of the cell. The maximum drift space is 3.5 mm and the minimum is 2 mm. The distance between two adjacent wires is 1.5 mm. The prototype has 12 sense wires. Both the drift lines and the equipotentials are shown.

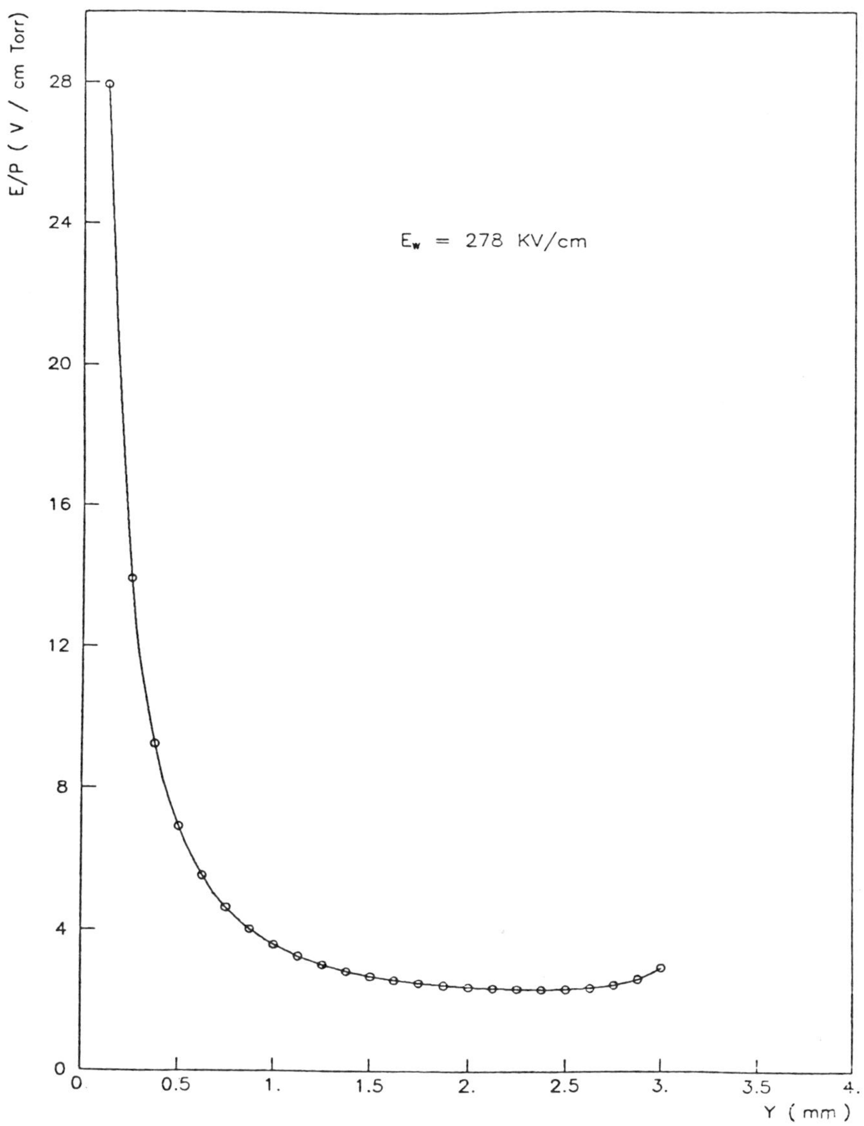

Figure 2. Behaviour of the electric field as a function of the distance from the sense wire.

2. Description

The constraints on the design of our vertex detector come from the characteristics of HERA. The maximum drift time is set at 500 ns or 5 beam crossings. It is imperative to have a space resolution of less than 50 μm, to be able to resolve tracks in jets (double track resolution better than 1 mm) and to withstand both the 2T magnetic field and a very high rate. [2]

A "cool" gas offers the advantages of being insensitive to the magnetic field, i.e. the geometry of the cell will be simple, and of a better spatial resolution than a "warm" gas, since the effect of ionization fluctuations will be depressed roughly by the ratio of the drift velocities in the drift space and in the amplification region. In order to achieve this result it is essential to shape the field so that it remains constant for as long a space as possible, rising to the gain field only near the wire surface. To do this we must optimize the three voltages on the sense, field and potential wires. Fig. 1 and Fig. 2 show the shape of the cell, its dimensions and the electric field.

From preliminary studies done with a little chamber, [3] we hope to achieve a space resolution of about 30 μm. Among "cool" gases, we have chosen DME because of its superior discharge – quenching capability and stopping power (150 total ions/cm).

The double track resolution is given by the pulse width multiplyed by the drift velocity. For a pulse width of about 100 ns and a drift velocity of 5 μm/ns (at $E_{drift}=2$ kV/cm) we should have .5 mm.

3. Design

The device will be built radially mounting alumina plates of the appropriate height (few cm depending on the azimuthal angle) on a cylinder cohaxial with the beam pipe. On these alumina plates all the necessary resistor chains, decoupling resistors and condensors will be surface mounted. Pd/Ag fingers will be used to solder the gold−plated W wires on. The left−right ambiguity will be resolved using information from the Central Tracking Detector.

The final design requiring that the alumina plates be very close to the beam pipe and that all the safety rules be respected will be relatively complicated, however the prototype with 3 full size cells, can be built with relative ease to prove the soundness of the conception.

At the time of writing, in fact, a cell has been built and shows no sign of electrostatic instabillity.

The electronics consists of Avantek amplifiers, preceded by a first amplifying stage made with one microwave transistor. 6 channels will fit in a 4−cm high alumina plate. At time of writing we have obtained the first samples which work without problems.

4. Time scale

We foresee the first bench test in March, followed by beam tests in april. Final results on the protortype will be available by june 1987.

References

[1] F. Villa Nucl. Instr. and Meth. 217 (1983) 273

[2] ZEUS Technical Proposal March 1985

[3] G. Bari et al. Nucl. Instr. and Meth. A251 (1986) 292

Discussion C. DEL PAPA

A.M. WALENTA

The original idea of the TEC for tracking was to make a centroid measurement in drift direction . You consider only leading edge measurement. Would a centroid measurement improve the resolution further?

C. DEL PAPA

So far as I understand centroid measurement is really useful only for long drift distances. Therefore I doubt that it would improve the resolution.

B.R. KUMAR

Is it clear that 30 μm resolution will be sufficient to detect short−lived particles at HERA?

C. DEL PAPA

From our Monte Carlo the answer is yes.

G. LUTZ

Do you have multi−hit electronics and what is your double track resolution?

C. DEL PAPA

We can definetely use multi hit electronics. The double track resolution will be in the order of half a mm.

P. WEILHAMMER

What is the multiple scattering contribution of beam pipe and inner wall of the TEC

C. DEL PAPA

Final design is not yet done, in general we expect equal contribution to the ~15 μm impact parameter error from measurement and multiple scattering error at momenta of a few GeV.

OPERATION OF MULTIDRIFT TUBES WITH DIMETHYL ETHER

R. Bouclier, G. Charpak, J. Garabatos, B. Guérard, J.C. Santiard, N. Solomey* and F. Sauli

CERN
Geneva, Switzerland

1. INTRODUCTION

The most difficult region to instrument in a detector for collider experiments is the one close to the vacuum tubes, where the density of tracks and the background rate are very high. This is, moreover, the most inaccessible region of the set-up; a major breakdown there has the worst consequences for the whole operation. One way of improving the reliability of a system based on gaseous wire detectors is to increase its modularity and redundancy; such an approach was originally taken by a group at SLAC with the development of cheap, mass-produced individual drift tubes or 'straws' that can be assembled to constitute a complete detector.[1] Extending this concept further, we have built several multidrift modules that have many independent drift cells packed within a thin carbon-fibre supporting tube, of hexagonal shape, which also contains the gas. Each sensitive cell consists of an anode wire centred in a hexagonal drift structure with cathode wires at the corners. Our first prototypes (called Model 1) had 128 anodes per module, with an anode-to-cathode distance (cell radius) of 1.15 mm; the method of construction and the main experimental results obtained with such a design have already been described.[2,3] Because of the problems encountered in the manufacturing of such a high-density detector, we have redesigned the detector as Model 2, slightly increasing the cell radius to 1.45 mm (1.27 mm between wire planes); this has simplified wiring and soldering, and greatly increased the reliability of the detector, at the expense of a decrease in the information density (70 cells per tube) and a slightly longer resolution time. All results described in this note refer to Model 2. When a charged particle traverses the tube, measurements of the drift-times on all wires provide the transverse coordinates,

* Permanent address: Ohio State University, Columbus, Ohio, USA.

perpendicular to the wires; the longitudinal coordinates are provided by a measurement of the charge ratio on the two sides of each wire. On an average, ten pairs of coordinates are recorded for each track in a single tube. Several tubes can be closely packed, for example in a cylindrical geometry, to instrument the central region in a collider detector; in the case of failure, individual modules may be disconnected or replaced.

2. CONSTRUCTION OF THE MULTIDRIFT TUBE

The method of construction of a module has been described in Ref. 3, and will only be shortly summarized here. Wires are strung between two high-quality fibre-glass plates of hexagonal shape (Fig. 1), 4 mm thick and 30 mm in diameter, through a set of holes of 600 μm diameter drilled on a high-precision numerical boring machine. A 50 μm thick copper sheet is glued on one side of the plate before drilling; all pins holding the cathode wires, once inserted and kept under tension, make contact with this common electrode that distributes the high voltage. As can be seen in the figure, a short, larger diameter (1.2 mm) bore around the holes destined to receive the anodes ensures the insulation of these pins from the cathodes. Wires are soldered to metal pins inserted in the holes of the supporting plates; these pins are precision-machined brass cylinders, 0.8 mm in diameter,

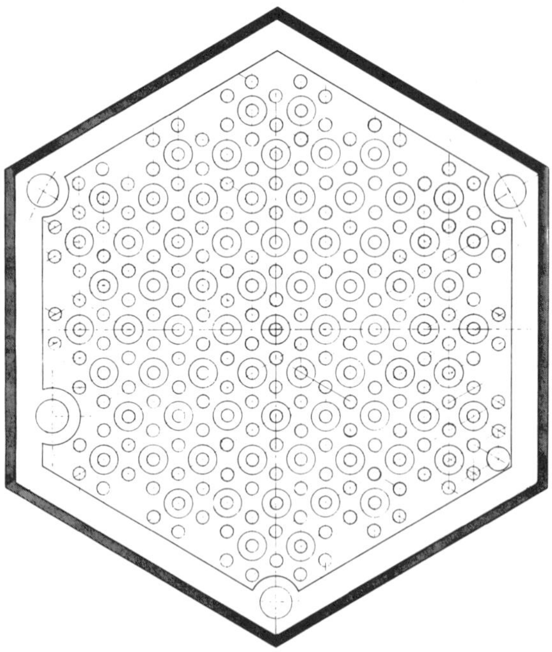

Fig. 1 Schematics of the drilling geometry of the end plates. Small circles represent holes for inserting the wire-holding pins; around the anodes, a short, larger-diameter drill ensures insulation against the metallized surface of the plate.

reducing to 0.6 mm at the tip in the section inserted into the plate (Fig. 2). A V-shaped groove, cut over the full length of the cylinder, receives the wire before soldering. Because of the different diameters of the corresponding wires, anode and cathode pins have grooves with different depth to ensure proper centring; the cathode pins are also shorter, and are completely covered by epoxy in the last phase of manufacturing. A bore on the edge receives an insulating tube for the gas inlet; the high voltage is provided to the interconnecting copper plate through a miniature pin soldered on the circuit. We have used 25 μm diameter gold-plated tungsten for the anodes and 70 μm copper–beryllium wires for the cathodes, tensioned at 60 and 120 g respectively.

To assemble the module, the two end plates are inserted on temporary brackets, rigidly mounted on a connecting horizontal bar. Wires are individually pulled through facing holes in the end plates, and kept under tension by a pulley and weight system. The appropriate brass pins are inserted on each side, with the wire resting face up in the grooves; wires are then soldered with a low-melting-point alloy and cut at both ends. The assembly proceeds in layers, starting from the lowest. Once the wiring is completed, the fibre-glass plates are fastened to a coaxial support fixed to three threaded pins on the corners, insulated from the copper layer (they will be used later to fasten the connector); on one side, the support consists of a long rod on which the supporting tube has been

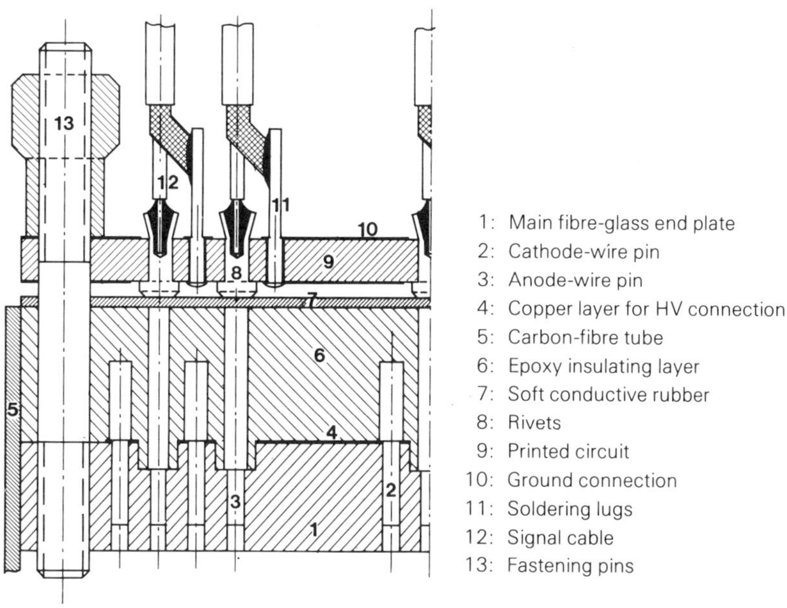

Fig. 2 Cross-section of the end-plate assembly. Once inserted and wired, cathode pins are completely embedded in epoxy, the common contact being provided by a pin (not shown) soldered to the copperized upper layer of the plate. The system of signal connection, which makes use of an anisotropic conducting sheath, is also shown.

previously inserted. The temporary brackets are then removed, all wires being kept under tension by the coaxial mounting, and the carbon-fibre tube is slid over the wires to its final position; epoxy is poured over the plates so as to completely cover and insulate the cathode pins and at the same time to seal the module. The carbon-fibre tube used for the prototypes is 80 cm long and 550 μm thick, with an outer diameter of 32 mm; it is composed of carbon and epoxy in almost equal proportions. Figure 3 shows a close view of one end of a completed module: one can recognize the top of the anode pins, the tube that provides the gas inlet, the pin for the high-voltage connection, and the three threaded rods used for fastening the signal connector. A contact to the anode wires is available at both ends of the module, to allow the determination of the longitudinal coordinate through a measurement of the current ratio on the resistive wire. At present, the anodes are realized with gold-plated tungsten wires that can be soldered to the pins; because of the low resistivity of such wire, the localization properties in current division are rather modest. We are currently developing an alternative method of mounting, using fast-setting conductive epoxy instead of soldering, which should allow the use of resistive wires of any material.

Because of the high density of output contacts, we have developed a special connector to bring the signals to the electronics boards. After several attempts, using various kinds of mechanical spring-loaded and conductive pad contacts that had rather

Fig. 3 One end of a finished module. Contacts with the anode pins are visible, as well as the tube for the gas and the pin for the high-voltage supply. The three threaded pins are used for assembly and to hold the signal connector in place.

poor reliability, we have adopted the solution sketched in Fig. 2. It consists in inserting an anisotropic conducting membrane between the anode pins and the printed-circuit matrix to which the cables are soldered. The membrane is a thin (0.3 mm) rubbery sheet with thousands of short metal wires, perpendicular to the surface, and insulated from each other. With our surface of contact (0.5 mm^2) the serial resistivity is typically between 10 and 50 Ω; while tolerable for time-recording electronics, such a spread is rather annoying for analogue recording as used in the current-division method. We are investigating alternative connection methods; one that seems promising consists in directly soldering the signal cables to the pins or gluing them with conductive epoxy, during the manufacturing process. This also has the advantage of reducing the amount of material in the end plates of the detector.

3. EXPERIMENTAL SET-UP

Prototype tubes with the described geometry have been installed for testing in a minimum-ionizing particle beam at CERN, together with a set of three high-accuracy drift chambers.[4] For the beam tests, the time reference was provided by a scintillation counter coincidence. Originally intended to provide a positional reference to study the localization properties of the multidrift tube, the chambers were in fact only used for gross track selection in the analysis, because of the very good quality of the results obtained within the tube itself.

As mentioned above, all anode wires (70 in each module) can be read out on both sides of the tube; for the measurements described here, however, we have only recorded the drift-time on all wires from one side, using a system of discrete electronics elements connected to the anodes with short (50 cm) miniature coaxial cables. For the measurement of the longitudinal coordinate through current division, four anodes only were equipped with charge- and time-recording circuits on both sides; in this case, we have used a low impedance amplifier terminating the coaxial input cable, followed by a splitter providing both the analogue and the shaped signals for charge and time measurements. Discrimination and shaping were realized with a home-made circuit[5] whose threshold could be regulated from 10^{-2} pC up, and having typically 3 ns slewing for a 40 dB variation in the input signal. The shaped output from the discriminators is sent to logical receivers in the counting room via flat twisted-pair cables, and the drift-time measured with multichannel time-to-digital converters; similarly, the analogue information from the current division was sent via coaxial cables to a set of analog-to-digital converters. Data were recorded with a small on-line computer, and analysed off-line.

4. EXPERIMENTAL RESULTS USING DYMETHYL ETHER

The first measurements confirmed the relative ease of operating the counter in many gas mixtures; the results realized with argon–hydrocarbon mixtures at atmospheric pressure, together with a theoretical analysis of the various sources of dispersion, indicated that an overall intrinsic limit in localization accuracy of around 100 μm r.m.s. had to be expected, mostly determined by the primary ionization statistics, i.e. the large average distance between primary ionizing interactions (300 μm in argon). Better localizations can be obtained only by increasing the gas density (using xenon or increasing the pressure, both rather inconvenient), or finding a gas in which the primary ionization density is higher. This is indeed expected to be the case of dense gases with low atomic number; dimethyl ether [DME: $(CH_3)_2O$], with a density at STP of 2.34×10^{-3} g/cm^3, is expected to have twice the primary ionization density of argon (indeed, higher than xenon). Originally proposed for use in gaseous counters because of the very low electron diffusion and drift velocity at high fields[6] DME proved also to be impervious to radiation damage, and indeed holds one of the world records of endurance in a multiwire proportional chamber (MWPC) under avalanche conditions.[7] Very good localization accuracies (30 μm and better) have been obtained in drift chambers operated with this gas.[8] Because of the very high electric field in most of our drift region, due to the close wire spacing, we did not expect to profit from a reduction in electron diffusion, but the other advantages remain. Provided straight from the bottle, DME is also a rather convenient gas to use since it does not require mixing; its vapour pressure at STP of more than 4 bars makes it easy also to operate at pressures larger than atmospheric.

The counting characteristics of the tube operated with DME at 1 bar are illustrated in Fig. 4, which shows both the efficiency in detecting minimum-ionizing particles and the single wire noise (without particles) as a function of applied voltage. Both measurements were realized with a discrimination threshold of around 5×10^{-2} pC at the wires. As compared to similar measurements realized with conventional argon-hydrocarbon mixtures (see for example Ref. 3), one can observe that full efficiency is reached at higher voltages (2.1 kV instead of 1.5 kV), with a very wide efficiency plateau (more than 500 V; the upper limit was indeed determined more by natural caution than by any sign of instability, as monitored through the singles counting rate). Dimethyl

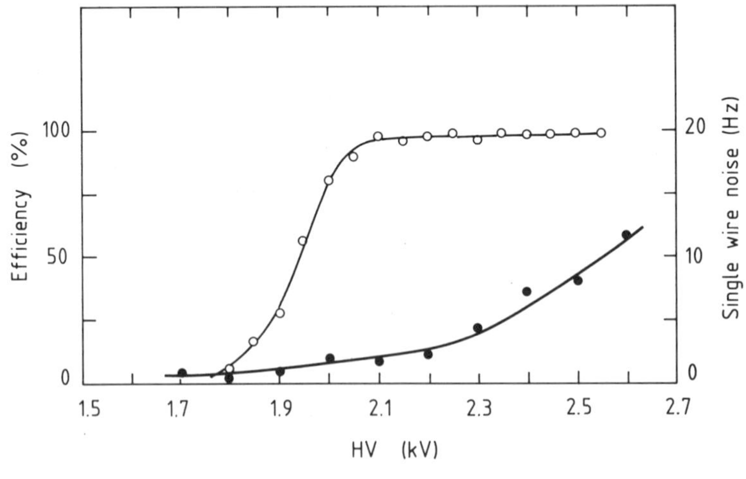

Fig. 4 Efficiency plateau for minimum-ionizing electrons (o) and singles (•) counting rate on one wire without radiation, measured in a fully instrumented module with DME gas filling.

ether appears to be a gas with remarkable quenching properties, a fact confirmed by other authors, who reported reaching the limited streamer operation in other chamber geometries.[1,9] Presumably because of the small drift space, we did not observe any sign of electron capture on impurities, a problem encountered by most authors using DME in drift chambers. Most of the beam data were recorded operating the tubes at 2.2 kV, conveniently above the beginning of the plateau; all results described here refer to this condition.

As indicated above, we recorded the drift-time on all wires, the time reference being provided by scintillation counters. The origin of the measured coordinate for each channel, or time zero, was deduced off-line from data histogramming the drift-time spectra for individual wires; after subtraction of this quantity and multiplication by a slope constant (also specific for each channel and determined from a previous electronic calibration), we obtained the drift-time as measured event per event by the hit wires.

It appeared soon in the data analysis that the assumption of a constant drift velocity resulted in rather poor localization accuracy. Data are unfortunately not available to allow computation of the space–time correlation in the very large range of fields across a cell in our gases; we have therefore directly measured the correlation with the following set-up. A special multidrift detector with a reduced number of cells (12) was built within a transparent glass tube; two openings facing the wires, and closed with ultraviolet (UV) transparent quartz windows, allowed a laser beam to traverse the structure. Using a pulsed N_2 laser, collimated and focused within the detector by UV lenses, we were able to produce an ionization trail, simulating charged-particle tracks, at a known and variable distance from the anodes; the width of the trail was below 100 μm, thus allowing very accurate measurements of the average drift-time. The space–time correlation measured in this way for several gases is shown in Fig. 5; the estimated error in positioning is around 10 μm for each point. Note that, because of the symmetric behaviour of the measured correlation on the two sides of the wire (not shown in the figure), both the time zero and the wire position reference can be deduced from the data themselves. The curve corresponding to the 'magic' gas mixture used in the measurements described in Ref. 3 (full dots) is indeed rather linear (except very close to the wire), as was deduced from the data analysis; for DME instead there is a definite change of slope at about half of the drift distance. This is unfortunate, since it implies a different influence of various electronic dispersions (e.g. the time binning) on the resulting accuracy. We are currently investigating if addition of small percentages of other gases to DME can straighten the correlation.

The straight-track finding proceeds then as follows. A first-level linear fit is attempted using only the hit-wire location (disregarding the actual value of the drift-time). On the perpendicular to the fitted track from the hit wires, the shorter distance is selected using the drift-time measurement (this solves the right–left ambiguity intrinsic in the detector); points having a distance from the initial track exceeding a given tolerance (2 mm) are discarded as background. The fit is then repeated with the new set of coordinates. The procedure converges in two or three steps to provide the best estimate of the track, as shown in the example of Fig. 6.

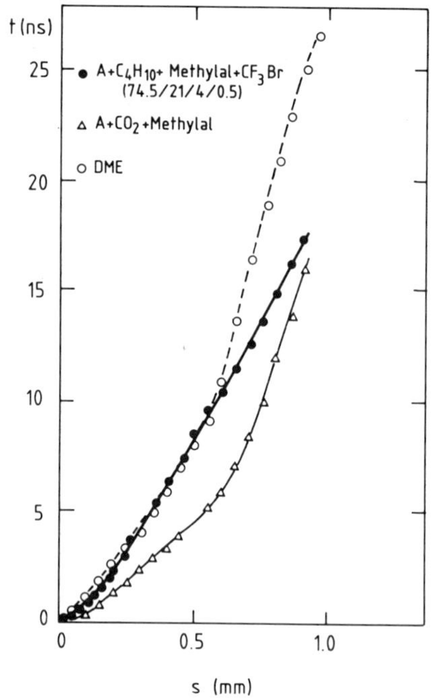

Fig. 5 Space–time correlation in various gases, measured in a dedicated drift tube with a pulsed laser beam.

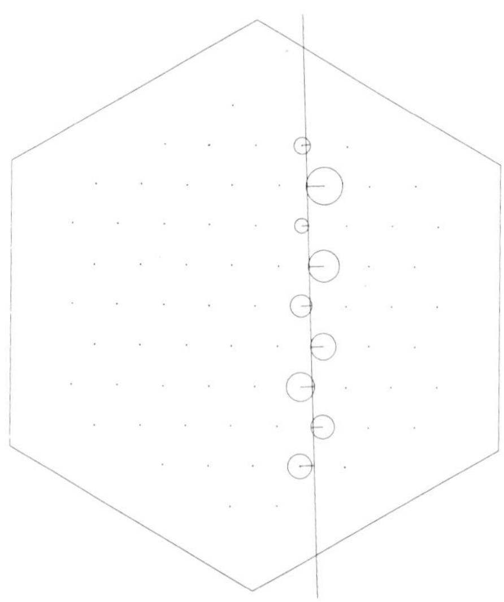

Fig. 6 Example of a single-track reconstructed event. The circles represent the measured drift-time on the hit wires.

The distribution of the difference between fitted measured coordinates on a wire is shown in Fig. 7; it has a standard deviation of 70 μm. This does not represent the true localization accuracy of the wire, since other wires contribute also to the dispersion; it is, however, rather close to the intrinsic accuracy because of the large number of points on each track (8 on an average). One can compute the average true single wire resolution σ from the expression

$$\sigma^2 = \Sigma(r_{meas} - r_{fitted})^2/(N - 2) ,$$

which takes into account, for each track, the number N of hit wires participating in the fit with two degrees of freedom. The distribution of σ for a typical run is shown in Fig. 8 and has a most probable value of 80 μm. This includes all statistical and systematic sources of error, such as wire positioning and errors in the time zero determination.

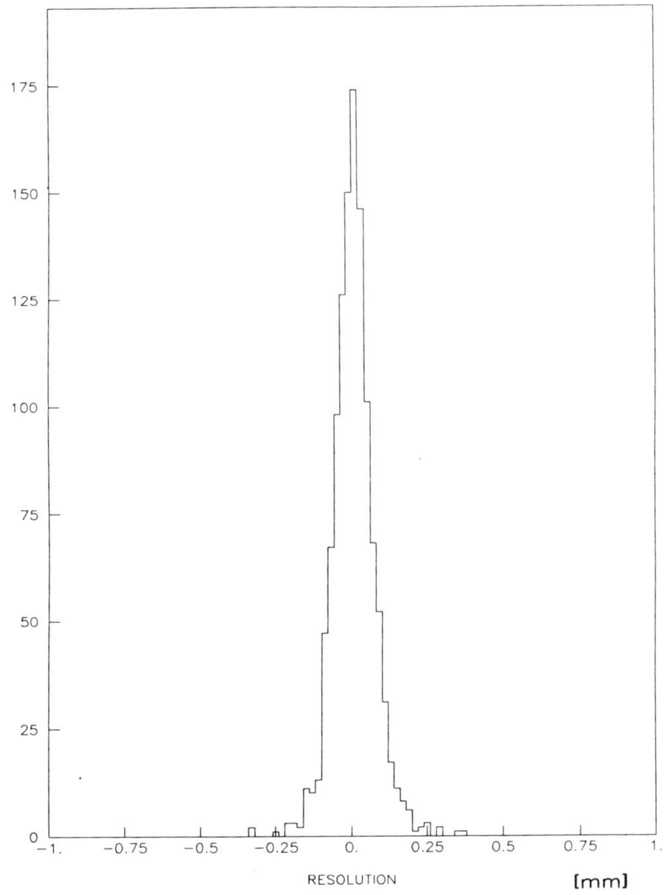

Fig. 7 Distribution of the difference between fitted and measured coordinates on a wire for a typical run; it has a r.m.s. of 70 μm.

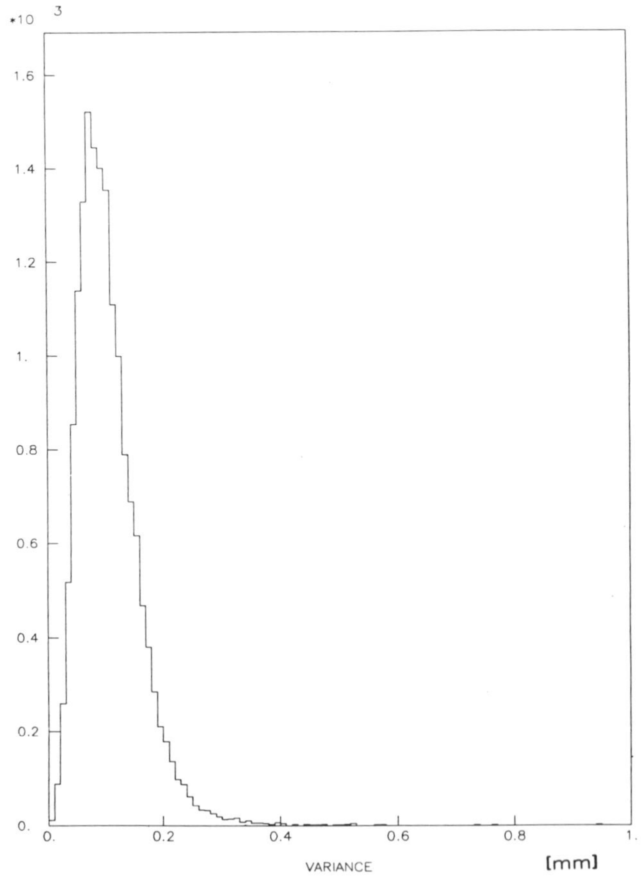

Fig. 8 Distribution of the overall variance from the fit σ, computed from the expression given in the text. Its most probable value, 80 μm, corresponds to the average intrinsic resolution of the wires.

For a reduced number of wires (four) we have recorded both charge and time to attempt longitudinal localization using the current-division method. The quality of the results was rather modest, as expected, because of the low resistivity of the gold-plated tungsten wire used in the prototype detector (180 Ω/cm). Since we expect a great improvement in the next models with the use of a resistive wire, not much effort has been invested in the present measurement. Figures 9a and b show the pulse-height spectra for minimum ionizing particles as recorded from the two sides of an anode wire, for a beam positioned at around 10 cm from one end of the module; Fig. 9c is a plot of the corresponding localization accuracy as deduced from the charge ratio, without any correction and including the beam width (10 mm); it has a r.m.s. of 30 mm (3.7% of the wire length). Other authors have reported,[10] in very similar conditions but using a resistive wire, localization accuracies better than 0.4%.

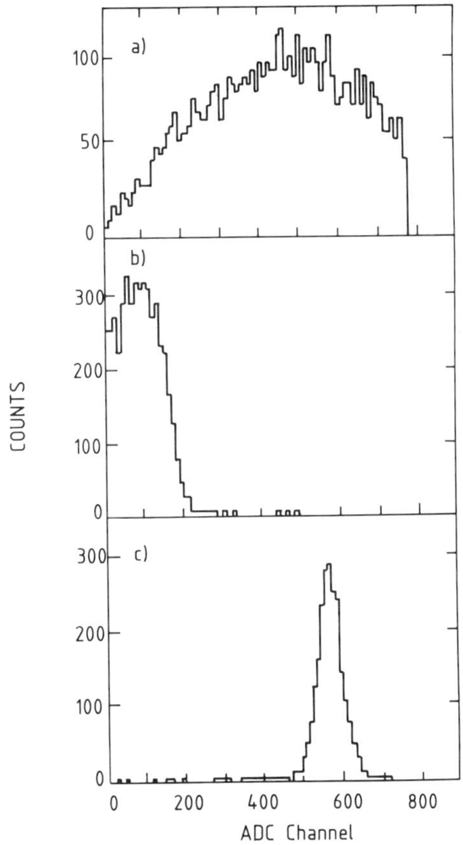

Fig. 9 Charge distribution for a collimated beam of minimum-ionizing particles, recorded on the right (a) and left (b) sides of an anode wire. c) Resulting localization accuracy distribution, deduced from the ratio of charges. It has a standard deviation of 30 mm, which includes the beam size (10 mm).

5. FUTURE DEVELOPMENTS

We are continuing the development work on the multidrift tube with the aim of improving its performance and reliability. A way of improving the transverse localization accuracy is to increase the operating pressure; as mentioned, DME has a vapour pressure of 4 bars at room temperature: however, we expect already a substantial gain at 2 bars, because of the corresponding reduction of the average distance of primary ionization clusters. Tubes were indeed mechanically tested up to 12 bars, without major problems (except a deformation of the carbon-fibre envelope); however, the general layout of the output connectors and the insulation of the high-voltage pins have to be improved to allow safe operation at higher voltages.

We are also testing alternative ways of holding the wires to the pins that would allow the use of resistive wires (these have very poor soldering characteristics); fast-hardening conductive epoxies seem very promising for this purpose. In order to reduce the material in the end plates of the modules, we are also considering a direct connection of the signal miniature cables to the anode pins, by gluing or soldering: this will also improve the reliability and decrease the serial resistance of the contacts.

The success of this approach to the realization of a high-rate, high-resolution detector will depend, of course, on the finding of cheap and reliable construction techniques, as well as on the development of high-density electronics.

REFERENCES

1. H. Nelson, Results from the MAC vertex detector, these proceedings.
2. R. Bouclier, G. Charpak, L. Ropelewski, J.C. Santiard, N. Solomey and F. Sauli, IEEE Trans. Nucl. Sci. NS-33:169 (1986).
3. R. Bouclier, G. Charpak, W. Gao, Ph. Miné, A. Peisert, J.C. Santiard, F. Sauli and N. Solomey, Nucl. Instrum. Methods A252:373 (1986).
4. A. Breskin, G. Charpak, F. Sauli, M. Atkinson and G. Schultz, Nucl. Instrum. Methods 124:283 (1975).
5. J.C. Santiard, A low threshold hybrid amplifier–discriminator, CERN EP Internal Report 82–04 (1982).
6. F. Villa, Nucl. Instrum. Methods 217:273 (1983).
7. S. Majewski, Proc. Workshop on Radiation Damage to Wire Chambers, LBL–21170 (1986), p. 239.
8. G. Bari et al., Nucl. Instrum. Methods A251:292 (1986).
9. F. Villa, personal communication at this meeting.
10. C.C. Young et al., IEEE Trans. Nucl. Sci. NS–33:176 (1986).

Discussion F. SAULI

B.R. KUMAR

Can you estimate the cost in mass production?

F. SAULI

The material cost for one tube (down to, but not including the electronics) is about 150 $. It takes, with the present method of construction, two weeks to one technician to manufacture a tube, which is by far more expensive. It is clear that to build hundred of tubes one will have to build some hardware for automating the mounting, e.g. to stretch wires and so on.

M. ATAC

Could you make a comment on the pulse height distribution across the cells from the center to outer layers?

F. SAULI

For most of the runs we used an amplifier + discriminator to record time, so I do not know the pulse hight distribution (but I know that efficiency is above 98% on each wire). We have one file of data, recorded to do current division, from which in principle I can reconstruct pulse height as a function of drift distance. But this analysis has not been done yet.

RESULTS FROM THE MAC[1] VERTEX CHAMBER

Harry N. Nelson

SLAC and Stanford University
Stanford, CA USA

INTRODUCTION

In the summer of 1983, the MAC and Mark-II collaborations gave evidence for bottom hadron lifetimes or 1-2 ps;[2] based on data accumulated at the PEP e^+e^- storage ring at SLAC. The importance and fundamental nature of this measurement[3] prompted the MAC collaboration to propose, in autumn 1983, the addition of a high precision vertex chamber to the MAC detector. The design utilized thin walled cylindrical drift tubes made of aluminized mylar. The device was installed in autumn 1984, immediately provided useful data, and continued to operate until March, 1986, when MAC data taking ceased. It has achieved the highest spatial resolution –35 μm– of any drift chamber used at colliders for physics results. In this paper, we describe the research and development of this chamber, its performance, and the B-lifetime result obtained with it.

The success of this chamber has helped convince many collaborations, including the Mark-III at SPEAR, CLEO at CESR, the upgraded TPC at PEP, AMY at TRISTAN, and JETSET at LEAR to build or propose similar devices. Indeed several features of this type of chamber are well matched to the probable conditions at the ELOISATRON or SSC: the small cell size is suited to the high rate and dense tracking environment anticipated; the simplicity and ruggedness makes a system of 10^5 channels credible. It should be noted that the MAC vertex chamber has operated only 4.6 cm from colliding beams, and has suffered *no* degradation due to radiation damage; this augurs well for survival in the more extreme conditions of the ELOISATRON or SSC.

Motivations and Strategy

Our initial motivation for using the thin walled mylar tubes[4,6] (often referred to as 'straws' because they are made in a spiral wrap, much like paper drinking straws) was simplicity and robustness. Problems of field wires and drift field mapping are eliminated. If a sense wire breaks, it is physically and electronically isolated from the other sense wires, so does not harm the rest of the chamber. However, there are other advantages to the use of straws. We have obtained 20 μm spatial resolution in beam tests with them, and 35 μm in our vertex chamber in use for physics results. Their small size is not only useful for the high rate environment, but aids pattern recognition in the dense regions of multitrack events. Their large, flat cathode makes straws less susceptible to some ills of radiation damage, including whisker growth and cathode coating.

Our Monte Carlo studies[7] indicated that the spacing of ionization clusters, and not diffusion, dominates the spatial resolution for drift distances as short as those in straws. This is in contrast to jet type cells, and indicated that the best strategy for high spatial resolution is to trigger the timing discriminator at the least possible ionization, instead of using the centroid in time of ionization arriving at the sense wire. To achieve a low threshold, we decided to operate the chamber at the highest possible gas gain, even in the 'self quenching streamer' mode if possible. We also decided to operate the chamber at high pressure, to reduce the spacing between ionization clusters.

Of course, a disadvantage of high gas gain is that the chamber might age quickly from beam-related radiation. We carefully designed our shielding and avoided this problem. Other disadvantages of straw chambers include: 1) the drift region close to the wire makes up a larger fraction of the active region than in jet type cells; this region *can*, but by no means *must*, have poor spatial resolution; 2) high gas gain is not linear with ionization, possibly impeding dE/dx measurements; 3) the radial drift field makes multihit resolution difficult; 4) the matter in the straw walls causes photon conversion and multiple coulomb scattering; and 5) very long straws are hard to keep straight. None if these disadvantages appear insurmountable.

We took great care in the machining of the endplates and the placement of our sense wires, but our strategy was always to survey the positions of the wires *in situ* with tracks from Bhabha scattering events. This has proved very successful. We have established the wire positions to 10 μm in this manner.

Gas Gain

Using a test straw of 6.9 mm diameter and 20 μm sense wire, we discovered that the aluminized mylar walls were photosensitive. Light from an incandescent bulb was shone on the straw wall, and caused ejection of photoelectrons, which we then used for the study of gas gain. Battistoni *et al.* have observed similar behavior in aluminum-walled streamer tubes.[5] We observed that the efficiency of the cathode for photoconversion was highly dependent on the history of the straw; higher efficiencies could be induced by several seconds of high voltage discharge in the straw.

We noticed that signals from the large amount of primary ionization from the 5.9 keV X-ray of an Fe^{55} source never went through the characteristic 'jump' from proportional to streamer mode. It is possible that electrostatic saturation limited the gain. In contrast, the signals from single electrons *did* go through the characteristic jump, as shown in Figure 1. In the streamer mode, single electron pulses were nearly as large as those from 5.9 keV X-rays. The stable streamer 'plateau' for single electrons lasted for 500 V in 4 atm absolute 50% Argon-50% Ethane. We found similar behavior in two gas mixtures with no organic additives: 49.5% Argon-49.5% CO_2-1.0% CH_4, with a 200 V plateau; and 50% Argon-40% CO_2-10% Xenon, with a very nice 900 V plateau. The limiting behavior at the high end of the plateau was photoelectric feedback from the cathode. Stable, large pulses from single electrons have been achieved in Magic Gas,[8] where the addition of freon suppresses feedback from the cathode; similar gas gains have been achieved in Dimethylether.[9]

Operation at high voltages where single electrons produce streamers would lead to the best spatial resolution for short drift distances, as discussed in the previous section. The noise requirements on the preamplifiers would not be stringent.

Radiation Damage

The primary obstacle to operation at high gas gain near colliding beams is damage induced by background radiation. We conducted studies of radiation damage in a test apparatus as similar as possible to our proposed vertex chamber. In particular, we conducted the tests at 4 atm absolute pressure, and did not circulate the gas; plastics such as teflon, PVC, and epoxy were used in the test vessel. More details of our tests appear in Ref. (10).

We found a variety of Argon-Hydrocarbon mixtures, including 50% Argon-50% C_2H_6, 70% Argon-30% C_4H_{10}, and 49.5% Argon-49.5% C_2H_6-1.0% H_2 failed reliably and repeatedly at 0.05 C/cm of accumulated charge. The straw would go into continuous discharge; replacement of the anode wire restored performance, clearly indicating anode damage. We opted not to add alcohol, as this damaged the glue in the straw.

We decided that 0.05 C/cm was unacceptable, so we turned to a non-organic quencher, CO_2. 50% Argon- 50% CO_2 showed no anode damage;

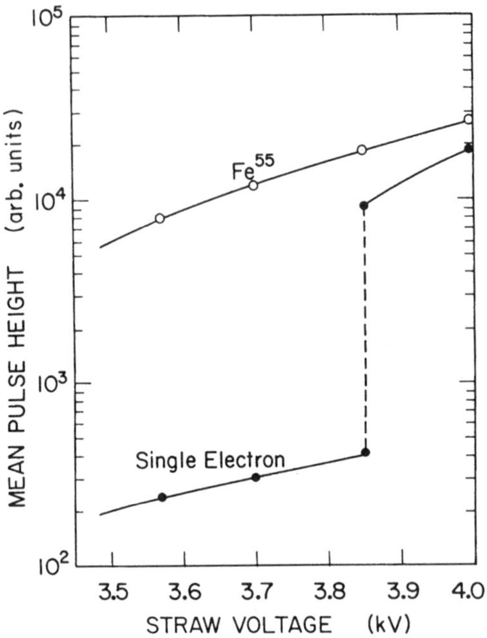

Figure 1 Gas Gain for 5.9 keV X-rays and single electrons. Data taken with 50% Argon- 50% Ethane at 4 atm. absolute, with a 20 μm sense wire and a straw of radius 6.9 mm. The signal from Fe^{55} X-rays never underwent the characteristic 'jump' into streamer mode, but the single electron signal did.

in fact, study of the Fe^{55} X-ray spectrum indicated the anode wire became cleaner up to 0.05 C/cm accumulated charge. However, the straw failed at 0.25 C/cm accumulated charge due to disappearance of aluminum on the cathode. We hypothesized that positively charged CO_2 fragments attacked the cathode. We added a small amount of CH_4 hoping charge transfer would occur, and that CH_4 species would cause less damage to the cathode. Only a slight improvement was observed; we settled on 49.5% Argon-49.5% CO_2-1.0% CH_4 as our operating gas. Use of Xenon rather than CH_4 led to a similar slight improvement in lifetime.

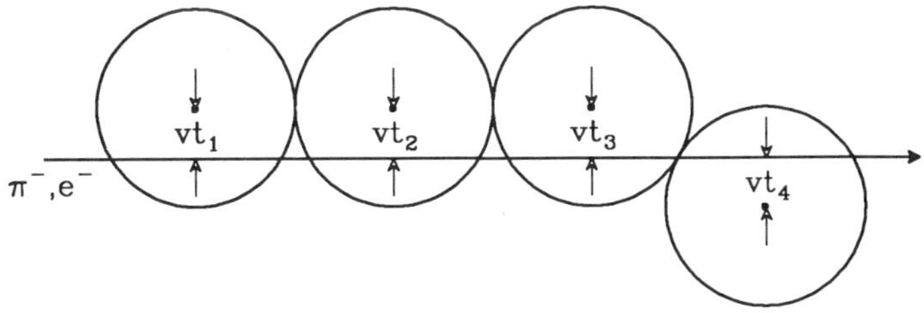

Spatial Resolution: $\Delta = t_2 - \frac{1}{2}(t_1 + t_3)$

$$\sigma_W = \sqrt{\tfrac{2}{3}}\, v\sigma_\Delta$$

Drift Velocity: $v \simeq 50\ \mu m/ns$

Figure 2 Straw Arrangement in Test Beam. Electrons and pions from were supplied in the SLAC Research Yard. v is the drift velocity, which was typically a constant $50\,\mu m/ns$ for the Argon-Hydrocarbon gases used. The spatial resolution, σ_W, was determined from the drift times in the first three straws. The fourth, staggered straw was useful for determining the time to distance relationship.

The aluminum layer on our mylar had a surface resistivity of 0.8Ω/square, indicating a thickness of $\approx 30\,nm$. A thicker layer or aluminum, or use of a different metal, may extend the liftime or straws.

Beam Test

We conducted a beam test[6] to determine the resolutions possible with a drift chamber made of straws. A variety of high voltages, discriminator thresholds, gases, and pressures were used. An unseparated beam of 8 GeV/c electrons and pions in the SLAC Research Yard was used. Four straws were arranged as shown in Figure 2. The simple geometry allowed a straightforward and systematic free determination of the spatial resolution. The spatial resolution as a function of drift distance is given in Figure 3, for a variety of discriminator thresholds. In the streamer mode at 4 kV, the spatial resolution was an average of 25 μm, *and did not degrade near the sense wire.* The data agree quite well with Monte-Carlo calculation. Figure 4 shows the spatial resolution measured for a variety of high voltages and gas pressures. Based on these test beam results, one sees that straws obtain spatial resolutions in the 25 μm regime.

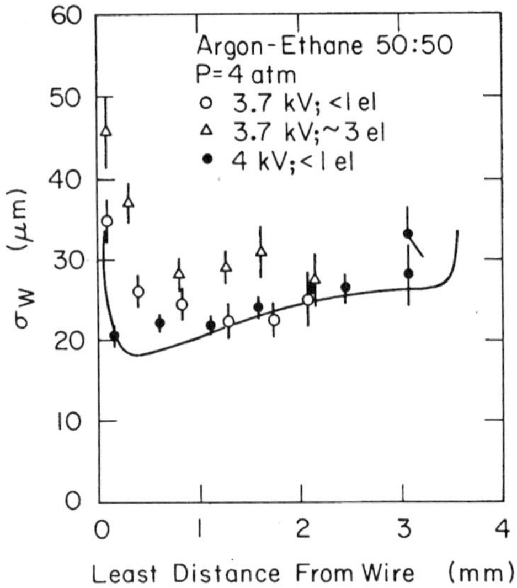

Figure 8 Spatial Resolution in Test Beam vs. Drift Distance for a Variety of Thresholds. Absolute calibration of the thresholds were performed with a source of single electrons produced via the photoelectric effect on the straw cathode. The streamer mode was fully efficient at 4 kV for single electrons. Our Monte-Carlo calculation is shown as well.

Design

Radiation Shielding. At the beginning of PEP operation, the MAC detector ran for a short time with a 3.15 cm radius beam pipe, and minimal shielding was employed. From this experience, we deduced that the primary radiation background came from off-energy electrons that were overfocussed by the 'low-beta' quadrupoles into the beam pipe and tracking chamber. Another important background source was synchrotron radiation, both from the 'soft bend' dipole and the low-beta quad.

We designed a close-in collimator to suppress these backgrounds. The collimator was a tantalum annulus of 2.8 cm inner radius and 0.8 cm thickness, placed 60 cm from the interaction region. Shielding of heavimet, a sintered tungsten material, surrounded the collimator to absorb electromagnetic showers. The location and dimensions of the collimator were chosen to allow sufficient room for this shielding, and to shield the vertex chamber from direct particle and photon trajectories from the soft bend and low-beta quad magnets. Figure 5 shows the shielding and these trajectories. A

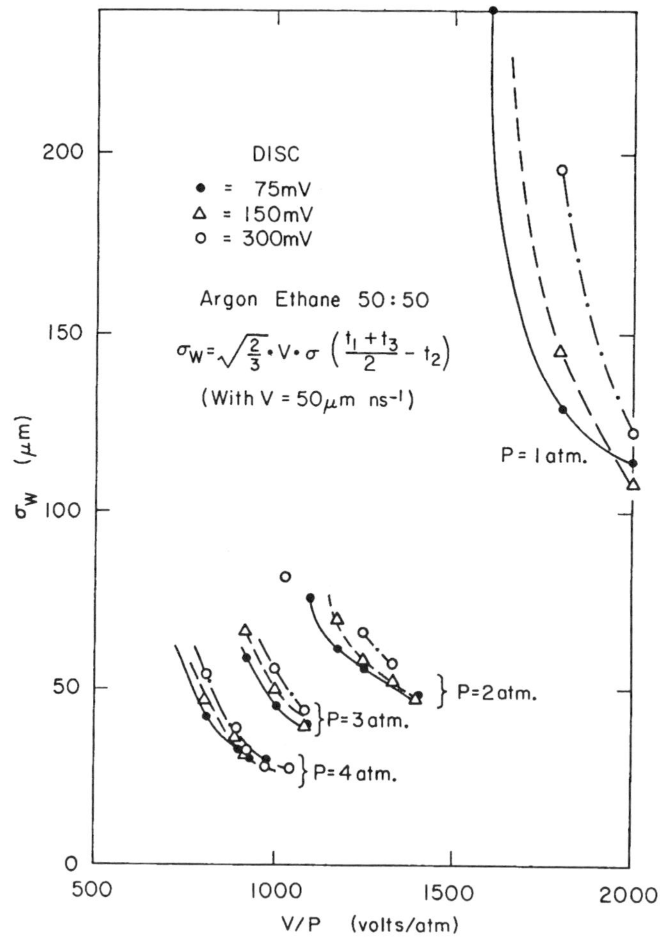

Figure 4 Spatial Resolution in Test Beam For a Variety of High Voltages, Pressures, and Discriminator Thresholds.

75 μm titanium liner, placed inside the beam-pipe to attenuate the ≈ 10 keV fluorescence of tantalum by a factor of ≈ 30, completed the shielding.

The heavimet occluded the low angle electron veto counters previously installed for the MAC search for $e^+e^- \to \gamma +$ *weakly interacting particles*.[11] To maintain this search, an annulus of BGO was placed between the heavimet and interaction region. The dense BGO was read out with compact photodiodes,[12] a well matched combination for such a cramped situation. This was the first application of this technology in an operating colliding beam experiment.

<u>Beampipe and Pressure Vessel.</u> The beampipe was 0.11 cm thick beryl-

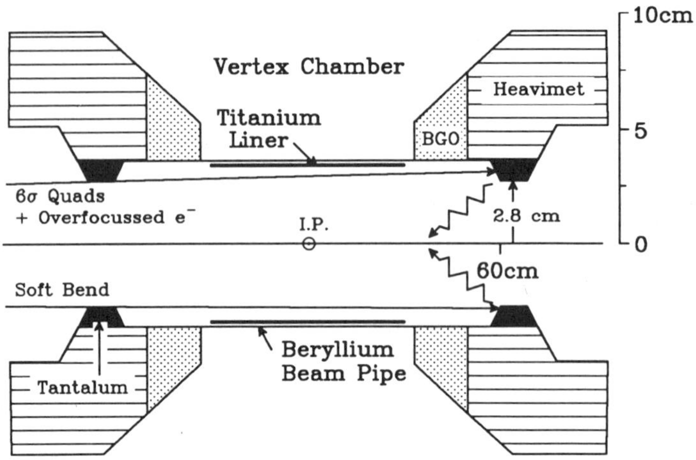

Figure 5 Shielding Configuration. The tantalum close-in collimators and surrounding heavimet shielding protected the vertex chamber from off-axis electrons, and off-energy electrons overfocussed by the nearby low-beta quadrupoles. The collimators also shadowed the vertex chamber from direct synchrotron radiation from the soft bend and low-beta quad. The titanium liner absorbed the $\approx$ 10 keV fluorescence from the tantalum. The remaining background were $\approx$ 25 keV synchrotron light from the low-beta quad, that was scattered off the tantalum.

lium, brazed to aluminum tubes at its ends. The aluminum tubes were in turn welded to the pressure vessel endplates, which were an explosion welded bimetal of aluminum and stainless steel. The remainder of the pipe was stainless steel. The outer wall of the pressure vessel was a 2.54 mm aluminum tube, also welded to the bimetal endplates. The outer wall carried the mechanical load of the beampipe and shielding. I-beams place vertically above and below the beampipe carried the load on either side of the vessel.

The pressure vessel was distinct from the vertex chamber. The coaxial signal cables from the chamber were potted in epoxy plugs that were captured in the endplates and sealed with O-rings.

Chamber. The straws were strung on a spool that consisted of a 0.076 cm thick, extruded beryllium tube, to which 1.6 cm thick aluminum endplates were glued. Holes were precision reamed in these endplates to accept the endplugs of the straws. The endplate layout is shown in Figure 6. There are six radial straw layers, arranged in three pairs. The inner layer of each pair is close packed in azimuth. The outer layer contains the same number

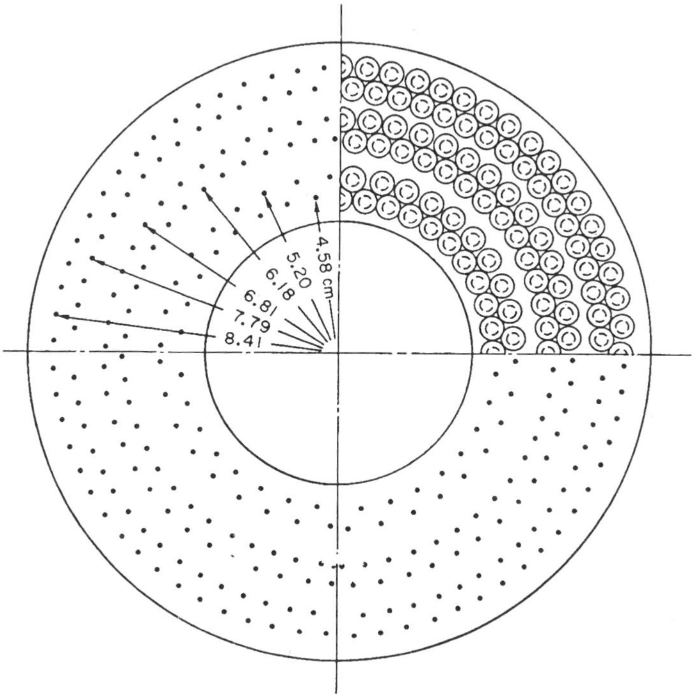

Figure 6 Spool Endplate Layout, Showing Placement of Straws in Radius and Azimuth.

of straws as the inner, but is staggered one half cell in azimuth. The stagger resolves the left-right ambiguity in the straws.

The straws themselves had an inner diameter of 6.9 mm. The wall consisted of a double wrap of 50 μm mylar, each layer coated on one side with 0.8Ω/square. surface resistivity aluminum. The conductive layers faced inward and outward. The pitch angle of the spiral wraps was 28°.

The end-fitting design is shown in Figure 7. An aluminum collar was glued with conductive epoxy to the straw inner face, trapping a Delrin (acetal) plastic plug. The plug performed several functions. First, it insulated the high voltage on the sense wire from the ground of the endplate and straws. For this reason, virgin Delrin, Delrin that was not reformed from old pieces, was used, to avoid cracks and crevices that lead to high voltage breakdown. Considerable trial and error lead to the shape of the plug inside the straw, in order to avoid edge breakdown at extremely high voltages. Second, the plug provided a means to apply tension once the straw was mounted between the endplates. The threaded tip of the plug allowed a nut to apply tension when tightened. Third, the plug provided a seat for the feedthrough that held and positioned the sense wire. Last, the plug provided a conduit for gas

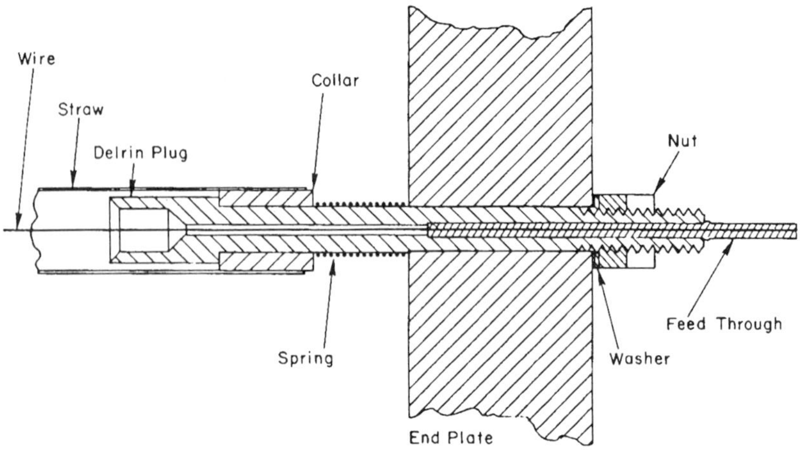

Figure 7 End-fitting Design for Straws. See text for more details. The collar contacted the endplate on one end of the straw; on the other, the spring made electrical contact.

flow through the straw, via two channels cut lengthwise on its outer radius. The 30 μm gold-tungsten sense wire was threaded through the 100 μm hole of a tempered stainless steel tube, or feed through, which was crimped to capture the wire.

Signal Processing. Figure 8 shows the electronic layout of the vertex chamber and associated electronics. The sense wires were operated at positive high voltage, and one end was terminated through a standoff capacitor and a resistor. The other end was connected to 19 meters of RG179B, a 75Ω, teflon insulated, miniature coaxial cable. This coaxial cable provided high voltage, and carried signals to preamps that were located away from the detector.

We used the LABEN 5242 preamplifier, a hybrid optimized for fast rise time and low noise. In retrospect, this preamp was better matched to sources with impedance higher than our signal cables.

Our discriminator circuit utilized the Plessey SP9687 dual comparator. Each discrimator channel used two comparators, one set to a low threshold, and one high, as shown in Figure 9. The discriminator output was fed to our time-to-voltage converters, or TVC's. These circuits demonstrated $< 300ps$ timing resolution in bench tests.

Gas System. The gas system was designed to maintain constant density in the chamber while allowing gas flow. Figure 10 shows the gas circuit of the final design. The gas pressure was maintained by mechanical regulators. These proved quite reliable, but were sensitive to ambient thermal variations,

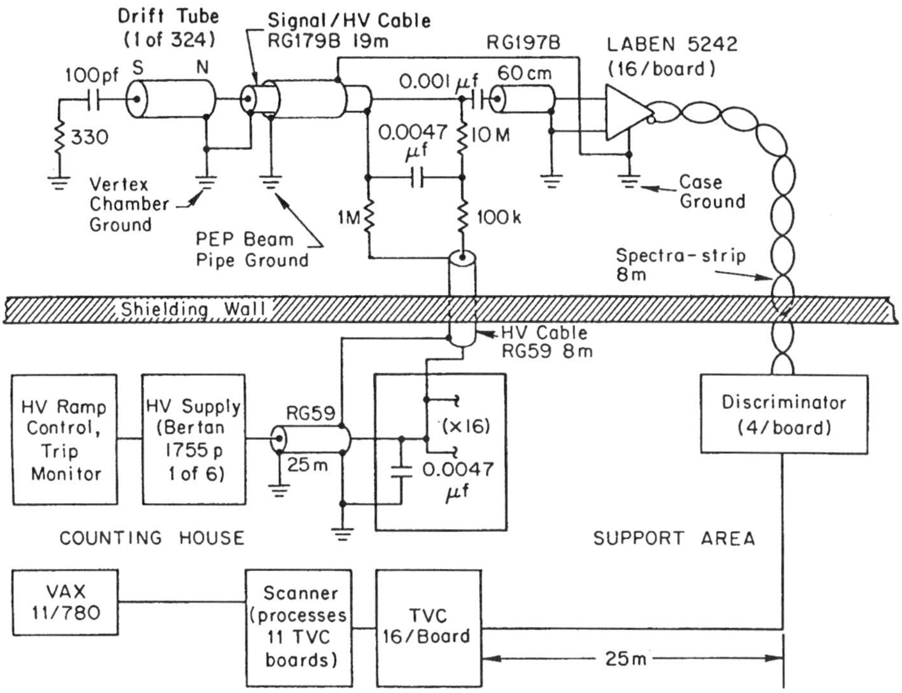

Figure 8 Electronic Layout of the Vertex Chamber. Straws are terminated in approximately their characteristic impedance at one end; signals go from the straw's other end, through 19 meters of coaxial cables to the preamplifiers, then to the discriminators and finally the TVC (time to voltage) drift time measuring system.

necessitating thermal isolation. Flow was maintained by a vacuum pump on the output of the system, which eliminated variation from atmospheric pressure fluctuation. The temperature of the chamber was maintained nearly constant by the cooling system of our solenoid; no effect of heating from the beam was observed.

This system maintained the gas density constant with an rms density of 0.8% over one week. The rms variation of the time to distance relationship contributed by this was a maximum of only 14 μm, at the straw wall. An electrical control system would seize control of the pressure if the density wandered outside $\pm 0.75\%$; this only occurred once in the operation of the chamber. Adjustment of the mechanical regulators for long term drift was made several times per week.

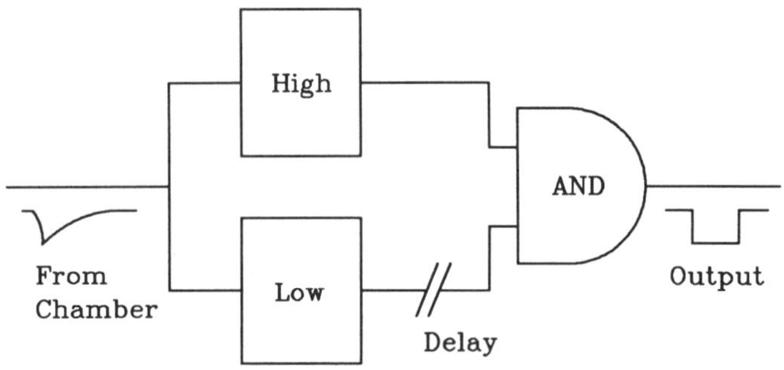

Figure 9 Principle of Discrimator Circuit. The low level output was delayed, so typically they reached the AND gate after the high level. Hence, the low level comparator controlled the timing of the output, but small pulses that did not fire the high level comparator were suppressed.

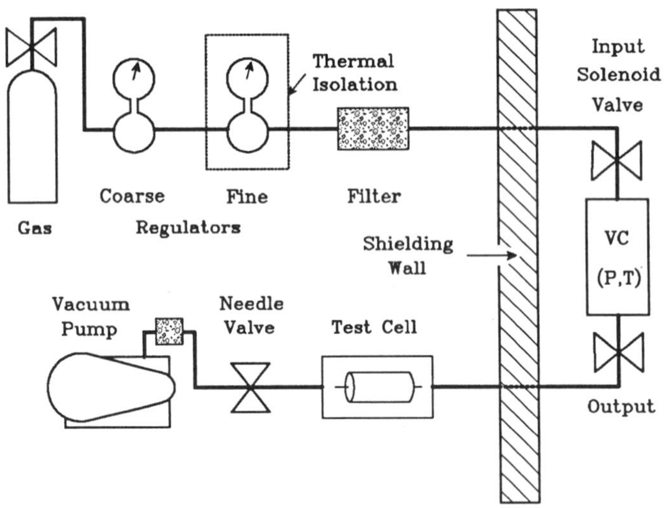

Figure 10 Gas System Circuit. Pressurized gas from cylinders flowed through Poly-Flo tubing to the vertex chamber, then back through a test chamber and a vacuum pump, which vented to the atmosphere. Absolute pressure was maintained by a pair of mechanical regulators on the cylinder output, and a needle valve on the output controlled the flow rate. Pressure and temperature were monitored in the vertex chamber (VC); if the density in the chamber drifted outside limits an electrical circuit could regulate the density via the solenoid valves.

Vertex Chamber Performance

Electronics. Refer to Figure 8 for the electronic layout of the vertex chamber and associated electronics. The electrical ground of the vertex chamber was insulated from the ground formed by the PEP beam pipe, to suppress pickup from the image currents of the beam itself. No pickup from this source was observed at the level of white noise from the preamplifiers. However, the long signal cables between the chamber and the preamplifiers proved susceptible to electromagnetic interference in the 1-100 MHz regime. Their susceptibility was reduced substantially by isolating both their center conductor and shield with 1-10 MΩ resistors from the high voltage supply cables. The dominant source of interference was the system for measuring drift times in the vertex chamber (VC), the time to voltage converters (TVCs). Moving these from 10 meters to 30 meters from the preamplifiers reduced the total pickup to approximately the level of the white noise of the preamps.

The long cables limited the 10-90% rise time at their output to 4 ns, and attenuated signals from the chamber by 15%. The output slew rate of the preamps was 90 mV/ns, yielding a 10-90% rise time of 8.0 ns for pulses from Fe^{55} sources in the chamber. The white noise of the preamps was 2.0 fC at their input with cables attached, 0.6 fC without. The low level of our discriminators was set at 6 fC input to the preamps, which corresponded to 30 mV at their output. Slewing and noise contributed less than 0.15 ns to our timing resolution.

TVC channels were calibrated with a CAMAC controlled delay generator. The delayed calibration pulse was injected at the preamplifier input, into half the channels at a time. For each channel, the delay was fit as a quadratic function of the voltage output from the TVC. The non-linearity contributed only 0.5% to the calibration. This calibration was performed 8 times over the 17 months of data taking.

The channel by channel pedestals had a standard deviation of 15 ns; this was governed by the variability of a CMOS switch in the TVC circuit. Over the data taking period these pedestals were stable to 1.3 ns. The slope term in the quadratic fit was far more reproducible from channel to channel, varying only 1.3%; this was governed by one resistor and one charging capacitor. The drift was only 0.5% in these slopes. Figure 11 shows the actual distributions of the offsets and slopes.

The calibration reproduced input delays to better than 0.5 ns. During data taking, four spare channels were fed fixed delays; these indicated that the system achieved a timing resolution over a typical two hour data taking run of 250 ps.

The combined effect of 1)slewing in the signal cables; 2)noise in the preamplifiers; and 3)resolution of the TVC system produces a timing resolution of < 300 ps. The maximum drift velocity in the vertex chamber itself

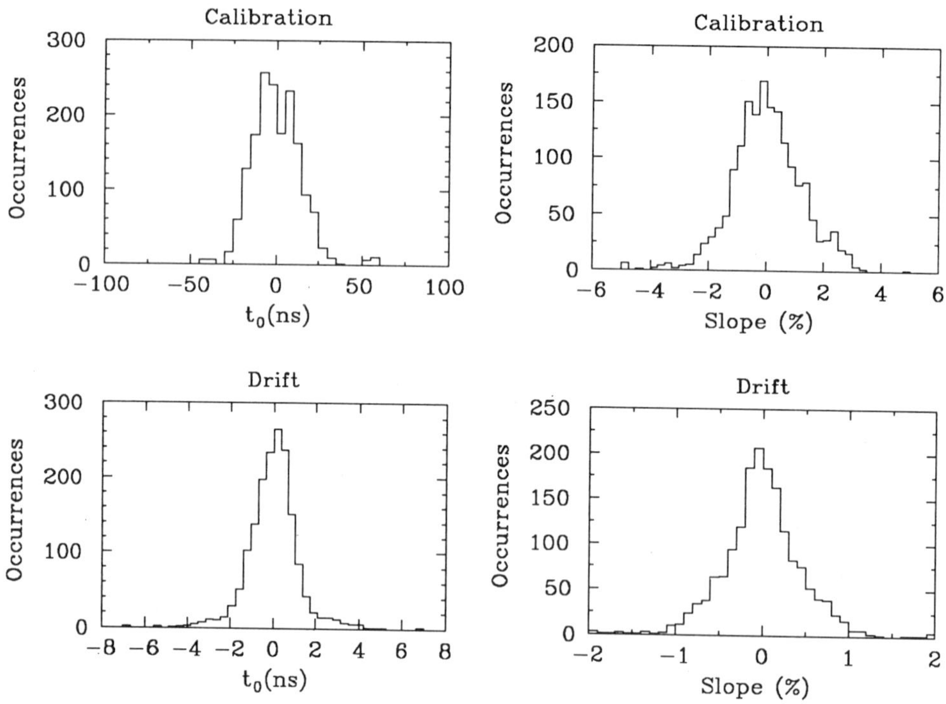

Figure 11 Fitted constants in the TVC system. The top plots are distributions of the timing pedestals and slopes. The bottom plots give the variability of these constants over the 8 recalibrations done during the 17 months of data taking.

was $100\,\mu$m/ns. Therefore the degradation in spatial resolution from purely electronic sources was $< 30\,\mu$m.

Operating Point. The high voltage and threshold at which the chamber was operated was a compromise between spatial resolution and radiation damage considerations. As discussed in earlier, for the short drift distances in our cells, triggering on the first arriving electron was desirable. The observed charge for various high voltages appears in Figure 12. At 3900 V, our threshold of 6 fC was very near the most probable signal for a single electron. Synchrotron X-rays were the dominant source of current drawn in the chamber. Approximately two random hits per beam crossing occurred in the chamber, causing a current of 3-5 nA/cm to be drawn in the innermost layer, for a high voltage of 3900 V. Higher voltages put the chamber in danger of accumulating more than 0.2 C/cm charge in the innermost layer, so we operated at 3900 V. In retrospect, we were slightly conservative; the innermost layer accumulated only 0.03 C/cm over the entire data taking period.

The two random hits per beam crossing interfered in no way with our

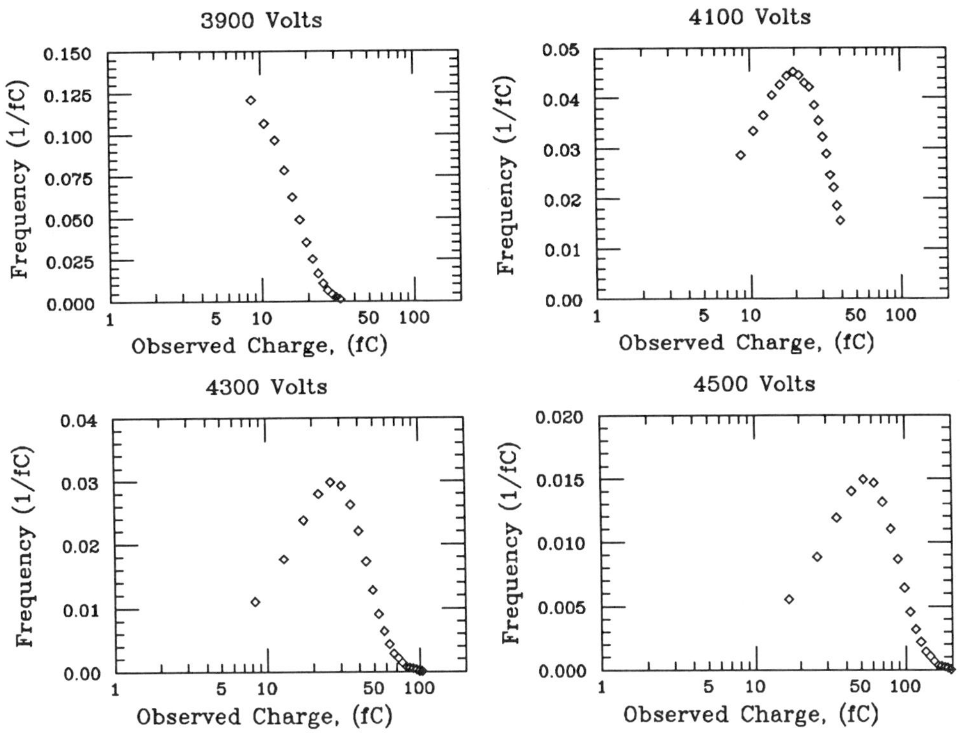

Figure 12 Single Electron Gain. The four plots are for different high voltages applied to the sense wire. These data were obtained with a test straw identical to those used in the vertex chamber. An incandescent light produced single electrons via the photoelectric effect on the straw cathode. Not shown are the very large charges due to self quenching streamer formation which were approximately 500 fC. Few single electrons initiated streamers at 3900 Volts, but most did at 4500 Volts. The spectra are not shown below 8 fC where amplifier noise begins to dominate. The gas mixture was 49.5% Argon, 49.5%CO_2, 1.0%CH_4, at an absolute pressure of 4 atm.

track finding procedure. Electronic cross talk did not limit our ability to maintain a low threshold: studies with Bhabha events indicated that electronic cross talk produced $< 0.1\%$ of the hits in five of the six layers. However, knock-on electrons did produce hits in the cells immediately neighboring a cell penetrated by a track 1-2% of the time.

Off–line calibrations

We performed off–line calibrations of wire positions, drift timing constants, the time to distance relationship, and drifts in these quantities. Bhabha scattering events proved very useful for this purpose due to their simple topology. The use of Bhabha scattering events is by no means mandatory. Any sample of tracks free of bias from pattern recognition could have been utilized. Bias can occur, for example, in dense regions of multihadron events, when two tracks pass through the same cell but only one is recorded.

Our calibration techniques require existing programs to link hits into tracks and to fit the track parameters. The programs used to do this are described in and reference (13).

Global Alignment. For a description of our tracking parameterization, see the appendix. The VC measured only the three parameters that describe the track in the x-y plane. Of these three, only two were well measured, δ_{0vc} and ϕ_{0vc}. Our existing central drift (CD) chamber extended from 12 cm to 45 cm in radius, and had stereo layers, so it measured all five helix parameters. Systematic comparison of δ_{0vc} and ϕ_{0vc} with δ_{0cd} and ϕ_{0cd} as a function of ϕ_{0cd}, θ, and z_0 allowed determination of the five relevant parameters that describe the relative global orientation of the VC and the central drift chamber.

These five parameters were measured 13 times over the course of data taking. Figure 13 shows the stability of the three most important over these measurements. The horizontal and vertical translations were stable to 10 μm.

Time to Distance Calibration. The gas used in the vertex chamber, 49.5% Argon, 49.5%CO_2, 1%CH_4, has a drift velocity approximately proportional to the applied electric field. In a cylindrical geometry, this implies the drift distance r and the drift time t are related by $r \approx \sqrt{t}$. This simple relationship provided an initial guess for our time to distance function.

We improved the initial guess with the following procedure. Tracks in Bhabha scattering events with 4-6 hits in the VC and 7-10 out of 10 possible hits in the CD were chosen. A VC hit waseliminated, and the remaining VC and CD hits were used in a new track fit. This new fit was used to predict the drift distance, r_{pred}, in the eliminated VC cell. Errors in the initial time to distance relationship tend to cancel in r_{pred}. We give a scatter plot of r_{pred} as a function of the measured drift time, t_{meas} in in Figure 14. The $r_{pred} \propto \sqrt{t_{meas}}$ character is evident. We fit data such as that in Figure 14 to obtain the improved time to distance relationship, $r(t)$, or its inverse, $t(r)$. Fitting r_{pred} as a function of t_{meas} or vice versa yield systematically different results for short drift distances, because no drift distances are defined as negative. Furthermore, the large slope of $\sqrt{t}$ near $t_{meas} = 0$ makes fitting r_{pred} as a function of t_{meas} difficult. Fitting t_{meas} as a function of r_{pred} proved to require fewer parameters and converged more easily. Cubic splines or orthogonal polynomials were used as fitting functions. The time to distance relationship converged to within 10 μm after one iteration of improvement. The relationship was evaluated five times over MAC data taking, usually after slight changes in gas mixture.

To account for drifts in timing, gas density, and electronics, run by run corrections were applied to the time to distance relationship. A typical run

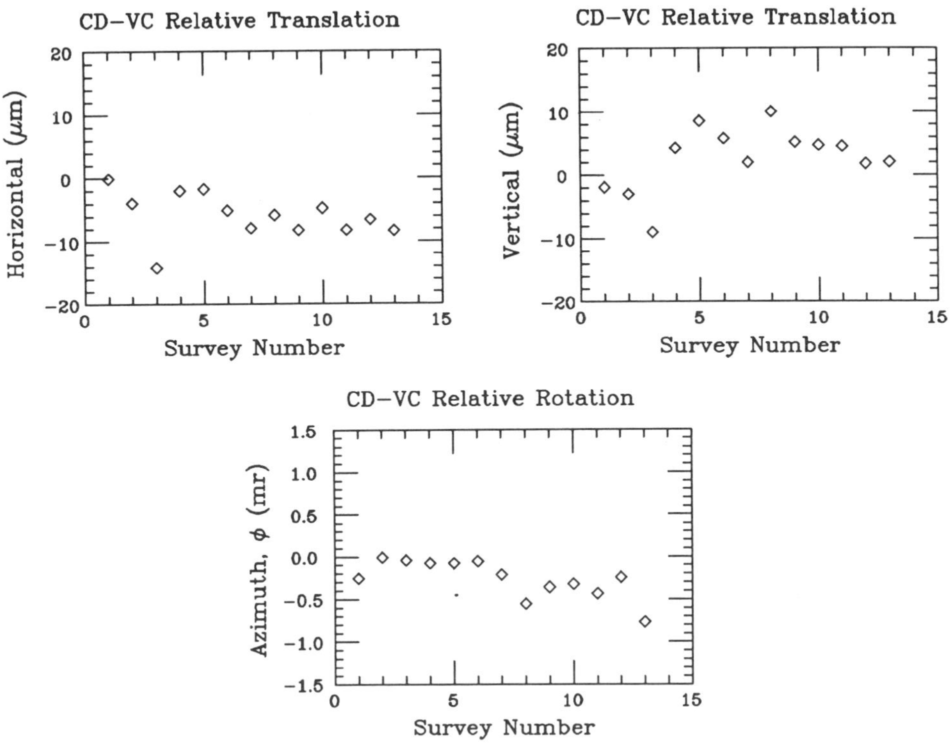

Figure 13 Drift of Global Alignment Parameters. The global alignment of the VC relative to the CD was done 13 times during the 17 months of data taking. The drift of the two translation parameters are shown in the upper plots, and the drift of the relative azimuth in the lower. The alignment was stable to 10 μm.

lasted 90 minutes and yielded 100 Bhabha scattering events for calibration. The drift time actually used in the time to distance relationship, t, was computed from

$$t = t_{0rr} + s_{rr} t_{meas}$$

where t_{0rr} and s_{rr} are the run by run timing offset and slope. These are found using a procedure very similar to that used to extract the time to distance relationship. Typically t_{0rr} could be measured to an accuracy of .2 ns and s_{rr} to 2 %. t_{0rr} remained stable for weeks, then would suddenly shift, usually due to a power failure or changes in PEP timing. The slope term was very stable, reflecting the stability of the gas density.

Channel by Channel Calibration. The timing constants and physical positions of each cell were calibrated offline. The procedure was similar to that used to obtain the time to distance calibration. The residuals $\delta r =$

$r_{pred} - r(t_{meas})$ and $\delta t = t(r_{pred}) - t_{meas}$ were fit for each cell with four constants:

1. An offset in time, t_0.
2. An offset in distance, in the azimuthal direction, d_0.
3. A slope in time, s.
4. A tilt in z, the axial direction.

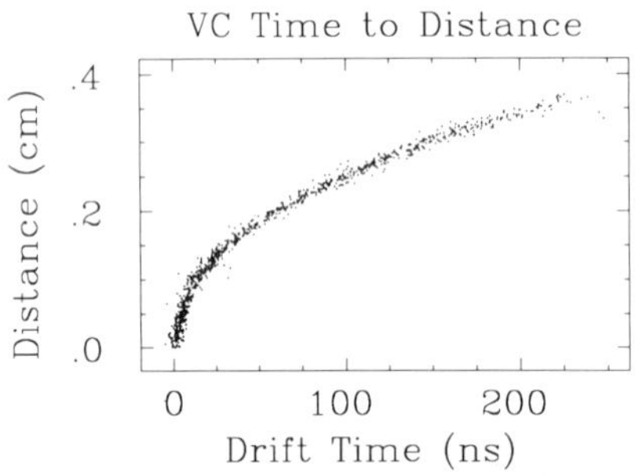

Figure 14 Time to Distance Relationship. The drift time is that measured in a cell of the VC, t_{meas}. The drift distance is obtained by dropping that cell from the track fit, then predicting the drift distance in the missing cell. In this manner the cell's 'pull' in the track fit is avoided. A fit is made to data like that above to obtain the functional time to distance relationship.

d_0 and t_0 effect δr with opposite signs on opposite sides of the sense wire. In a typical calibration, 100 runs were used, corresponding to $6\,\mathrm{pb}^{-1}$ of integrated luminosity. This yielded approximately 150 residuals in each of the 324 cells. The largest correction was d_0, appearing in Figure 15. The calibration was performed 13 times. The standard deviations of d_0 for each cell over the 13 measurements also appear in Figure 15, indicating a reproducibility of 10 μm.

The channel by channel calibrations were made in a single pass through Bhabha scattering data. The improvement in spatial resolution due to the calibration was then immediately known via the multivariate analog of the

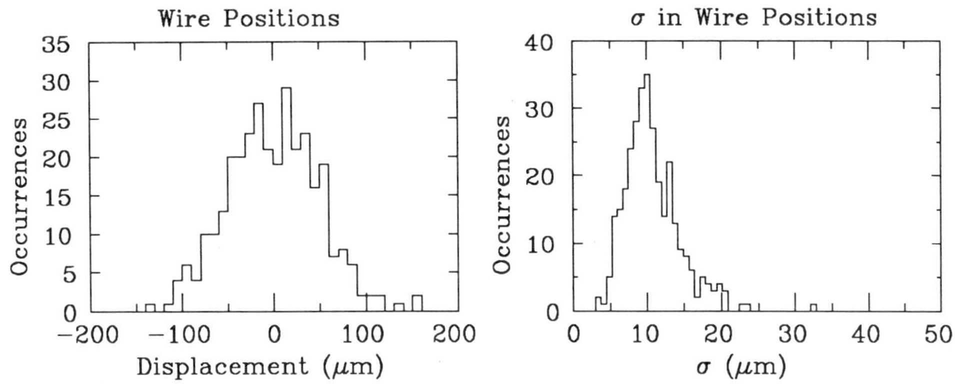

Figure 15 Sense Wire Positions. The histogram on the left gives the wire offsets in azimuth as obtained in the channel by channel calibration. The accuracy of wire placement was controlled by the size of the hole in the crimp pins that capture the sense wire, and the pins' tendency to bend slightly. This calibration was repeated 13 times; the histogram on the left is the standard deviation of each wire's offsets among the 13 calibrations. It indicates that the calibration procedure found the wire positions to $\approx 10\,\mu$m accuracy.

Table 1 Contributions of Survey Terms. These are the contributions to the *uncorrected* spatial resolution that may be attributed to the various survey corrections. The tilt in z indicates a deviation of the ends of the sense wires from their average position of $38\,\mu$m.

Correction	Contribution (μm)
t_0	45
d_0	49
s	20
tilt in z	11

relationship:

$$\sum_{i=1}^{N} (x_i - \overline{x})^2 = \sum_{i=1}^{N} x_i^2 - N\overline{x}^2$$

Use of orthogonal polynomials as fitting functions allowed straightforward isolation of the contribution of the various constants. Table 1 gives the importance of each of the constants.

The miscalibration of a channel in the vertex chamber 'pulls' the track fit and therefore can induce an apparent miscalibration in of channels in other layers. This coupling was not explicitly accounted for in our calibration procedure. The offsets in azimuth between layers tend to reduce the importance of this effect. Further, when we iterated the calibration procedure we found the constants to converge within one iteration to 10 μm of their value after three iterations.

Certain correlations between distant cells were found to occur. For example, the initial channel by channel calibration resulted in a slight uniform translation of all cells by an equal distance in the azimuthal direction. This created the appearance that the tracks in Bhabha scattering events were tangent to a circle centered on the geometric center of the chamber. The effect was eliminated by constraint of the two tracks in a Bhabha event to have equal but opposite distances of closest approach to the geometric center of the chamber. This constraint was made only for the track fits used to determine the calibration constants. Similar distortions in the mean acollinearity and momentum difference were observed and suppressed. In general, correlations between distant cells are not necessarily well constrained in our cell by cell calibration procedure. However, spurious correlations are easily removed by constrained fitting of tracks, or by constraint of the final survey constants.

Pattern Recognition

Algorithm. Track segments were obtained by independent pattern recognition programs in the CD and the VC. The CD algorithm was complex, due to the presence of stereo layers and non-negligible track curvature.[13] The VC algorithm was simpler, because tracks are approximately straight lines in the VC.

We regard the six layers of the VC as three double layers for the first step of pattern recognition. Because of the overlapping, regular geometry of the VC double layers, the drift distances r_1 and r_2 for hits produced by a radial track are related by:

$$r_1 + \frac{R_1}{R_2} r_2 \approx r_0 \qquad [3.1]$$

Here R_1 is the radius of layer 1 or the double layer, R_2 that of layer 2, and r_0 the straw radius. Pairs of hits in double layers were associated if they met [3.1] within 200 μm. Random associations were thereby suppressed, with little loss of real tracks. Most loss of real tracks occurred for very low momentum tracks or tracks from very long lived species, such as K^0s and Λs. The ambiguity of which side of the sense wire the track passed is resolved by [3.1] in most cases.

Associated pairs in the three double layers were then linked based on their agreement in azimuth. A straight line fit was made to the resulting set of hits and drift distances. The straight lines found in the VC were

Table 2 Vertex Chamber Efficiency.

Events	VC Efficiency
Bhabhas	0.994 ± 0.001
Taus	0.982 ± 0.003
Multihadrons	0.939 ± 0.001

then compared to and matched with tracks found in the CD. Unmatched CD tracks were extrapolated into the VC and a search made for VC hits. A weighted, least squares fit was performed to both the VC and CD hits and drift distances. VC hits were assigned weights based on the measured resolution as a function of drift distance (Figure 19). Multiple scattering was accounted for by allowing partially constrained kinks at the beam pipe, the boundary between the VC and CD, and at fixed locations within the chambers. A final search was made for hits along the track fit, and .1% of the VC and CD hits were dropped due to large residuals.

Considerable effort was expended scanning graphical representations of events and interrogating the pattern recognition programs to tune its decisions. After some optimization of cuts, results at least as good as those from scanning were achieved.

Efficiency. Since the CD and VC are in one sense redundant devices, it was possible to measure the efficiency of the VC by using a sample of tracks well measured in the CD. Sources of inefficiency include dead cells, overlap of particle trajectories, and inadequacy of the track finding algorithm for particles with low momentum or very large impact parameter.

We define well measured tracks in the CD by four requirements: 1)ten hits from the CD must be linked to the track, 2)$z_{0cd} \leq 5$ cm., 3)$\theta \leq 20°$, and 4)$\hat{\chi} \equiv \sqrt{2\chi^2} - \sqrt{2n_D - 1} \leq 4$. Table 2 gives the efficiency for well measured CD tracks to link to ≥ 3 VC hits. The Bhabha efficiency was determined from 2 pb^{-1} of integrated luminosity, tau lepton and multihadron efficiencies from 94 pb^{-1}.

Track overlap was a source of inefficiency in the VC in multihadron events. If two tracks were separated by less than the diameter of a straw, 6.9 mm, they could leave only one hit in a layer of the VC. The staggering of VC layers usually prevented this phenomenon from happening in every layer. Double-hit electronics probably could not eliminate it, because the drift field for straws is radial. Figure 16 shows the efficiency for well measured CD tracks to link to ≥ 3 VC hits as a function of VC track separation. The VC track separation is the minimum distance in the azimuthal direction from the well measured track to the nearest CD track, within the VC boundary.

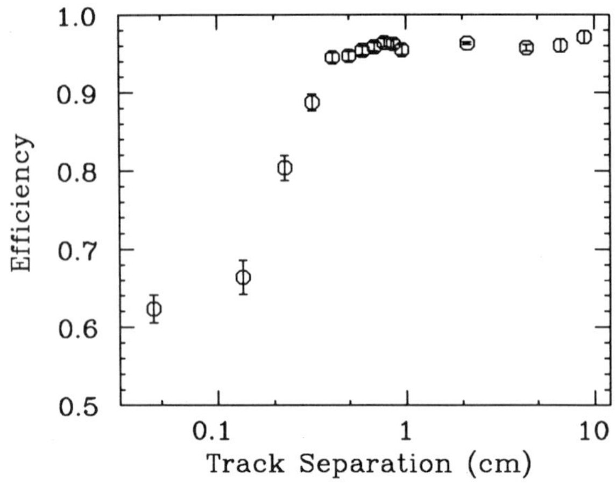

Figure 16 Effect of Track Overlap on VC Efficiency.

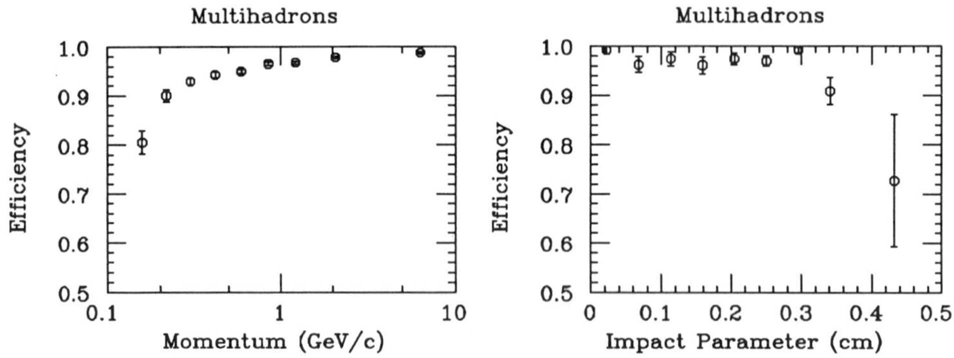

Figure 17 VC Efficiency for Various Momenta and Impact Parameters.

The efficiency was near 50% for small separation, indicating one track of the overlapping pair was found in the VC. For large separations, the efficiency was better than 96%.

The VC track finding algorithm was most efficient for straight tracks coming from the beam centroid. The efficiency for CD tracks with VC track separation greater than one straw diameter is shown versus momentum in Figure 17. The efficiency was $\geq 90\%$ above $200\,\mathrm{MeV}/c$. Tracks softer than $500\,\mathrm{MeV}/c$ had an error in impact parameter due to multiple scattering in the

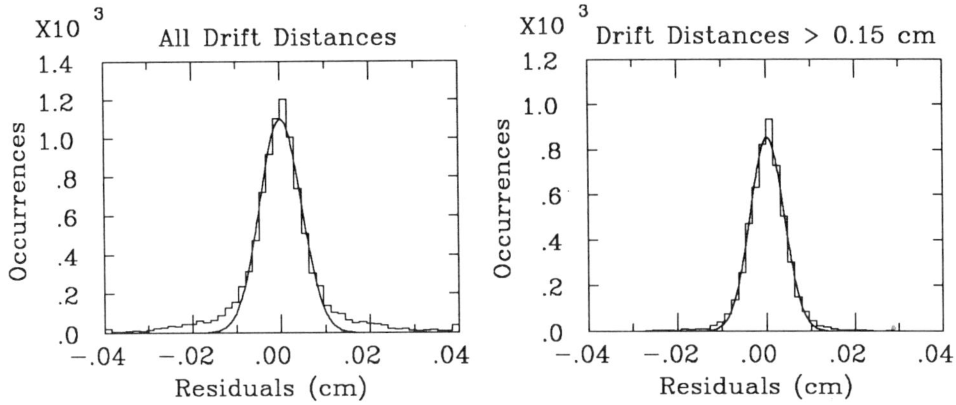

Figure 18 VC Residuals. The hit was *removed* from the track fit in these data. Residuals for all drift distances are on the left, where the fitted gaussian has $\sigma = 45\,\mu\text{m}$. On the right, only data from distances $> 0.15\,\text{cm}$; the fitted gaussian has $\sigma = 38\,\mu\text{m}$. As discussed in the text, these data *overestimate* the spatial resolution of the chamber.

beam pipe equal to their error due to chamber resolution. Since the multiple scattering error scales as $1/p$, tracks with $p < 200\,\text{MeV}/c$ contribute little information to lifetime measurements. Figure 17 also exhibits the efficiency for the same tracks versus impact parameter itself. Typical impact parameters of particles from B-meson decay are $300\,\mu\text{m}$, and the VC efficiency is flat and $\geq 96\%$ to some ten times this value, 3 mm. We conclude that linking efficiency effects lifetime measurements very little.

Well measured tracks in the VC can be used to measure the efficiency of the CD. Defining a well measured track as one with six hits in the VC, 0.932 ± 0.001 are matched with tracks having six or more hits in the CD, and 0.798 ± 0.001 are matched to tracks having seven hits or more.

Observed Resolutions

Spatial resolutions were measured using residuals from the track fit. If the VC hit used to measure the residual is left in the track fit, the resulting residual distribution is systematically narrower than the true residual distribution, due to the pull of the hit in the fit.[14] Conversely, if the hit is removed from the fit the distribution is wider then the true distribution, due to interpolation error in the track fit. The size of the bias is almost equal in these two cases, but opposite in sign. Figure 18 displays the residuals from Bhabha scattering events, where VC hits were removed from the fit. The non-gaussian tails arise from the variation of resolution with drift distance. The resolution as a function of drift distance is shown in Figure 19. σ as a function of drift distance both for the hit left in the fit, and removed. For most drift distances, the resolution is 35 μm. Near the wire the resolution

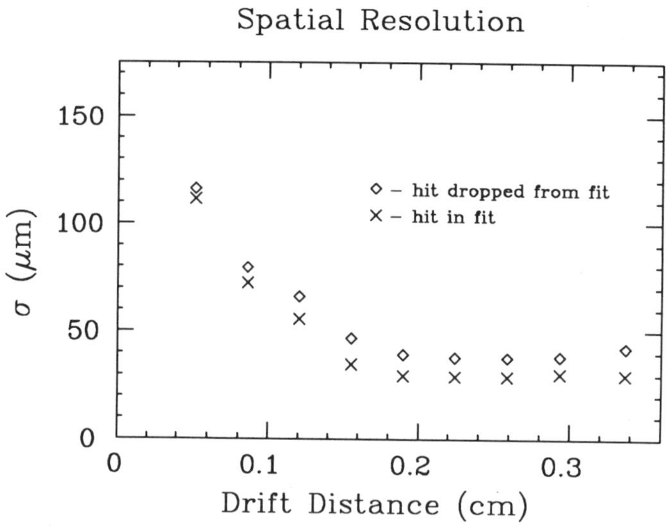

Figure 19 VC Spatial Resolution. The fitted σ of the residual distribution is shown as a function of drift distance. Both cases of hit left in the track fit and removed are shown. The true σ is approximately half way between these two extremes.

degrades, most probably due to the dispersion in pulse heights in our gas mixture, which is underquenched. Estimates of degradation due to spacing of ionization clusters, and due to timing resolution, indicate that these do not provide the degradation of resolution near the wire. The mean resolution is 45 μm.

The figure of merit for lifetime measurements is σ_v, the resolution in extrapolated distance of closest approach to the geometric center of the chamber. σ_v depends on the position resolutions of both the VC and the CD, the radii of their sense wires, and the amount of matter that causes multiple scattering in the beam pipe:

$$\sigma_v^2 \approx \frac{\overline{\sigma}^2}{N}(1. + f_1^2(\overline{\rho}) + f_2^2(\overline{\rho})) + \frac{\sigma_p^2}{p^2} \qquad [3.2]$$

The first term arises from the non-zero measurement resolution in the drift chambers, the second from multiple coulomb scattering. Here $f_1(x)$ and $f_2(x)$ represent the errors in extrapolation due to the slope and curvature in the fit, and are polynomials of order 1 and 2 respectively:

$$f_1(x) = x$$

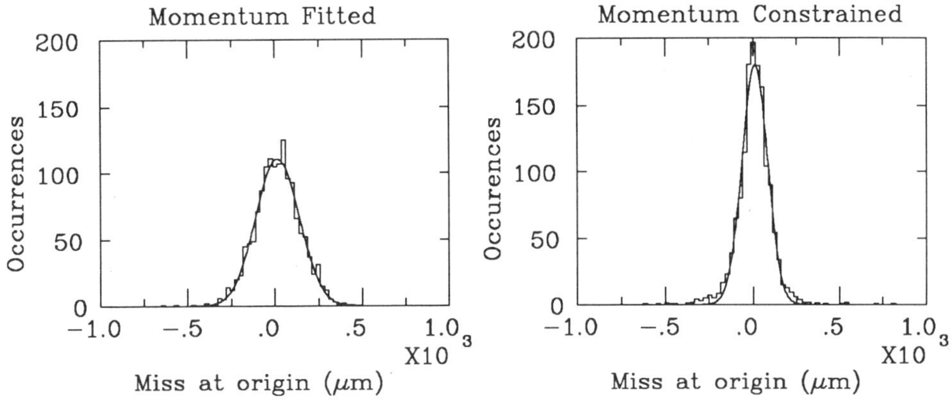

Figure 20 'Miss' distance in Bhabha Scattering Events. These data show the sum of the δ_0s fort the two tracks in Bhabha events. On the left, the momenta were allowed to vary in the track fit. The distribution is very gaussian, with $\sigma = 124\,\mu$m. On the right, the track momenta were constrained to beam energy, and the distribution narrowed to $\sigma = 74\,\mu$m. This indicates the importance of the term arising from track *curvature* in [3.2].

$$f_2(x) = \frac{x^2 + \gamma_{1r}x + 1}{\gamma_{2r} + \gamma_{1r}^2 + 2}$$

$\overline{\rho}$, the dimensionless mean extrapolation distance, is $\overline{\rho} \equiv \overline{r}/\sigma_r$; given the radii $\{r_1, \ldots, r_N\}$ of the sensing layers, and their spatial resolutions $\{\sigma_1, \ldots, \sigma_N\}$,

$$\overline{r} \equiv \frac{\overline{\sigma}^2}{N}\sum_{i=1}^{N}\frac{r_i}{\sigma_i^2} \quad , \quad \frac{1}{\overline{\sigma}^2} \equiv \frac{1}{N}\sum_{i=1}^{N}\frac{1}{\sigma_i^2} \quad , \quad \sigma_r^2 \equiv \frac{\overline{\sigma}^2}{N}\sum_{i=1}^{N}\frac{1}{\sigma_i^2}(r_i - \overline{r})^2 \quad ,$$

and γ_{1r} and γ_{2r} are the coefficients of skew and curtosis defined in analogy with σ_r. For the combination of CD and VC, $\overline{r} = 8.6\,\text{cm}, \sigma_r = 7.4\,\text{cm}, \gamma_{1r} = 3.5$, and$\gamma_{2r} = 15$. $f_1^2(\overline{\rho}) = 1.4$, while $f_2^2(\overline{\rho}) = 12$, indicating that uncertainty in the curvature in the fit dominates the extrapolation error. This uncertainty arises primarily from the small maximum radius, 45 cm, of the MAC CD; the vertex resolution decreases as the *square* of this radius. The first term in [3.2] predicts a vertex resolution of $72\,\mu$m for $45\,\mu$m resolution in the VC. In the second term σ_p depends on the angle of the track with respect to the beam axis, but was typically $65\,\mu$m- GeV/c.

The resolution in impact parameter was measured with Bhabha scattering events. Ideally, the opposing electron tracks would have equal and opposite impact parameters. Figure 20 shows the sum of the measured impact parameters. The distribution is very gaussian, with fitted σ of 124μm. This indicates a resolution $\sigma_v = 124/\sqrt{2} = 87\,\mu$m in impact parameter. Constraint of the momentum leads to $\sigma_v = 52\,\mu$m.

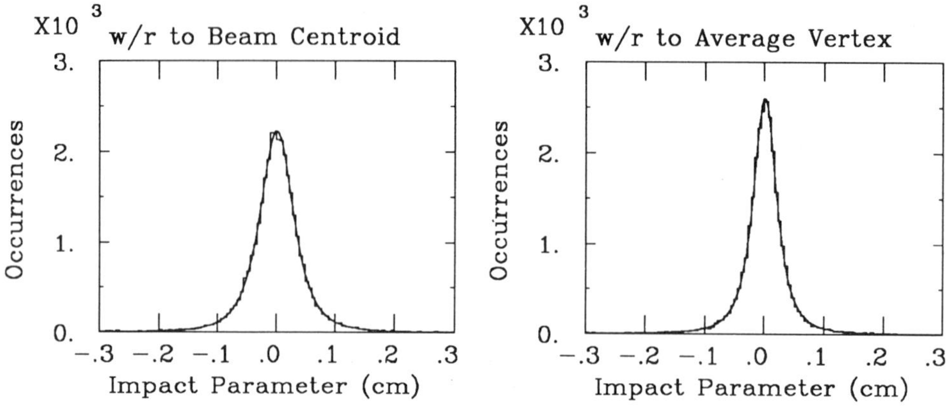

Figure 21 Vertex Resolution in Multihadron Events. These data show the distance of closest approach of tracks with respect to two estimates of the e^+e^- annihilation point. In the plot on the left, the estimate is the beam centroid, as determined in Bhabha scattering events. The fit is to a Student's t distribution, with $\sigma = 306\,\mu\text{m}$ and asymptotic power 1.9 describing the tails. In the plot on the right, the 'average vertex' was used; the fit yields $\sigma = 210\,\mu\text{m}$ and power 1.5.

The effective spatial resolution in the VC degraded to $\approx 70\,\mu\text{m}$ in multihadron events. The primary source of this degradation was the misassociation of hits with tracks in the dense environment of a multihadron event. Hits in adjacent cells due to knock-on electrons probably caused some confusion as well.

The resolution in extrapolated distance of closest approach, or impact parameter, in multihadron events is given in Figure 21. Tracks with: 1) 3 or more hits in the VC; 2)7 or more hits in the CD; and 3) momenta $> 500\,\text{MeV}/c$ were used. The first plot makes use of the beam centroid, as determined from Bhabha scattering events on a run-by-run basis. The resolution in this case is dominated by the size of the beam itself, which we measured to have $\sigma_x = 350\,\mu\text{m}$ and $\sigma_y = 70\,\mu\text{m}$. For the second plot, an improved estimate of the e^+e^- annihilation location, the 'average vertex,' was made on an event by event basis. These data indicate a vertex resolution of $210\,\mu\text{m}$, of which $110\,\mu\text{m}$ was due to multiple coulomb scattering.

Figure 22 exhibits the tracks and VC hits in a typical multihadron event. This event is in no way unusual in complexity or lack of noise.

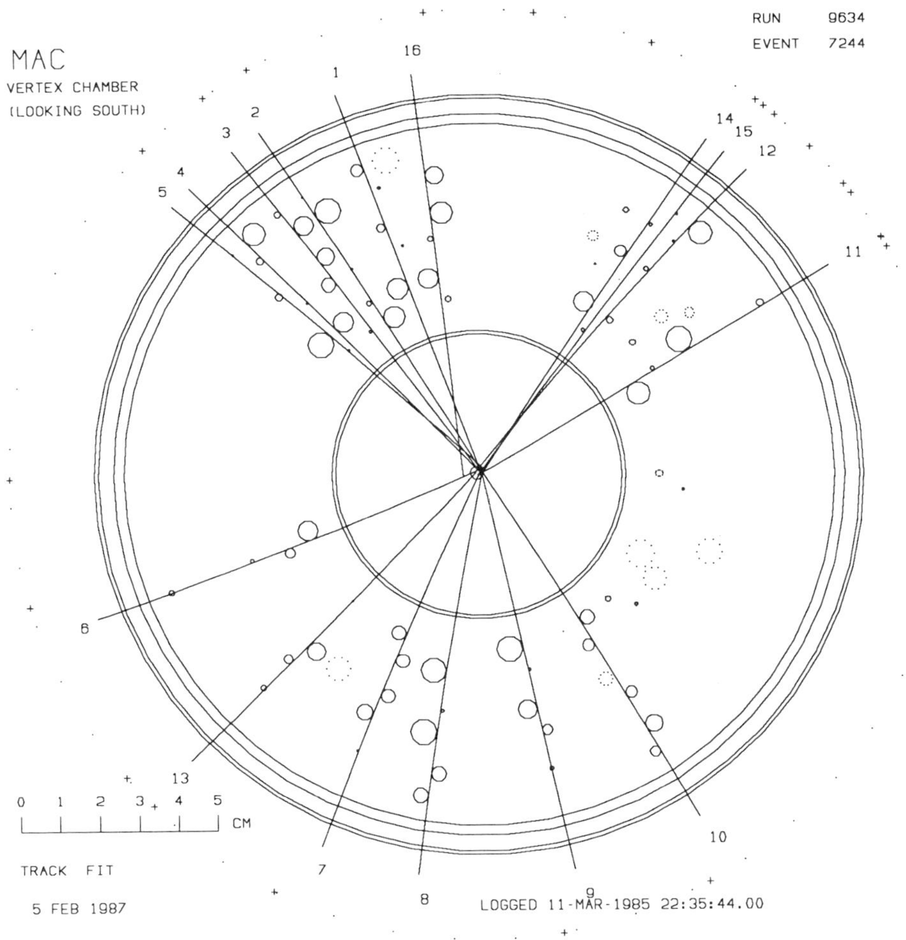

Figure 22 A Typical Multihadron Event in the VC. Each small circle is the contour of constant drift distance found from the drift time measured in that cell. Solid circles indicate the hit was linked to a track; dotted indicate it was not. Tracks should be tangent to the small circles. The innermost small circles represent the beam pipe, the outermost the outer chamber wall and inner wall of the CD.

B-LIFETIME MEASUREMENT

Event selection We collected $94\,\mathrm{pb}^{-1}$ of integrated luminosity, all at $\sqrt{s} = 29\,\mathrm{GeV}$, with the vertex chamber in place. This provided a sample of 44,000 multihadron events. We enrich a multihadron subsample in events from $b\bar{b}$ production by requiring a lepton with $p > 2\,\mathrm{GeV}$ and $p_\perp > 1.5\,\mathrm{GeV}$, where $p_\perp$ was taken with respect to the calorimetric thrust axis.[16] The thrust axis is required to be $> 30°$ from the beam direction, and the thrust is required to exceed 0.72 to ensure that the axis is well measured.

A detailed description of the MAC detector can be found elsewhere.[13,15] Substantial improvements in the MAC muon identification have been made. MAC features an excellent system of external drift chambers to measure the exit direction of muons from the calorimeters. The steel in the hadronic calorimeters is toroidally magnetized, yielding a muon momentum measurement in addition to that provided by the central tracking inside our solenoid. We used the redundancy of the MAC systems to understand and improve our efficiencies and misidentification probabilities. The resulting $b\bar{b}$-enriched subsample tagged with muons contains 117 events for data taken with the VC. We have also reanalyzed the 101,000 multihadron events from 220 pb^{-1} of integrated luminosity accumulated prior to installation of the VC, and find 336 muon-tagged events.

We perform electron identification with the central electromagnetic calorimeter of MAC, and find 35 electron-tagged events in the data taken with the VC, and 74 in the data taken prior to VC installation. The electron sample is smaller than the muon sample due to smaller acceptance, and due to lower efficiency within that acceptance. A cut to reduce background from 2γ processes has been introduced that was not present in MAC's earlier B-lifetime analyses.

We estimate from Monte-Carlo and detector simulation studies that the events in the lepton tagged sample are 60% from $b\bar{b}$-production, 22% from $c\bar{c}$-production, and 18% from light quark production.

Lifetime Measurement

We use all tracks of sufficient measurement quality to estimate the B-hadron lifetime; previously, only the lepton tracks were employed. The quality cuts are: 1)$p > 0.5$ GeV; 2)number of CD hits ≥ 7; and 3)number of VC hits ≥ 6, in data where the VC was present. In the data with the VC, 4.4 tracks/event pass these criteria; in previous data, 6.3 tracks/event pass.

The *signed* impact parameter of tracks with respect to a point that estimates the $b\bar{b}$ production location is used to measure lifetime. The situation is exhibited in Figure 24. Since the signature of a non-zero lifetime is an average *positive* impact parameter, the sign convention is crucial. We use the calorimetric thrust axis to estimate the line of travel of the decaying parent; we assign an impact parameter a positive sign if it appears that the parent flew along the thrust axis into the hemisphere containing the track. This sign is uncorrelated with the sign conventions a host of systematic errors would provide.

Previously, we used the beam centroid, as determined from Bhabha scattering events, as our estimate of the $b\bar{b}$ production location. However, after addition of the VC, our resolution at the vertex is smaller than the beam, which we measure to have a half-width of 350 μm in the horizontal and 70 μm in the vertical. So, we have devised a method to improve our estimate of

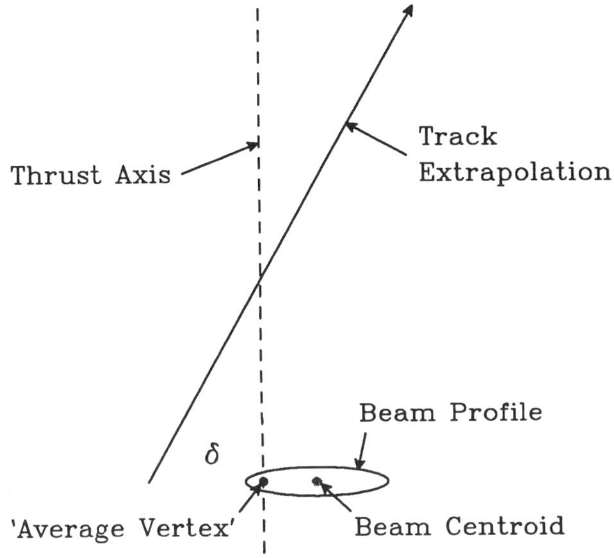

Figure 24 Impact Parameter and Sign Definitions. We use the thrust axis to estimate the parent direction; if it appears that the parent flew forward along the thrust axis, then decayed, we assign $\delta > 0$. This diagram is a projection in the plane perpendicular to the beam; if the thrust axis and track have no components along the beam direction, $\delta > 0$ in this diagram. Negative impact parameters result most often from extrapolation error, but can result from backward decays.

the $b\bar{b}$ production location on an event-by-event basis. Within each lepton-tagged event, we separate the N tracks that pass the quality cuts into $N-1$ tracks which we then use to make a better estimate of the $b\bar{b}$ production location, and a track for which we measure the impact parameter. The improved estimate of the $b\bar{b}$ production location, or average vertex, is that point which maximizes a likelihood function constructed to represent the assumption that the $N-1$ tracks came from a common point, and that point came from the beam. We assume the log-likelihood is parabolic, and take the weights from the covariance matrix of the track fits, and the horizontal and vertical beam size. We omit the track providing an impact parameter to avoid statistical correlations; we then loop over all quality tracks in the event, in turn omitting and defining a new average vertex for each track. The improvement yielded by the average vertex procedure was shown in Figure 23; the vertex resolution improves by some 50% compared to use of the beam centroid.

Tracks that are near the thrust axis in azimuth provide little information on non-zero parent flight paths, and are subject to sign errors induced

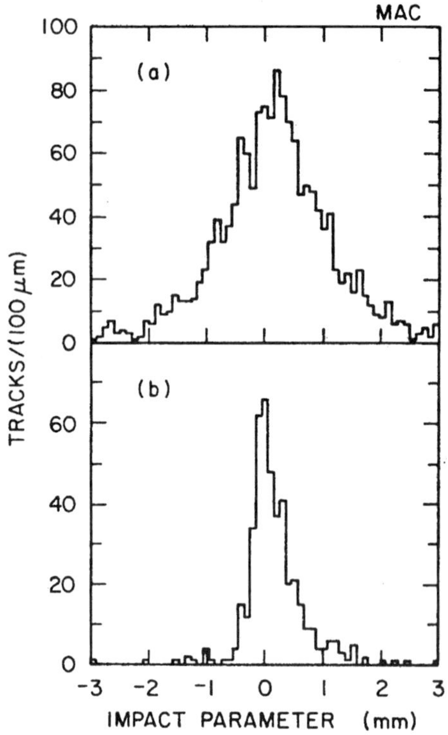

Figure 25 Signed Impact Parameter Distributions. (a) 1558 tracks in lepton tagged events fromt the 220 pb^{-1} taken prior to installation of the VC. (b) 441 tracks in lepton tagged events taken in the 94 pb^{-1} accumulated with the VC. Not just the lepton tracks, but all tracks passing criteria described in the text are included. Both distributions show clear offsets from 0, and the VC data show an exponential-like tail.

by the error in thrust axis direction. Therefore, we cut tracks closer than $0.2/|sin(\theta_t)|$ radians in azimuth to the thrust axis, where θ_t is the polar angle of the thrust axis with respect to the beam axis. This leaves 2.9 tracks/event in the data taken with the VC, or a total of 441 tracks; 3.8 tracks/event, or 1558 tracks remain in the data set without the VC. The impact parameter distributions of these data are given in Figure 25.

We use the trimmed mean of the impact-parameter distribution as a robust estimator of the lifetime![17] The trimmed mean is defined as the mean of a distribution where fractions f, in the tails, have been *symmetrically* removed. This estimator is robust in the sense that tails of the resolution function do not compromise the statistical precision of the estimator *and* in the sense that the trimmed mean, for a given lifetime, is insensitive to the shape of the distribution function. The statistical efficiency of the trimmed mean is be very near to that provided by a maximum likelihood fit, but the systematic error do to ignorance of the resolution function is substantially less than for a maximum likelihood fit. We find good statistical efficiency for $f = 0.1$, although the efficiency is insensitive to f between 0.05 and 0.25. For the data taken with the VC, we find a trimmed mean of $129 \pm 19\,\mu$m, and $154 \pm 26\,\mu$m for previous data. The fraction f was 0.1 in both cases.

We determine lifetimes using our Monte Carlo event generator and detector simulation. Because $b\bar{b}$ events have higher multiplicity than light quark events, the *track* fractions differ from the *event* fractions: we find that 70% of the tracks come from $b\bar{b}$ production; 16% from $c\bar{c}$ production, and 14% from light quark production.

The B lifetime determined from the data taken with the VC is $\tau_b = 1.35 \pm 0.30(stat)$ps. Data taken prior to VC installation gives $\tau_b = 1.24 \pm 0.29(stat)$ ps. All electron tagged data yield $\tau_b = 0.92 \pm 0.35(stat)$ps, while all muon tagged yield $\tau_b = 1.30 \pm 0.25(stat)$ ps. The combined result is $\tau_b = 1.29 \pm 0.20(stat)$ ps.

The dominant systematic error arises from uncertainty in how many tracks are produced in the primary hadronization of the b-quark; this arises primarily from the uncertainty in the mean fraction of the beam energy retained by the B-hadron. Taking this fraction to be 0.78 ± 0.05[18] indicates a ±10% uncertainty in the lifetime. This uncertainty is dominantly *multiplicative*, it does not degrade the *statistical* significance of the result. Other systematic uncertainties are discussed in (19). Our final combined result is $\tau_b = [1.29 \pm 0.20(\text{stat.}) \pm 0.07(\text{syst.})\,\text{ps}] \times (1.00 \pm 0.15)$, where the last factor is the systematic multiplicative uncertainty. This value is in agreement with the world average from PEP and PETRA detectors, $\tau_b = 1.26 \pm 0.16$ ps, where statistical and systematic errors have been combined in quadrature.

CONCLUSIONS

The MAC vertex chamber, built of aluminized mylar proportional tubes, or straws, has achieved a spatial resolution of $35\,\mu$m, operated reliably in a region very close to colliding beams, and provided good data used to measure the B-hadron lifetime. The motivations and design considerations for this chamber have been reviewed, as has its performance and use for physics results. The attributes of this type of chamber make it an attractive option for use in high energy hadron colliders, such as the ELOISATRON.

ACKNOWLEDGEMENTS

I wish to specially thank F. Villa, G. Charpak, and A. Zichichi for organizing this interesting and informative workshop, and the Ettore Majorana Center for wonderful hospitality. I wish to thank E. Askeland, J. Broeder, N. Erickson, J. Escalera, N. Palmer, the late C. T. Pulliam, J. Schroeder and D. Wright for their contributions to the vertex chamber design and construction. This work was supported in part by the U. S. Department of Energy under contract numbers DE-AC02-81ER40025 (CU), DE-AC03-76SF00515

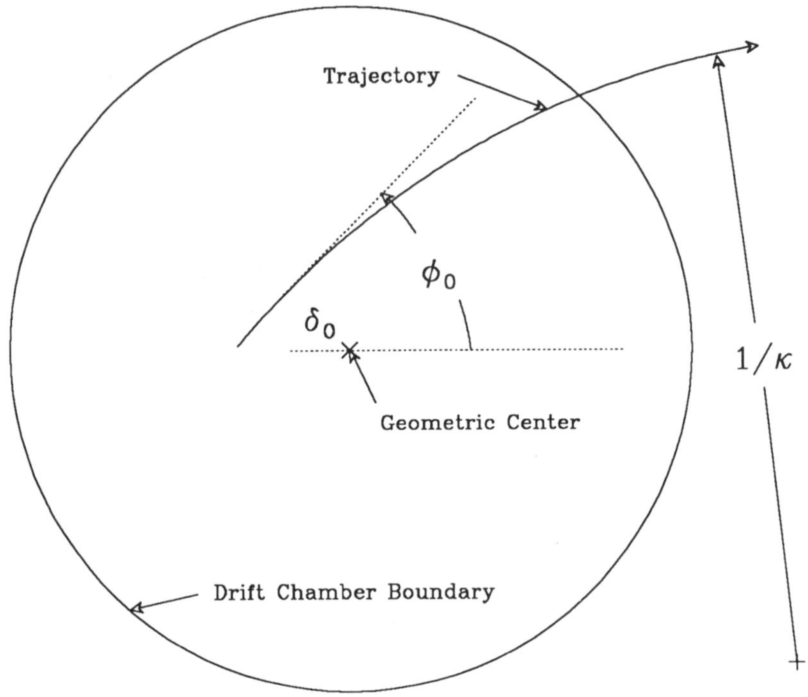

Figure 26 Parameters Describing a Particle Trajectory. The figure is in the plane perpendicular to the beam. The outer circle schematically represents a boundary in the drift chamber system. δ_0 is given a sign based upon the cross product of a vector pointing along the particle trajectory taken with a vector from the geometric center to the perihelion of the trajectory.

(SLAC), and DE-AC02-76ER00881 (UW); by the National Science Foundation under grant numbers NSF-PHY82-15133 (UH), NSF-PHY82-15413 and NSF-PHY82-15414 (NU), and NSF-PHY83-08135 (UU); and by the Istituto Nazionale di Fisica Nucleare.

Appendix: Tracking Parameterization

Five parameters describe the helical trajectory of charged particles in the axial magnetic field of MAC. Three of these describe the circular projection of the trajectory in the plane perpendicular to the beam axis: κ, the reciprocal of the radius of curvature; δ_0 the (signed) distance of closest approach of the circle to the geometric center of the MAC coordinate system; and ϕ_0, the azimuth of the trajectory at closest approach. The sign of δ_0 is determined from the cross product of the direction of the particle trajectory at closest approach taken with the direction from the geometric center to the point of closest approach. Figure 26 exhibits the the definition of κ, δ_0, and ϕ_0. The remaining two parameters describe the particle trajectory out of the plane perpendicular to the beam. θ is the angle with respect to the beam axis of the trajectory, and z_0 its intercept with the axis, viewed along the azimuth containing the distance of closest approach.

In the MAC track fitting routines, multiple coulomb scattering is described by 'kinks' at the radii of matter concentrations, such as the boundary between the VC and CD. Several kink angles are fit, but with a constraint by the r.m.s. angle predicted by multiple coulomb scattering. This procedure quite naturally incorporates errors from multiple coulomb scattering into the covariance matrix of $\kappa, \delta_0, \phi_0, \theta$, and z_0.

REFERENCES

1) The MAC collaboration consists of: W. W. Ash, H. R. Band, T. Camporesi, G. B. Chadwick, M. C. Delfino, R. De Sangro, W. T. Ford, M. W. Gettner, G. P. Goderre, D. E. Groom, R. B. Hurst, J. R. Johnson, K. H. Lau, T. L. Lavine, R. E. Leedy, T. Maruyama, R. L. Messner, J. H. Moromisato, L. J. Moss, F. Muller, H. N. Nelson, I. Peruzzi, M. Piccolo, R. Prepost, J. Pyrlik, N. Qi, A. L. Read, Jr., D. M. Ritson, F. Ronga, L. J. Rosenberg, W. D. Shambroom, J. C. Sleeman, J. G. Smith, J. P. Venuti, P. G. Verdini, E. von Goeler, H. B. Wald, R. Weinstein, D. E. Wiser, and R. W. Zdarko, *University of Colorado, INFN-Laboratori Nazionali di Frascati, University of Houston, Northeastern University, Stanford Linear Accelerator Center, Stanford University, University of Utah, and University of Wisconsin.*
2) E. Fernandez *et al.*, Phys. Rev. Lett. **51**, 1022 (1983); N. S. Lockyer *et al.*, Phys. Rev. Lett. **51**, 1316 (1983).
3) M. Kobayashi and K. Maskawa, Prog. Theor. Phys. **49**, 652 (1973); P. H. Ginsparg, S. L. Glashow, and M. B. Wise, Phys. Rev. Lett. , **50**, 1415, (1983) and **51**, 1395, (1983).
4) D. Rust in *Proceedings of the Third International Conference on Instrumentation for Colliding Beam Physics, Novosibirsk, 1984*, (USSR Academy of Sciences, Siberian Division, Institute of Nuclear Physics, Novosibirsk, 1984); P. Baringer, C.K. Jung, H.O. Ogren, D.R. Rust, Nucl. Instrum. Meth. **A254**, 542 (1987).
5) G. Battistoni *et al.*, Nucl. Instrum. Meth. **A235**, 91 (1985).
6) E. Fernandez *et al.*, SLAC-PUB-3390, Aug. 1984, in *Proceedings of the Twelfth SLAC Summer Institute on Particle Physics, Stanford, 1984* edited by P.M. McDonough (Stanford Linear Accelerator Center, Stanford, 1984); W. Ash *et al.*, SLAC-PUB-4311, Apr. 1987, submitted to Nucl. Instrum. Meth.
7) G. Chadwick and F. Muller, MAC Internal Memo No. 683: this study employed the model described in J. Va'Vra, Nucl. Instrum. Meth. **225**, 445 (1984).
8) M. Breidenbach, F. Sauli, and R. Tirler, Nucl. Instrum. Meth. **108**, 23 (1973).

9) G.L. Godfrey "Proportional Tube Lifetimes (Magic Gas, A-CO_2, DME)," in *Workshop on Radiation Damage in Wire Chambers, Berkeley, CA*, edited by J. Kadyk (Lawrence Berkeley Laboratory, Berkeley, CA, 1986).
10) H.N. Nelson "Lifetime Tests for MAC Vertex Chamber," SLAC-PUB-4017, Jul. 1986, in *Workshop on Radiation Damage in Wire Chambers, Berkeley, CA*, edited by J. Kadyk (Lawrence Berkeley Laboratory, Berkeley, CA, 1986).
11) W.T. Ford *et al.*, Phys. Rev. **D33**, 3472 (1986).
12) D.E. Groom, Nucl. Instrum. Meth. **219**, 141 (1984).
13) E. Fernandez *et al.*, SLAC-PUB-4048 (in preparation), and to be published.
14) F. Solmitz, Ann. Rev. Nucl. Sci. **14**, 375 (1964).
15) E. Fernandez *et al.*, Phys. Rev. D **31**, 1066 (1985).
16) E. Fernandez *et al.*, Phys. Rev. Lett. **50**, 2054 (1983).
17) W. T. Eadie *et al.*, *Statistical Methods in Experimental Physics*, American Elsevier (1971).
18) This value and error result from our own analysis of world data. See E. H. Thorndike, Ann. Rev. Nucl. Sci. **35**, 195 (1985); S. Bethke, Z. Phys. C **29**, 175 (1985).
19) W.W. Ash *et al.*, Phys. Rev. Lett. **58**, 640 (1987).

Discussion H. NELSON

P. WEILHAMMER

What makes the difference in statistics in the data samples taken for determination of B lifetime with and without straw chambers?

H. NELSON

Most of the data have been taken without the straw chamber (220 pb^{-1}vs 94 pb^{-1}). About 85% of all tracks used in the lifetime measurement from the data sample with the straw chamber and central detector together are identified in the straw chamber.

B.R. KUMAR

Is it possible to get a B lifetime measurement without using thrust information, just by plotting the impact parameter of high p_T leptons?

H. NELSON

We use thrust information to get the sign of the impact parameter; in this way we measure the B lifetime through the displacement of the impact parameter distribution. I can't think of how to get the sign without thrust – like information. Without the sign, one might be able to derive a lifetime from the shape of the distribution of absolute values of impact parameters, but probably not at our current level of precision.

WIRE CHAMBER AGING AND WIRE MATERIAL

Muzaffer Atac

Fermi National Accelerator Laboratory*
P.O. Box 500
Batavia, Illinois

Abstract

Recent studies on wire chamber aging has shown surprising results [1,2] indicating that anode wire material may have a strong influence on the useful life-time of a chamber or counter. The aging process discussed here is related to changes in the surface characteristics of the anode wires. Those changes may be due to an oxidation process involving molecules produced from the dissociation of organic quenching agents during the avalanche process. It is observed that after some running time the anode wire surface is coated with a nonconductive film, which drops the electric field around the anode wire, reducing the gain as the aging process continues.

Aging due to polymerization and other processes have been known since the early years of Geiger and proportional counter operation. Interesting phenomena found here shows that the aging process is not only dependent on the gas mixture, it is strongly related to the anode material. Results

*Operated by Universities Research Association under Contract with the United States Department of Energy.

reported earlier[1] showed that chambers with gold plated tungsten anode wire using Argon-Ethane-Ethyl alcohol gas mixtures performed well for a very long time. They showed no appreciable loss of gain after an accumulated charge of one to two Coulombs per centimeter of wire length. There was no evidence of polymerization.

Aging was accelerated when the gold plated tungsten anode wire was replaced by Stablohm-800 resistive alloy (75% Ni + 20% Cr + Al + Cu) wire using the same gas in the same chamber. However, the gas mixture of Argon-ethane-isopropyl alcohol provides a very long lifetime for a drift chamber made of Stablohm-800 anode wire. The details of the findings are discussed below.

Resistive Alloy as Anode Wire

A small chamber (Figure 1) was used for this test. The anode wires were Stablohm-800 resistive wire (50 μm diameter), the field shaping wires were 150 μm stainless steel and the cathode planes were 25 μm thick aluminum strips on 100 μm thick of G-10 sheets. The purpose of these choices was to use all the materials in the same configuration as in the radial wire drift chambers[3] of the CDF. The resistive wire is used for determining track coordinates along the wire by current division.

Equal portions of argon and ethane gas were bubbled through ethyl alcohol at -7° C, adding 1.1% vapor in solution. Using an intense Sr^{90} β-source (collimated to illuminate 6mm) the wire current drawn from the HV-power supply was monitored for the following tests. The count rate obtained from the wire approached $2x10^{5}$ per second and showed a 0.4 μA current draw. A pulse height spectrum of Fe^{55} was taken by illuminating various parts of the chamber. The spectrum obtained at the illuminated spot was compared to the spectra obtained from the adjacent wire for 5.9 keV peak as far as 6 cm away from the spot. Wire aging appeared after the first day of running time. Figure 2 shows the drop in the pulse height as a function of

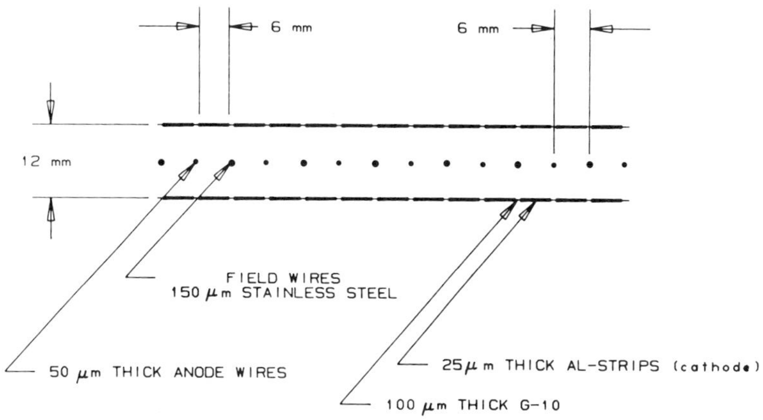

Figure 1 Cross section of the test chamber.

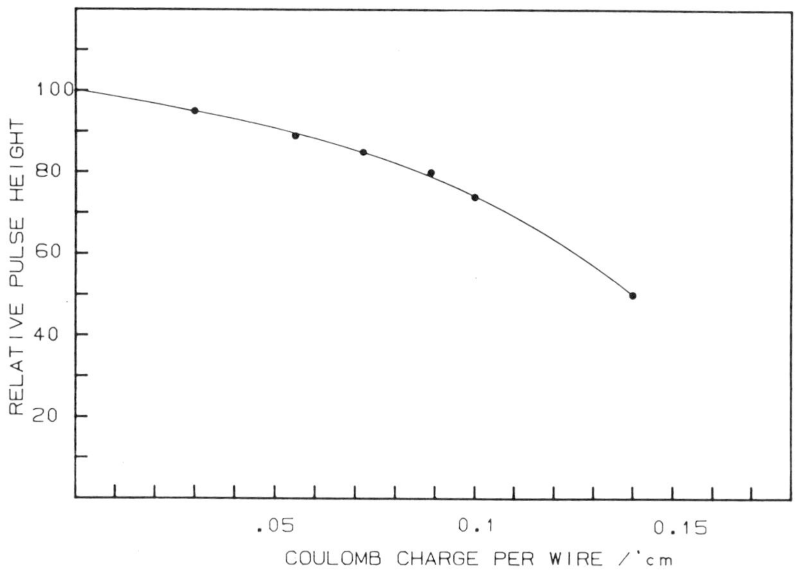

Figure 2 Pulse height of 5.9 keV line from the Fe^{55} source as a function of integrated charge indicating the aging process. Stablohm wire with 50% A/50% C_2H_6 gas bubbling through CH_3CH_2OH at -7°C, adding ~1.1% vapor.

the accumulated charge. Aging is reflected in the pulse height resolution as seen in Figure 3. The results were alarming since earlier tests showed argon-ethane gas needed ethanol as an additive quenching agent to prevent breakdowns, glow discharges and fast aging.

50% A/50% C_2H_6 gas mixture has been popular due to its excellent drift velocity saturation. It provides good spatial resolutions at atmospheric pressures. The test results obtained earlier with argon-ethane-ethanol using gold plated tungsten wires showed very little or no aging up to 1-2 Coulomb charge,[1] per wire per cm. The difference is related to the material of the anode wire. We needed a remedy and this remedy was replacing ethanol vapor with isopropyl alcohol vapor.

Test with Argon-Ethane-Isopropyl Alcohol Gas Mixture Using Stablohm-800 Wire

Using the same chamber with the Ni-Cr-Al-Cu alloyed wires, the above tests were carried out under the identical conditions after replacing ethyl alcohol with isopropyl alcohol. The temperature of the alcohol bath was -5° C, adding 0.25% vapor to the 50% A/50% C_2H_6 gas mixture. This gas mixture resulted in about a factor of 50 longer life-time for the same amount of gain drop. The gain drop was 4% after an accumulated charge of 0.6 Coulomb. Figure 4(a) shows the pulse height spectra at the beginning of the runs and after accumulating 0.6 Coulomb charge per cm per wire. Pulse height resolution is unaffected.

The same gas mixture using 50 μm gold plated tungsten wire showed no aging up to 2 Coulomb per centimeter of wire length. There was no gain drop, no change in pulse height resolution. No residue could be seen on the exposed region of the wire. The test was stopped to carry out the following experiments. Figure 4(b) shows the pulse height spectra obtained with the Fe^{55} source on the aged spot.

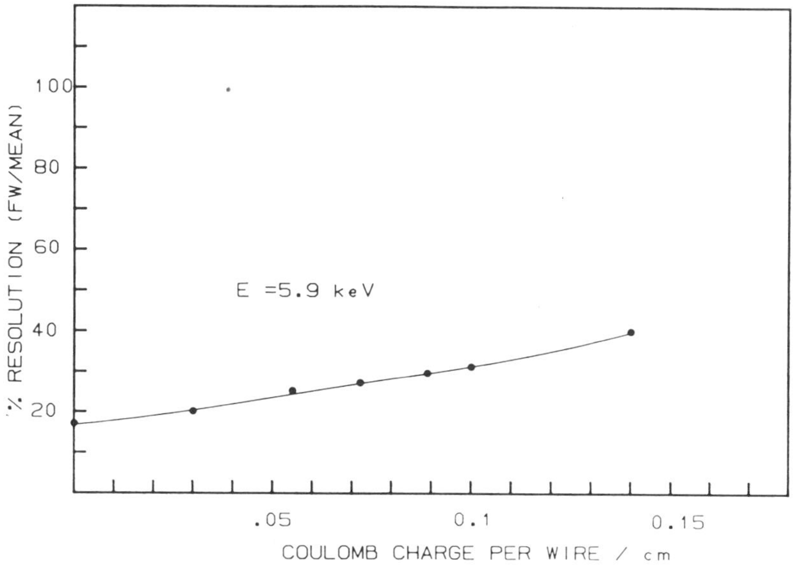

Figure 3 Pulse height resolution, FWHM, is degraded as the Stablohm wire is aged.

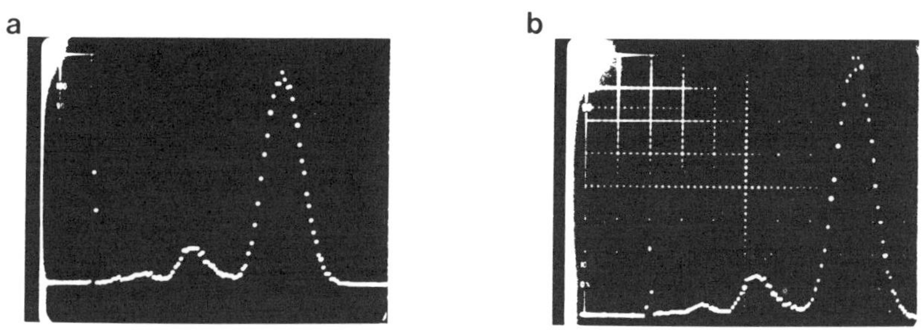

Figure 4 Aging is greatly reduced with isopropyl alcohol (2-propanol). About 0.25% isopropanol vapor added to the A-C_2H_6 gas mixture instead of ethanol.
Figure 4(a) shows the Fe^{55} spectrum at the beginning and Figure 4(b) after accumulation of 0.6 Coulomb/cm wire. The drop in the gain is only 5%.

Stablohm Wire with A-C_2H_6-CH_3OH

To gather more clues on the aging process of the resistive alloy, a 50% A/50% C_2H_6 gas mixture was bubbled through methanol at -5° C, adding about 2% vapor to the mixture. The wire aged very rapidly. After four hours of running with 0.4 μA average current, (corresponding to 0.008 Coulomb per centimeter) wire pulse height resolution was destroyed and average pulse height dropped (Figure 5). The aging was faster with methanol than ethanol mixture possibly due to higher concentration of alcohol (2% methanol) mixing. We will see further that the dissociation products of ethanol and methanol may be responsible for this aging. We also see a peak appearing below 5.9 keV peak; this will be explained further in the paper. Ethanol and methanol vapors are effective in preventing polymerization, but the dissociation products of these are reacting with this particular wire. Two remaining questions are: which components of the Stablohm wire are sensitive to dissociation products of the ethanol and methanol, and why aren't they reacting with the dissociation products of isopropanol. To answer the first question, the following aging tests were done on aluminum, nickel, copper, and chrome.

Tests with Aluminum Wire

Hard aluminum (Al 5056) showed very fast aging with argon-ethane-methanol mixture. A double 5.9 keV peak appeared after 0.001 C/cm on the 6 mm portion of the wire. The height of the lower peak was increased after 0.003 C charge as seen in Figure 6. One possible explanation for the double peaks is that the aged spot is coated with fairly uniform layers of Al_2O_3 which is highly resistive. The resistive layer may charge up, and polarize under the electric field; thus reducing the gas gain in that region. The lower peak is from the aged region. The scattered X-rays from the collimator result in the higher peak. Further results given will help resolve this issue. A remarkable point is that both peaks show good resolutions indicating a uniform aging of the anode wire. The color of the aged surface of the wire was like white chalk.

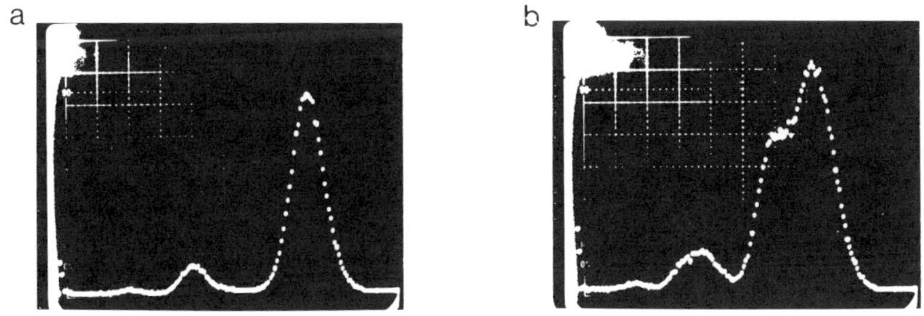

Figure 5 The Stablohm wire with the addition of 2% Methanol, CH_3OH, to the A-C_2H_6 gas. Very rapid aging is observed. Spectrum (a) is taken before the aging test and spectrum (b) is after only 0.008 Coulomb/cm wire.

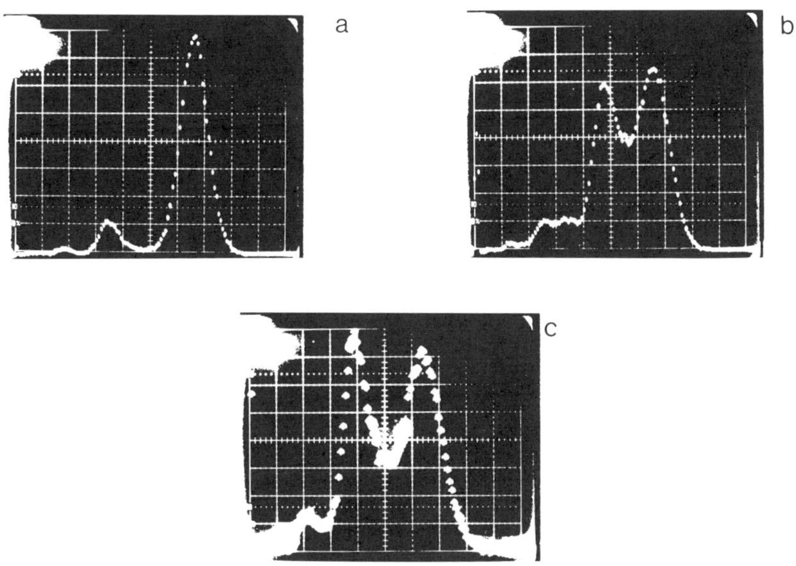

Figure 6 Aluminum wire (Al 5056) with the 2% ethanol added A-C_2H_6 gas mixture.
Figure 6(a) the pulse height spectrum before aging,
Figure (b) after 0.001 C charge per cm,
Figure (c) after 0.003 C.

Nickel Wire

Nickel wire also aged but not as fast as the aluminum wire did. There was a 16% gain drop after 0.05 C charge per cm per wire with the argon-ethane-methanol mixture. Figure 7 shows again a lower peak appearing on the shoulder of the 5.9 keV peak. The surface of the aged spot had red and blue colors, depending on the intensity of the aging, indicating surface oxidation. NiO is an insulator[3] but not as good an insulator as Al_2O_3. Aged surfaces of Ni and Al wires were examined using α-scattering by Dr. S. Majewski (University of Florida). The spectral analysis showed a measurable amount of oxygen on both surfaces.

Neither with nickel nor with aluminum was any substantial residue seen on the aged sections of the wires. Scanning electron microscope pictures were taken at Argonne National Laboratory by B. Tani. Figures 8(a) and (b) show these pictures on the clean and aged parts of the aluminum and nickel wires respectively. There is no visible increase in the wire thickness but there may be some evidence of corrosion on both wires. X-ray analysis was taken during the scanning process of the aluminum and nickel wires. Figures 9(a) and (b) show that there are no detectable elements above aluminum mass like Si and Br which have been found in the polymerization processes. Only the expected impurities in the nickel or aluminum were detectable.

Copper, Stainless Steel, and Platinum

These three wires showed no measurable aging after 0.15 C/cm per wire using the argon-ethane-methanol gas mixture. Because gold, platinum and stainless steel-304 are resistive to corrosion, the dissociation products of methanol and ethanol must be corrosive to aluminum, nickel and their alloys. It is not clear why the copper wire did not show fast aging. One possible explanation is that aluminum, nickel and copper are all oxidized in the process of aging. The gas gain drops with aluminum and nickel because Al_2O_3 and NiO are not conductive, thus dropping the electric field. CuO is

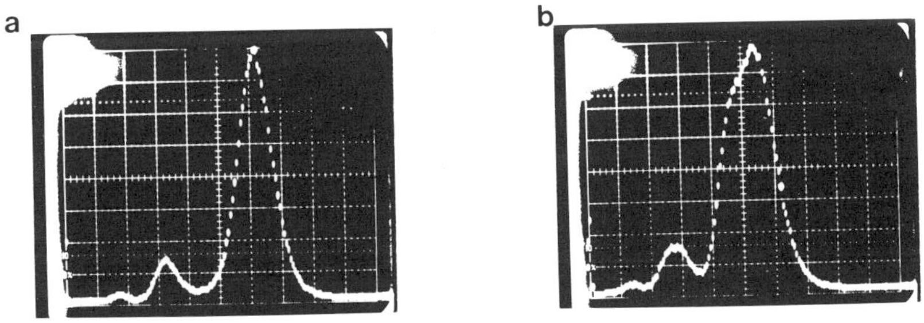

Figure 7 Nickel is aged slower than aluminum.
Figure 7(a) before the run,
Figure 7(b) after 0.05 C/cm.

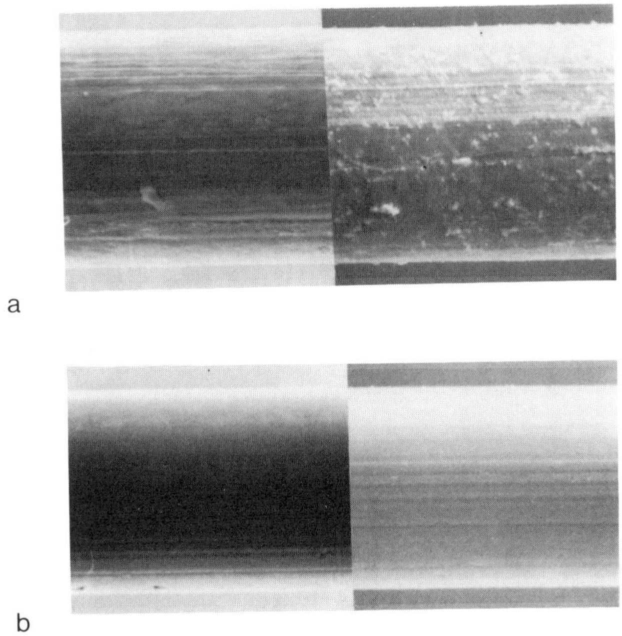

Figure 8 Scanning electron microscope pictures.
Figure 8(a) nickel wire surface,
Figure 8(b) of the aluminum. Right hand slides on the aged sections and left hand sides from the sections which are 2cm away from the aged portions. There is no observable increase in diameter of the wires. Appearance of corrosion may be observable.

a semiconductor; therefore, the electric field on the wire is not affected. Gold, platinum and stainless steel are not oxidized; their surfaces stay clean and show no measurable aging.

Isopropyl Alcohol

Why does the argon-ethane-isopropyl alcohol gas mixture not cause aging on any of the above mentioned wires? The answer may be that isopropyl alcohol gas dissociates differently than ethanol and methanol do. In the first stage of dissociation[5], isopropyl alcohol will loose two of its hydrogens and become acetone:

$$CH_3CHOHCH_3 + \gamma \text{ -------> } \underset{\text{acetone}}{CH_3COCH_3} + H_2$$

As the gas flows through the chamber, the water and the acetone molecules are vented out of the chamber. Experimental evidence confirms this hypothesis. An oxidized copper wire was used as an anode in the argon-ethane-isopropyl alcohol gas mixture until 0.1 C charge was collected from the collimated regions using the same method mentioned previously. Then the wire surface was examined. The oxidation was greatly reduced especially in the region of the source collimation.

We know that ethane dissociation products polymerize. Therefore, most of the UV quenching must be done by the isopropanol molecules for the above mentioned reactions to be useful in greatly improved life-times.

Several of the experiments repeatedly showed that equal portions of argon and ethane with about 0.25% isopropyl alcohol gas mixture provides very long life-times in drift chambers. An important question is how good a drift chamber gas is for electron drift velocity saturation. Figure 10 shows electron drift velocities as a function of isopropyl alcohol temperatures. For this the single cell drift chamber[6] was used with a nitrogen UV-laser. The

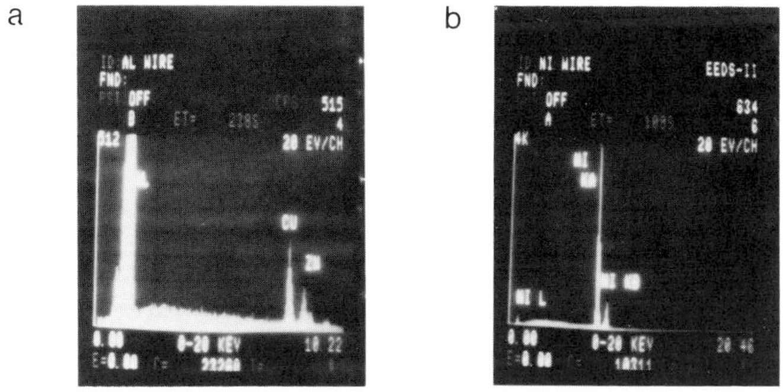

Figure 9 X-ray spectra of aluminum and nickel wire surfaces from the aged sections taken during the scanning electron microscope analysis. There is no substantial element observable above aluminum mass. Only the expected elements in the wires are seen.

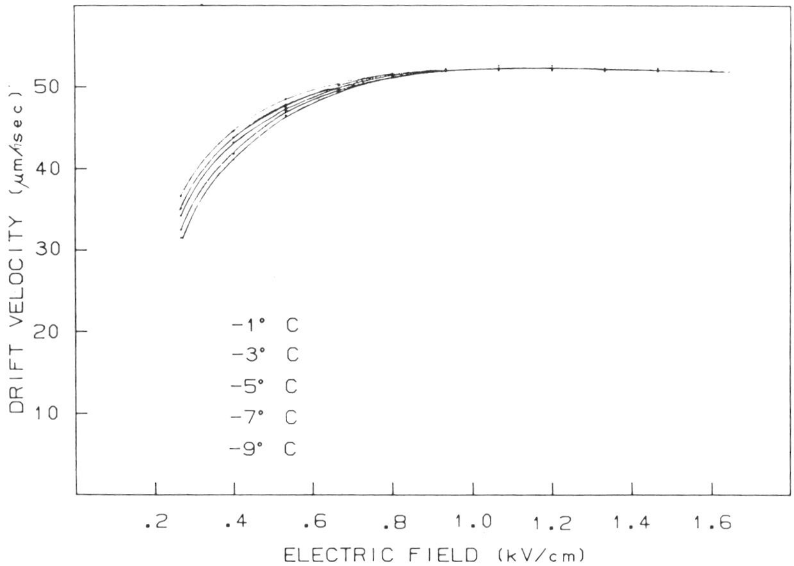

Figure 10 Drift velocity as a function of electric field at various temperatures of 2-propanol (isopropyl alcohol) as the 50% A/50% C_2H_6 is bubbled through. The drift velocity is well saturated above 900 v/cm and is independent of the alcohol vapor concentration above this field value.

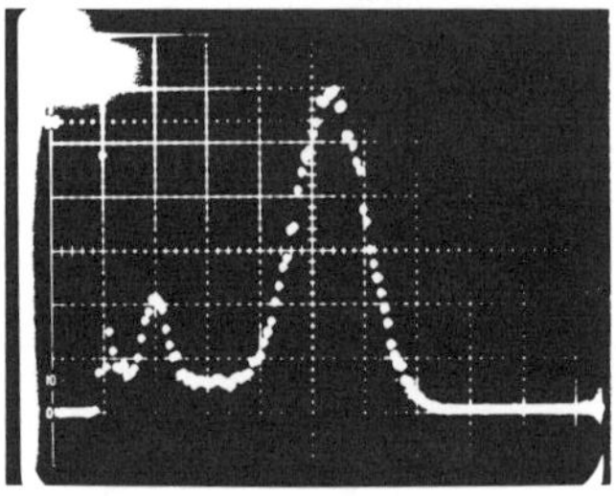

11(A)

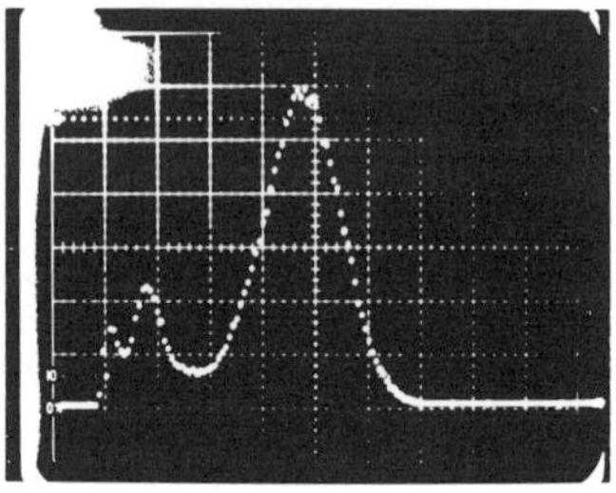

11(b)

Figure 11 12% gain drop after 0.2 Coulomb/cm-wire with a gold plated tungsten wire or Stablohm wire when a 12.5 mm aluminum tube was used.

Figure 11(A) shows the pulse height spectrum of the Fe^{55} source. 5.9 keV line and aluminum K-edge line are seen before aging.

Figure 11(b) shows the spectrum after the aging.

UV-beam was collimated to 400 μm and was precisely positioned to several cm distances to provide precise measurements. The drift velocity is saturated above 900v/cm, independent of the alcohol temperatures.

In the cases of methanol and ethanol it is possible that formaldehydes (CH_2O) and formic acid (CH_2O_2), and acetaldehyde (C_2H_4O) and acetic acid ($C_2H_4O_2$) result as dissociation products which can react with aluminum and nickel but not with gold, platinum, and stainless steel.

Aging of Cathode by Carbon Dioxide

Some tests were done using highly pure CO_2, a brass tube with gold plated tungsten wire and/or Stablohm wire and also using an aluminum tube with gold plated tungsten and Stablohm wires. With or without isopropyl alcohol vapor in CO_2, there was no observable aging of the anode wires up to an accumulated charge of 0.5 Coulomb per cm wire, and there was no observable drop in gas gain with brass tube (diameter = 12.5 mm), but there was 12% drop in gain when the same size aluminum tube was used (see figure 11). The anode wire was perfectly clean, indicating that the aging in this case occurs on the surface of the aluminum tubes. As CO_2 molecules dissociate, the inner surface of the aluminum tube is oxidized. There must be sufficiently thick film of aluminum oxide formed on the inner surface to drop the electric field in that region of the wire where the collimated β-source was kept. One would expect much faster aging with aluminum cathode or field wires instead of an aluminum tube which has much larger surface area to be oxidized relative to wires.

CONCLUSIONS

1. Glow discharge results in fast aging of wires with A-C_2H_6, A-i-C_4H_{10} (very high discharge current), etc. Polymers form with radicals like Si, Br, S, ...

2. Addition of CH_3OH, CH_3CH_2OH, $CH_3CHOHCH_3$ may prevent glow discharge and formation of polymers.

3. Dissociation products of C_2H_6, CH_3OH and CH_3CH_2OH are corrosive, thus oxidation of wire surface occurs with corrosive elements. Au, Pt and stainless steel wires show very slow aging, alloys with Al and Ni age fast.

4. A-C_2H_6-$CH_3CHOHCH_3$ found to be best for extending life-times of wires. No appreciable aging of Stablohm wires up to ~1 Coulomb/cm, no trace of aging of Au-W wire up to 2 Coul/cm.

   Reduction and cleaning of wire surface observed. Pulse height resolution was improved using long time oxidized Cu-anode. J. Kadyk (LBL) observed increase in pulse height with an aged stablohm wire.

5. Aging was observed on the aluminum cathode surface with no observable aging on the anode wire when CO_2 gas was used.

ACKNOWLEDGMENTS

The author would like to express his appreciations to Drs. J. Huth, M. Mishina and A. Tollestrup for their useful discussions, to W. Coleman for his assistance during the tests, and to A. Atac for making the drawings.

REFERENCES

1. M. Atac, IEEE Trans. on Nuclear Science, col. NS-31, No.1 (1984) 99.

2. J. Kadyk, Proc. of the Workshop on Radiation Damage to Wire Chambers.

   M. Atac, Proc. of the Workshop on Radiation Damage to Wire Chambers.

3. M. Atac, T. Hessing and F. Feyzi; IEEE Trans. on Nuclear Science, vol. 33, No.1 (1986) 189.

4. O. Kubaschewski and B.E. Hopkins; Oxidation of Metals and Alloys.

5. Private Communication with Dr. R. Schuller, Notre Dame University.

6. M. Atac, E. Coleman, T. Hessing, R.W. Kadel, M. Pratt and R.L. Wagner; Nuclear Instr. and Meth. in Phys. Research, 1986 (To be published).

Discussion M. ATAC

H. NELSON

You have suggested the dissociation of isopropanol into acetone and hydrogen occurs in the avalanche, and that the acetone is important to keep the wire clean and/or cleaning the wire. Have you tried adding acetone alone? Perhaps trace amounts are sufficient.

M. ATAC

We have not tried adding acetone. It attacks plastic insulators in the chamber.

G. CHARPAK

Methylal and isopropyl alcohol were used at the time of wire chamber to slow down the ageing. What seems to me to be new in your work is the connection made between specific elements in the wire (Al) with specific products of gas decomposition. If the production of H^2 is responsible for the curing process by alcohol, then one understands why people claim to have solved the problem of ageing with alcohol when the ageing agent is an oxyde deposed in the wire.

D. RITSON

Your point that exist gas mixtures that are self–cleaning is very important. Another example of a self–cleaning well quenched gas is 45% argon, 45% CO_2, 10% xenon (no organics). The anode becomes cleaner with time. Shouldn't we go for self–cleaning configurations in high radiation environments?

M.ATAC

I see nothing wrong with your gas mixture provided that you do not use aluminium wires for anode and aluminium cathode surface since the dissociation products of CO_2 is oxidizing aluminium. This will drop the gas gain.

J. KIRKBY

There is another reducing gas which should be investigated for use as a cleaning additive to drift chamber gases: Carbon monoxide. This may have some attractive features in view of its close relationship with CO_2.

M. ATAC

CO is a problem for safety. It could be a good UV quencher but it may not be a good addition for drift chambers.

CHARM PHOTOPRODUCTION USING A SILICON VERTEX DETECTOR

B. R. Kumar

University of Toronto
Toronto, Ontario, Canada

Representing The Tagged Photon Spectrometer Collaboration †

University of California, Santa Barbara, California, USA
Carleton University, Ottawa, Ontario, Canada
Centro Brasileiro de Pesquisas Fisicas, Rio de Janeiro, Brasil
University of Colorado, Boulder, Colorado, USA
Fermi National Accelerator Laboratory, Batavia, Illinois, USA
National Research Council, Ottawa, Ontario, Canada
Universidad de Saõ Paolo, Saõ Paolo, Brasil
University of Toronto, Toronto, Ontario, Canada

ABSTRACT

The use of a silicon microstrip vertex detector (SMD) in a high energy photoproduction experiment is described. The experiment, E691, used the Tagged Photon Spectrometer at Fermilab. The reconstruction of tracks and vertices using the SMD is described, and the method used for extracting clean charm signals is discussed. Measurements of D^0,D^+ and D_s^+ lifetimes based on 30% of the total data sample are presented. Some performance characteristics of the SMD system are given. The changes being made for a future experiment are outlined.

INTRODUCTION

Since the discovery of the D meson more than 10 years ago, many experiments have tried to take advantage of the high absolute rates of charm production in fixed target collisions. However, despite valiant efforts, success has been limited, principally because the non-charm background is much higher than in e^+e^- collisions. Hence the observed charm signal to background was either too poor, or the size of the signals extracted from the data too small, to make a major contribution to charm spectroscopy or charm decay phenomenology, which thus became the preserve of e^+e^- experiments.

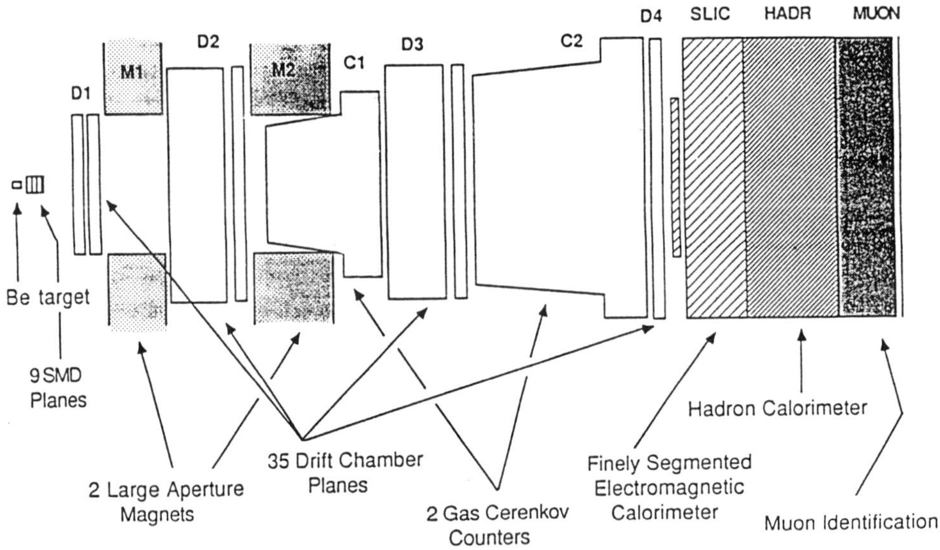

Fig. 1. The Tagged Photon Spectrometer.

This situation is now changing. The latest generation of fixed-target experiments has been able to make an impact on this field through the use of high precision vertex detectors. In particular the charm photoproduction experiment described here, experiment E691 at Fermilab, has obtained large charm samples of several thousand D^0 and D^+ decays on low backgrounds. This has permitted measurements of charmed meson lifetimes with unprecedented accuracy, and will allow detailed investigations of many other topics in charm physics.

In this paper the experiment is described with the emphasis on some key features, in particular the Silicon Microstrip Detector (SMD) and associated electronics. The reconstruction of tracks and vertices using the SMD and the method used for extracting clean charm signals is discussed in some detail. The first physics results of this experiment, namely the measurement of the lifetimes of the D^0, D^+ and D_s^+, are briefly treated. (Throughout this paper, charge conjugate states are implicitly included). Some performance characteristics of the SMD system are given and the changes being made for a future hadron-beam experiment are outlined.

EXPERIMENT E691

The experiment

A schematic of the Tagged Photon Spectrometer used in Fermilab experiment E691 is shown in Fig. 1.

An earlier version of the spectrometer has been described elsewhere [1]. The two magnets together provide a p_T kick of 0.5 GeV/c. The 9 SMD planes and the first drift chamber station allow a straight line measurement of the initial track trajectory with potentially 17 points. The 35 drift chamber planes arranged in 3 views (0° +20.5°, -20.5°) contain 6400 sense wires. A fast track measured in the SMDs and all 4 drift chamber stations has a momentum resolution $\Delta p/p$ of

0.01% p. The two threshold Cerenkov counters provide π/K separation in the range 6-37 GeV/c.

The tagged photon beam is obtained in the usual 3-stage process. Typically an 800 GeV proton beam of intensity 2.5 $10^{12}p$/22-sec spill is extracted from the Tevatron and on being dumped produces a 260 GeV electron beam of 6.5 $10^{7}e$/spill which is transported to the experimental area and put through a 20% W radiator, resulting in a bremsstrahlung photon beam of 1.3 10^{7} γ above 80 GeV per spill. The photon energy is tagged on an event-by-event basis to about 7% by detecting the scattered electron and subtracting its energy from that of the incident electron, and correcting for multiple bremsstrahlung. The measured photon spectrum spans the range 80-220 GeV with a mean of 120 GeV. On striking the 5 cm long Be target the tagged photons create about 200 hadronic interactions/sec, on top of an electromagnetically produced background from all photons of about 3 10^{5} $e^{+}e^{-}$ pairs reaching the detectors per second. The contamination of hadronic events by $e^{+}e^{-}$ tracks from beam photon conversions occurs at the level of only a few percent. The size of the beam spot (σ ~0.8 cm) is well matched to the rectangular cross section of the target (1.27 x 1.90 cm).

Some features which contributed to the success of the experiment were:

- The presence of the high precision vertex detector.

As described below, the SMDs allowed the identification of charm decays and a strong suppression of background.

- The use of a photon beam.

Photons are relatively favourable for charm production: the charm photoproduction cross section is estimated to be about 0.4% of the total photoproduction cross section at our energies – this fraction is some 5-10 times higher than in hadroproduction.

- The upgrades made to the Tagged Photon Spectrometer.

After its use in a previous experiment (E516) the following changes were made to the TPS :

  - The two Cerenkov counters were improved by installing new mirrors and by implementing finer segmentation.
  - More planes were added to the second and third drift chamber (DC) stations, and the first DC station was moved from inside the first magnet to just upstream of it.
  - The field in the second magnet was increased by a factor 1.5 to improve the momentum resolution.
  - The electron beam-line was modified to obtain increased electron acceptance.

- The use of an open and efficient trigger.

The trigger required a total transverse energy E_T, as measured by the electromagnetic and hadron calorimeters, of > 2.2 GeV. This requirement suppressed the hadronic interaction rate by a factor 3.2 while accepting about 75% of the charmed events.

- The advantages of the Tevatron.

The combination of proton and electron energies of 800 GeV and 260 GeV made possible by the advent of the Tevatron, together with the increased acceptance of the electron beam-line, gave almost twice the number of electrons per proton as the previous combination of 400 GeV protons and 170 GeV electrons. Furthermore, the Tevatron duty cycle of 35% (one 22 second spill every minute) compared very favourably with the pre-Tevatron situation of a 1-second spill every 10 seconds.

- High data acquisition rate.

The data was recorded onto tape at a rate of about 100 events/sec, with each event consisting of about 1600 16-bit words. Typically 2000 events/spill were recorded with a dead-time of 30%.

- Excellent computing facilities.

The mainframe Cybers at Fermilab provided powerful computing with reasonable turnaround. The use of the Advanced Computer Program facility dramatically removed the computing bottleneck which reconstructing 100 million events would have become.

The data sample

In a 6 month run (March-August 1985) about 100 million triggers were recorded on 2000 data tapes. Of this 90% were taken with the transverse energy trigger while the remaining 10% required simply a hadronic interaction with no E_T condition. Another 10 million events were taken in a subsidiary experiment with a closed geometry configuration to measure the A-dependence of ψ photoproduction.

From the measured charm enrichment of 2.4 provided by the trigger we estimate that the recorded data contain about 1 million charm events. For a favourable charm decay mode such as $D^0 \to K^-\pi^+$, the product of the relevant acceptance factors for usefully identifying the charm decay (the fraction of events for which the decay products are well measured in the spectrometer, the fraction of kaons from the decay which are in the momentum region suitable for Cerenkov identification, and the fraction of decays surviving the vertexing cuts) is about 0.1; we expect to obtain about 4500 such events in the experiment. For higher multiplicity modes, the overall efficiency drops by about 0.65 for each additional charged pion. For channels containing a K^0 or π^0, the efficiency is considerably lower but will still allow useful samples of such modes to be obtained.

Computing

The computing time required for reconstructing one event through the charged track reconstruction and through the calorimetry and vertexing is about 0.9 sec on the Cyber 175. In 7 months of processing, we were able to reconstruct 15 million events on the Fermilab Cybers. Extrapolation of this rate gives a period of three years before the full data sample would be available for physics analysis. This virtually unacceptable delay was fortunately avoided by our succesful use of the Fermilab Advanced Computer Program (ACP). In its first application in large scale data reduction, the ACP processed a further 15 million events through the charged track reconstruction in just 4 weeks. The ACP [2] is a collection of processors configured such that each processor or node reconstructs events independently. The heart of each processor is a commercial 32-bit microprocessor. The 53 node system with which production was started has the computing power of 5.5 Cyber 175's or 35 VAX 11/780's and the full system of 140 nodes is expected to be available soon. The reconstruction code running on the processors is derived from essentially standard Fortran code. Several university groups are now in the process of setting up their own ACP system.

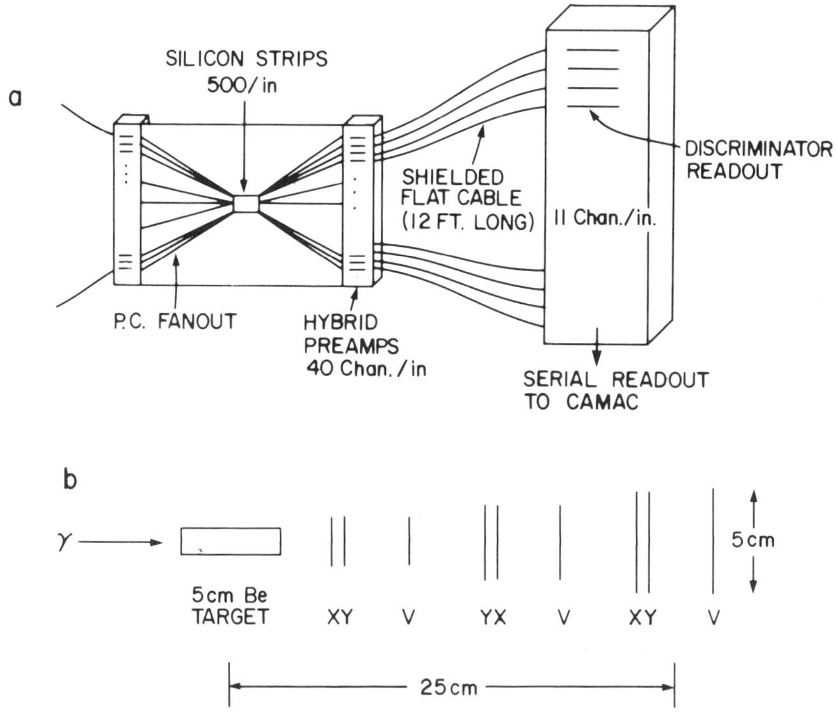

Fig. 2. (a) Schematic of an SMD plane and readout; (b) SMD Layout.

THE SILICON MICROSTRIP DETECTOR

A schematic of one SMD plane and its readout is shown in Fig. 2a. The configuration of the 9 SMD planes with respect to the 5 cm Be target is shown in Fig. 2b.

Each Si plane is 300 μm thick with strips of pitch 50 μm. The 9 planes are grouped into 3 sets of triplets measuring 3 views each: X(0°),Y(90°),V(-20.5°). The 3 upstream planes, manufactured by Enertec-Schlumberger (France) have an active area of 2.5 x 2.5 cm^2 and have 512 strips each. The 6 downstream planes, manufactured by Micron Semiconductor (United Kingdom) have an active area of 5 x 5 cm^2 and have 1000 strips each [3]. The edges of the middle triplet were not instrumented, resulting in a total of 6840 active channels for the system. The triplet assemblies are mounted on two granite beams to facilitate alignment. The system was operated in a cooled environment with nitrogen flowing onto the detectors. All the detectors are ion-implanted passivated devices made by the planar process [4].

At the operating reverse-bias voltage of about 100V the detectors are fully depleted [5]. The charge deposited by the tracks passing through the detector (a minimum ionizing particle deposits 25000 electrons or 4 fC) is collected (without

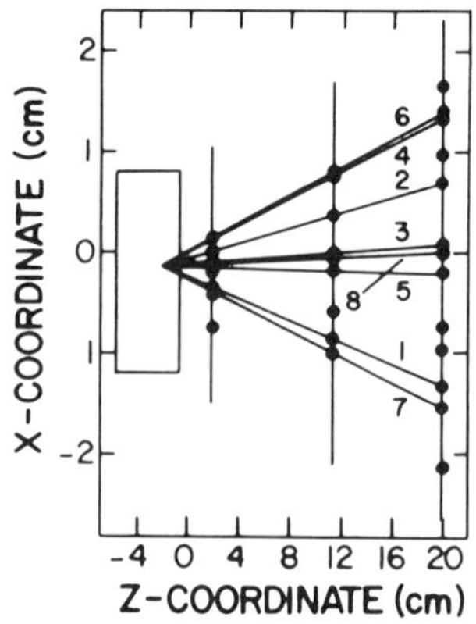

Fig. 3. SMD hit pattern of a typical charm event. Tracks 1 and 2 are from a $D^0 \rightarrow K^-\pi^+$ decay.

internal amplification) by the electric field and propagates from each strip via a wire-bond onto the printed circuit traces laid down on a G-10 board and into the preamplifiers which are directly connected to the fanout.

The preamplifier used was the 4-channel MSD2 card manufactured by Laben (Italy). The characteristics of these hybrid current-sensitive devices are low noise and fast processing time [6]. The high packing density of 40 channels/inch required for their mounting onto the detector assembly was made possible by their low power dissipation of 15 mW/channel. Their current gain of 200 resulted in an output signal of about 1 mV for a minimum ionizing particle.

These small signals propagated along 12-foot-long flat shielded transmission-line cables with a nominal impedance of 28 ohms (chosen to match the input impedance of the next stage of the readout chain). The 1710 cables required were manufactured from commercial 9-conductor ribbons covered on both sides by a dielectric tape and aluminum foil, which was electroplated at both ends with copper through which alternate conductors were grounded, leaving four signal strips per cable.

The cables passed the preamplified analog signals to the readout electronics consisting of modified 8-channel Nanosystems (USA) S710/810 cards mounted in crates on either side of the detector axis. The original discriminator/delay/latch/shift-register circuit intended for use with MWPCs was modified by adding a 1-transisitor amplifier circuit per channel to provide additional amplification and to invert the positive signal from the SMDs; also a potentiometer in the new circuit enabled the discriminator threshold to be adjusted separately for each channel to an accuracy of about 5%. Although the signal to noise ratio at the discriminator input is fairly high (greater than 10) the thresholds must still be set more accurately than for a wire chamber. The thresholds were set, using a standard test pulse, to about 0.4 of the size of a signal from a minimum ionizing particle; this setting is such that it causes almost no loss in efficiency. Signals greater than this enabled a latch bit to be set in the card's shift register. To allow time for the high-level trigger decision to be made, the discriminated signal was delayed by 750 ns before being latched; the delay for each channel had to be set separately by adjusting a potentiometer. The gate width at the latch input was 50 ns. The shift registers of the entire system were daisy-chained into 6 independent streams, each of which was read out serially by a Nanosystems Camac scanner at 5 MHz. The readout racks were cooled by fans to minimize the sensitivity of the electronics to the heat generated by the high power dissipation (250 mW/channel) of this old (c. 1975) readout system.

The pattern of hits in the SMDs for a typical charm event is shown in Fig. 3 for the 3 X planes. The picture shows two features which will be quantified later : the high efficiency of the SMDs and the low noise rate of 1 or 2 noise hits/plane/event.

The following comments about the SMD can be made:

- Since every strip is read out as on or off, with no analog output being recorded, the position resolution is given by the strip width/ $\sqrt{12}$, i.e. 14.4 μm. This number sets the resolution limit of the entire system.
- The 9 planes we used constituted a fairly minimal system, since there are only 3 planes per view with which to find straight lines. Furthermore the 3 Y views have only weak downstream support from the drift chambers, whose orientations are predominantly X-like. Considerations of finance and time precluded the use of more planes.
- The geometrical acceptance of the SMD system is about $\pm$100 mrad, which is reasonably well matched to the drift chamber system acceptance. Some higher angle tracks which miss the SMDs can be picked up in the DCs.
- Ideally a 3-view system would of course be oriented at 0°, 60° and 120°. In our case the fanout scheme constrained two of the views to be perpendicular (X and Y). A choice of 45° for the orientation of the third view (V) was rejected because of the acceptance loss at the corners of each plane characteristic of this scheme. Instead the V view was chosen to be X-like, at an angle of 20.5° to the X view, to follow the orientation of the drift chamber planes. While this made for easier pattern recognition, it resulted in poorer resolution in Y than in X.
- The cost of setting up the entire system of SMD plus readout was about US $200000. This was made up of per-channel costs roughly as follows: $10 for purchase of, and mechanical work on, the detectors, $8 for the preamplifiers, and $10 for the cables and readout.
- We benefitted greatly from the experience and development work of other groups working with silicon devices. The ACCMOR group at CERN were pioneers in the use of microstrip detectors. Our Enertec devices are based on the detectors designed by E. Heijne et al. [6] for use at the CERN Omega Spectrometer, while the Micron planes are a modified form of the detectors designed for the CERN NA14 experiment.

EVENT RECONSTRUCTION

Track finding

The presence of the SMDs enabled us to modify our track-finding algorithm, which now starts off in the SMDs rather than in the drift chambers (DC). This is a natural choice because the SMDs are superior to the DCs in cleanliness and efficiency, they are well suited for extrapolation to the DCs, and tracking in the SMDs is substantially faster than in the DCs. The track reconstruction proceeds as follows:

- Find all 1-view SMD track candidates. Combine 2 views to predict hits in the third view, and thus find SMD track candidates having 6 or more hits (out of 9 possible).
- Find SMD+DC track candidates by picking up reconstructed DC space hits ('triplets' made from the overlap of hits in the 3 DC views) downstream of the magnets. This is done by projecting the SMD trajectory in the non-bend plane to define a horizontal band in the bend plane and allowing each triplet lying within the band to establish a trajectory road.
- Do a momentum fit to the SMD+DC hits (up to 44 points per track).
- Require that most SMD hits on a track be unique (some hit sharing is still allowed).
- Pick up SMD-only tracks (these could be kinks from hyperon decays).
- Pick up DC-only tracks (these could be tracks from V^0 decays, tracks outside the SMD aperture, tracks created downstream of the SMD, or tracks missed in the SMD).

After this, a typical charm event with the decay $D^{*+} \rightarrow D^0\pi^+ \rightarrow (K^-\pi^+)\pi^+$ has 10.0 charged tracks reconstructed, of which 8.5 are SMD+DC tracks and 1.5 DC-only; of the former, 5.7 are faster tracks which go through both magnets and hence are very well measured, 1.9 are 1-magnet tracks and 0.9 are SMD-only tracks.

High hit density in the central region and insufficient hit redundancy cause two connected problems in the tracking:

A large number of 'phantom' track candidates are generated when using only the SMD hits : these spurious tracks are created from a few hits from each of several different real tracks. For example, before the link-up with the drift chambers is made, there are several times more phantom 8-hit tracks than real 8-hit tracks even after stringent χ^2 cuts. The phantom track problem is reduced to a tolerable level of a few percent in the best measured tracks by requiring a good link with the DCs and by requiring that most SMD hits on a candidate track have not already been used. The proportion of phantom tracks in the small SMD-only class is much higher. It is important for physics analysis to keep down the number of phantom tracks since they can cause an explosion of combinatorial background in mass plots. As an example, one event processed with an earlier program version had 16 reconstructed SMD+DC tracks which gave 26 $K3\pi$ combinations within 200 MeV of the D mass, all of which satisfied a tight D^*-D mass difference cut. When the criterion on the number of shared hits allowed was tightened up for the final program version, the same event had only 10 SMD+DC tracks giving only 3 entries into the corresponding plot.

The other problem is that the track clustering and confusion in the central region of the detector causes a loss of track-finding efficiency. The per-track efficiency for finding an SMD+DC track in the data is estimated to be about 92%, because some tracks are not found in the SMDs even though they pass through them. This loss of efficiency is really a multiple inefficiency in that when one track is not correctly identified, another one or more tracks will generally be lost too.

Track fitting

The 4 parameters describing the track initial straight line trajectory, namely the two intercepts x_0 and y_0 and the two slopes s_x and s_y, and the covariance matrix of these four quantities, is required for the vertex fitting. Since an accurate knowledge of this error matrix is critical, we chose for this purpose to use the track parameters

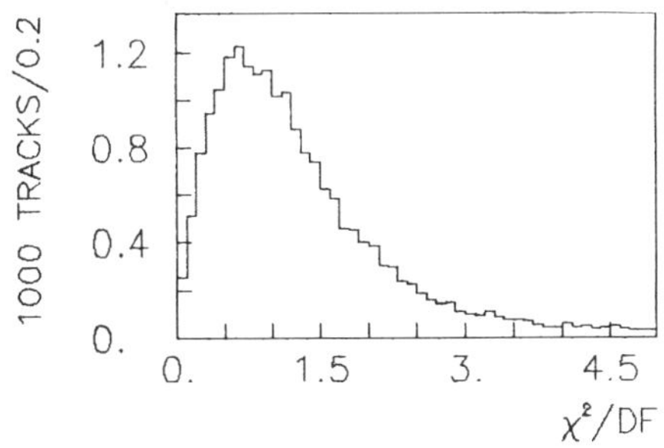

Fig. 4. The χ^2/Degree of Freedom distribution for straight-line track fits to SMD hits.

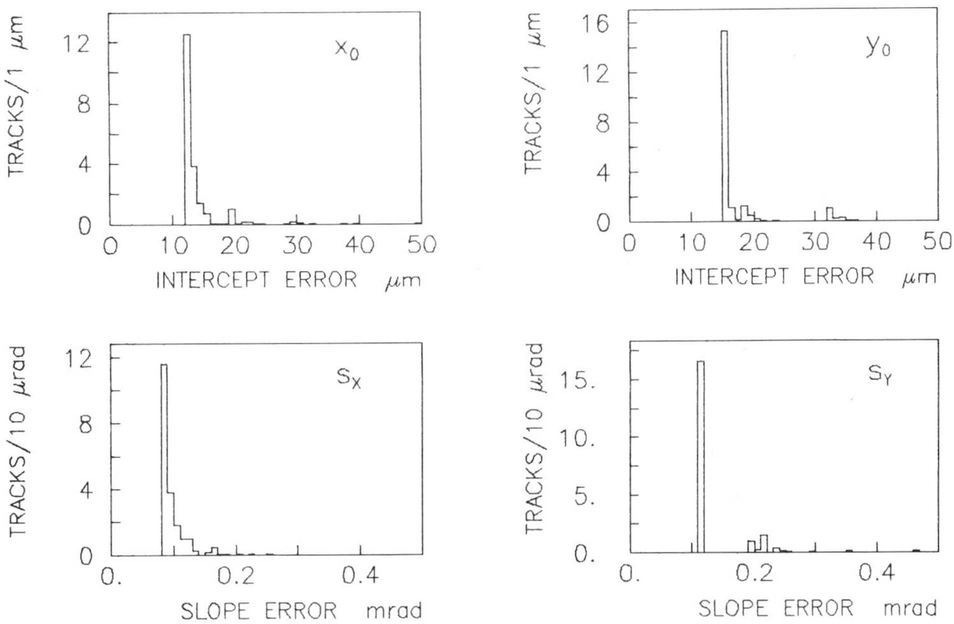

Fig. 5. Track intercept and slope errors from the track fit to SMD hits. The vertical scales are in units of 1000 events.

and errors as obtained from fitting the tracks to the SMD hits only, rather than to the entire trajectory.

The tracks were fitted to the SMD hits using the nominal strip resolution of 14.4 μm. Fig. 4 shows the χ^2/DF for all tracks in a sample of D^*-dominated events; the distribution follows the theoretically expected one.

The error distributions for the four track parameters are shown in Fig. 5; the tail at higher errors is caused by tracks with fewer than the maximum number of hits. Typical values of the errors are

$$\sigma(x_0) \sim 13\ \mu\text{m} \quad ; \quad \sigma(\text{y}_0) \sim 16\ \mu\text{m} \quad ; \quad \sigma(\text{s}_\text{x}) \sim 100\ \mu\text{rad} \quad ; \quad \sigma(\text{s}_\text{y}) \sim 130\ \mu\text{rad}.$$

(The intercept errors are evaluated at z=0 near the target exit). The size of these errors depends on the intrinsic strip resolution and on the number and position of the hits. It turns out that for our setup the track intercept error is about the same size as the strip resolution – recall the expressions for the intercept and slope errors for a straight-line fit to N points of measurement error δ spaced equally along a distance L, which approximates to our situation [8]:

$$\sigma(\text{intercept}) = \delta\frac{2}{\sqrt{N}} \quad \text{independent of } L \ ; \quad \sigma(\text{slope}) = \frac{\delta}{L}\sqrt{\frac{12}{N}}.$$

Before using the track-fit errors in the vertex fit, the contribution to the track errors from the track's multiple scattering in the target must be added in quadrature. For a track emerging from the middle of the target, which goes through 7% of a radiation length, these contributions are

$$\sigma_{\text{MS}}(\text{intercept}) = \frac{50}{p(\text{GeV})} \ \mu\text{m} \quad ; \quad \sigma_{\text{MS}}(\text{slope}) = \frac{3300}{p(\text{GeV})} \ \mu\text{rad}$$

and so the intercept error is significant for tracks of momentum below about 5 GeV/c.

The multiple scattering in the 9 Si planes, whose total thickness is 2.8% of a radiation length, can in principle be accounted for in the SMD track fit by increasing the errors on each measurement from the nominal 14.4 μm by an amount dependent on the track momentum and plane position [9]. However no improvement was found in the charm signals on doing this, probably because those tracks which are most susceptible to improvement are not reconstructed in the first place because they fail pattern recognition cuts. The shape of the χ^2/DF of Fig. 4 implies that, overall, the multiple scattering in the detector planes is not a serious problem for reconstructed tracks, and so this correction was not implemented.

Vertex fitting

Using the track parameters and the track errors, one can fit tracks to a vertex by either a simple or a constrained least-squares fit to find the vertex coordinates. The fit also gives the errors on the vertex positions; these errors will be essential for extracting the charm signal. Two independent vertexing algorithms were developed to check how sensitive the vertex finding was to the vertexing method:

One method builds up vertex candidates by starting with a 2-track vertex. The tracks in the event are added to the vertex one at a time and accepted if the resulting vertex gives an acceptable fit. If there is a track left over unfitted after all the tracks have been tested, it is used to make a new seed vertex and a different vertex is built up by accretion as before. All reasonable combinations are output in a vertex list which is later checked against any charm candidates in the event.

The other approach starts by identifying, from an effective mass calculation, a set of charm candidate tracks. All the other tracks in the event are fitted to a potential primary vertex. If the fit is bad, the tracks contributing the most to χ^2 are removed one by one until an acceptable fit is found.

The two approaches were found to give essentially the same result. The lifetime results presented here were obtained using the first algorithm while most of the other plots in this paper were obtained using the second program on a sample of 1300 events containing about 520 $D^{*+} \to D^0\pi^+ \to K^-\pi^+\pi^+$ events.

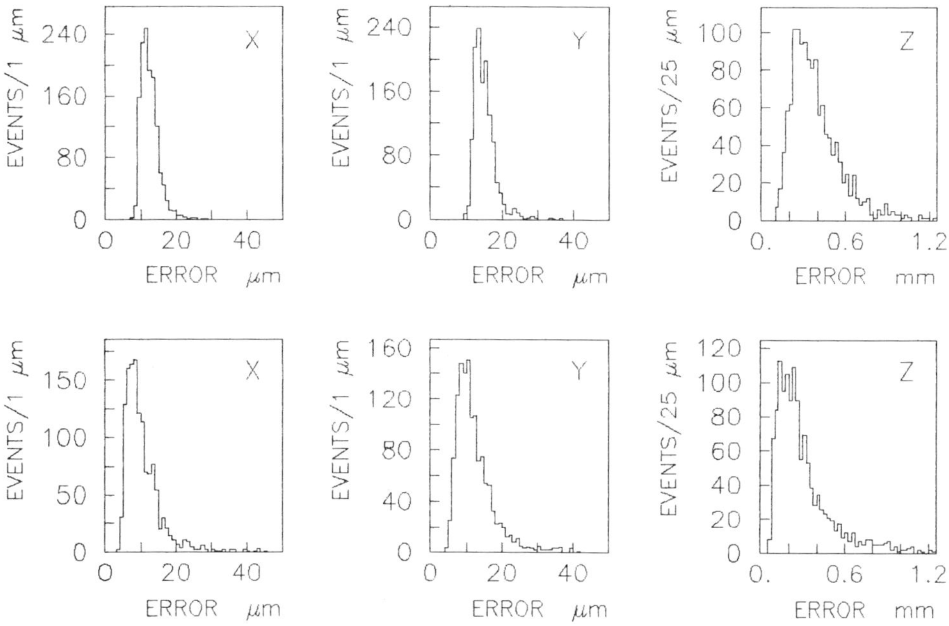

Fig. 6. Vertex position errors from the vertex fit to : (top row) the charm candidate vertex, and (bottom row) the main vertex, in a sample of D^*-dominated events.

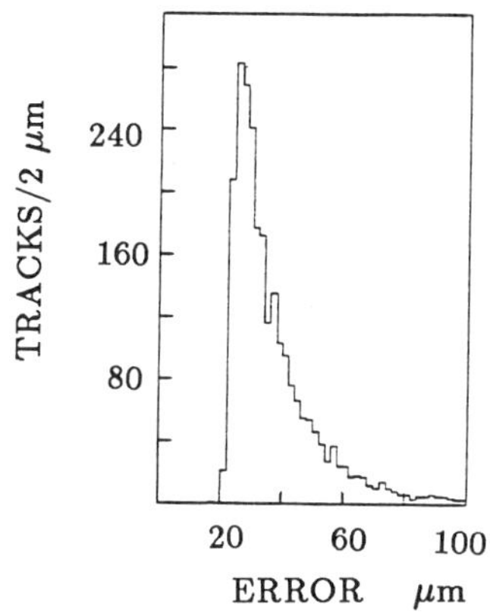

Fig. 7. The impact parameter resolution at the main vertex.

In our setup the vertex errors are heavily dominated by the error on the track intercept – the angular error contributes little. Fig. 6 shows the errors in the fitted vertex position for the charm candidate vertex and for the main vertex in the sample of D^* dominated events. Typical values for the charm vertex errors are :

$$\sigma(\mathrm{X}) = 12\ \mu\mathrm{m}\ ; \quad \sigma(\mathrm{Y}) = 14\ \mu\mathrm{m}\ ; \quad \sigma(\mathrm{Z}) = 300\ \mu\mathrm{m}$$

The transverse vertex error for the D^0 2-body decay turns out to be about the size

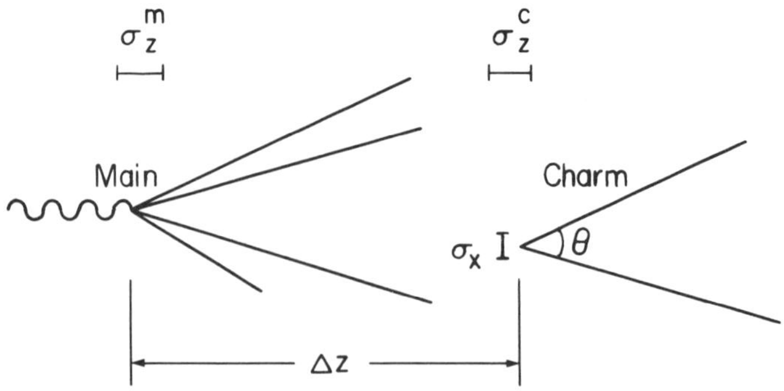

Fig. 8. Illustration of quantities used in the definition of the intervertex separation parameter N_σ.

of the SMD strip resolution. The errors on the main vertex are somewhat smaller, reflecting the higher number of tracks fitted. σ(Y) is slightly bigger than σ(X) because of the lower accuracy of the Y view mentioned above.

Finally Fig. 7 shows, for the same sample, the error on the impact parameter (or distance of closest approach) at the main vertex for all tracks fitted as belonging to the main vertex. From this, one can obtain the figure of merit often used for vertex detectors, namely the impact parameter resolution; in our case this is about 25 μm, a few microns of which comes from target multiple scattering.

EXTRACTION OF CHARM SIGNALS

Previous experience has shown that, without some form of signal enhancement, most charm modes are essentially unobservable in fixed target experiments because of the combinatorial background. Experience has also shown that many methods considered for improving the charm signal, such as kaon signatures, lepton signatures or multiplicity changes, are not particularly rewarding. The only clean separation of charm from background comes from the finite charm decay length. The charm mean lifetimes correspond to proper decay lengths of a few hundred microns, and the γ factor from the laboratory momenta of the charmed particles results in the actual decay distances being typically a few millimeters. Comparing this with our longitudinal vertex position resolution of about 300 μm shows that the obvious way to enhance the charm signal is to cut on N_σ, the significance of the intervertex separation measured in standard deviations :

$$N_\sigma = \frac{\Delta z}{\sigma_z}$$

where Δz is the distance between the charm vertex and the main vertex, and σ_z is the error on Δz, obtained from compounding the errors on the main vertex σ_z^m and on the charm vertex σ_z^c (see Fig. 8).

For a mode such as $D^+ \rightarrow K^-\pi^+\pi^+$, requiring $N_\sigma > 10$ results in a background suppression of about 300 while accepting 0.7 of the signal. The strong signal so obtained (see Fig. 14) would, without a vertex cut, have shown up as a barely visible peak of 3 σ significance.

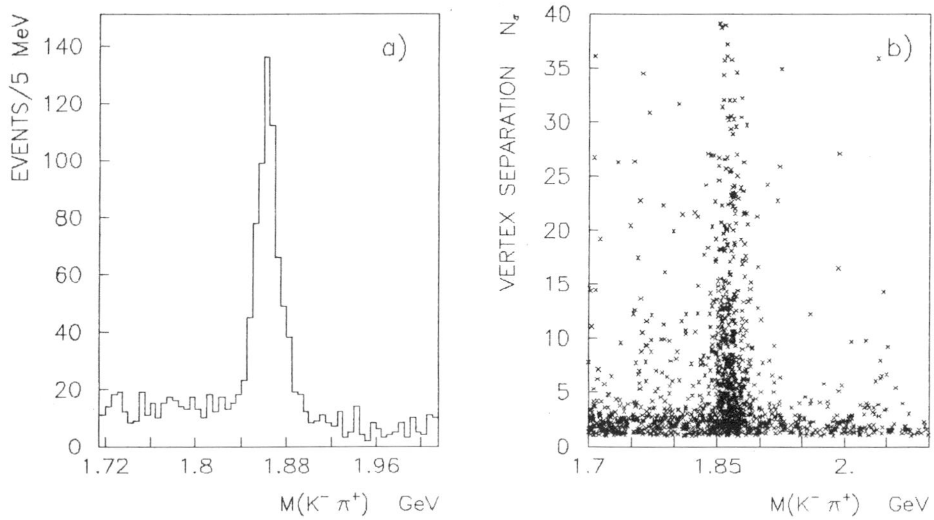

Fig. 9. (a) The $K^-\pi^+$ mass spectrum showing the D^0 signal, and (b) M($K^-\pi^+$) vs. N_σ, from the D^* sample. The condition N_σ >1 has already been imposed.

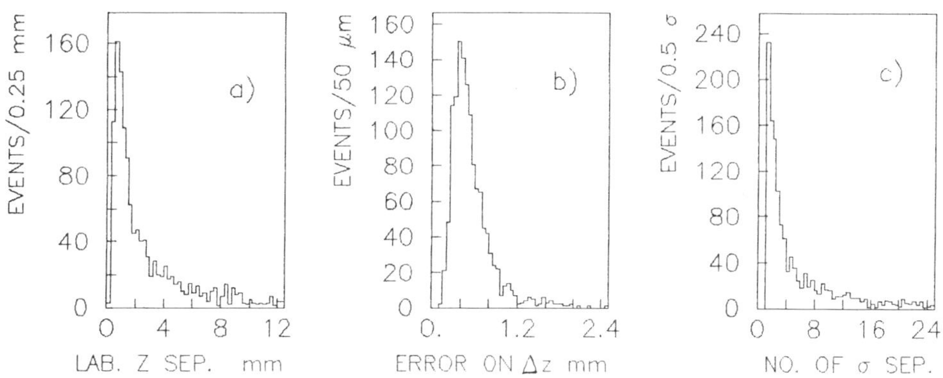

Fig. 10. Intervertex separation parameters for D^* sample. (a) Lab. separation Δz; (b) σ_z, the error on Δz; (c) $N_\sigma = \Delta z/\sigma_z$, the significance of the separation.

The cut on N_σ is almost independent of the charmed particle momentum because both the vertex separation Δz and the error on the charm vertex σ_z^c are proportional to γ: $\Delta z = \gamma vt$, where t is the decay proper time, while $\sigma_z^c \sim \sigma_x/\theta$, where σ_x is the transverse error on the charm decay vertex and θ is the decay opening angle. But since θ is given by $\theta \sim m/E = 1/\gamma$, where m and E are the decaying particle's mass and energy, σ_z^c is also proportional to γ. The γ factor does not cancel out entirely since σ_z^m does not depend on γ, but for $\gamma >$ about 20 (which most of our identified D's satisfy) N_σ is only weakly dependent on γ and is given approximately by

$$N_\sigma = 15t$$

where t is the charm proper lifetime in picoseconds. Thus for example a D^+ with a

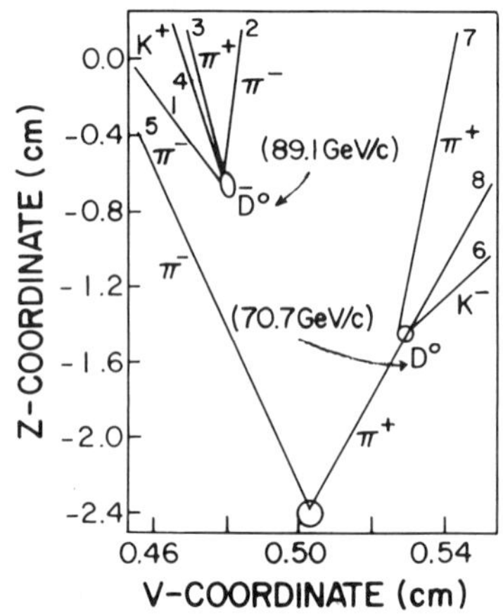

Fig. 11. Vertex region of an event containing a D^*. Tracks (6,7)8 are the $(K,\pi)\pi$ from the D^* while tracks 1,2,3 and 4 are from the accompanying $\overline{D}^0$ decay. The ellipses are the 1σ error contours from the vertex fit.

proper lifetime of 1 ps will generally give a vertex separation of about 15 σ.

It is important to note that N_σ is a signed quantity, since dz, the longitudinal separation of the main and charm candidate vertices, is just as likely to be negative as positive for the non-charm combinatorial background. The actual charm decays show up as a tail at high values on the gaussian shape of the N_σ distribution.

The effect of the N_σ cut can be seen in the sample of D^{*+} events decaying into $D^0\pi^+ \rightarrow (K^-\pi^+)\pi^+$ obtained from 30% of the data. In this case, because of the D^*-D mass difference selection, the D^0 signal is visible without a vertex cut, but is strongly cleaned up with little loss of signal by a weak requirement on N_σ. Fig. 9a shows the $M(K^-\pi^+)$ spectrum for events having a mass difference $M(K^-\pi^+\pi^+) - M(K^-\pi^+)$ in the range 0.144 to 0.147 GeV, and requiring $N_\sigma >$ 1. A very strong D^0 signal is evident. Fig. 10 gives the dz, σ_z and N_σ distributions for these events, showing respectively the actual decay length of a few mm, the typical error on this of about 500 microns, and the tail of high sigma events. Fig. 9b shows a scatter plot of the $K^-\pi^+$ mass against N_σ; it is clear from this plot that tightening the N_σ cut to 5 or 10 will produce a much cleaner signal, as the background is concentrated at low N_σ. There is a small irreducible background, which is caused partly by secondary interactions and spurious tracks, but which is also partly (perhaps up to a half) charm-associated, arising both from combining tracks from the second charm decay with main vertex tracks, and from feed-down from higher multiplicity charm decays (in this particular example mainly from $D^0 \rightarrow K^-\pi^+\pi^0$ decays populating the $K^-\pi^+$ mass region below the D^0).

Fig. 11 is a blowup of the vertex region of one of these D^* events, which shows in addition the decay of the second charmed particle in the event. Note that both decays occur within the target, as is usually the case.

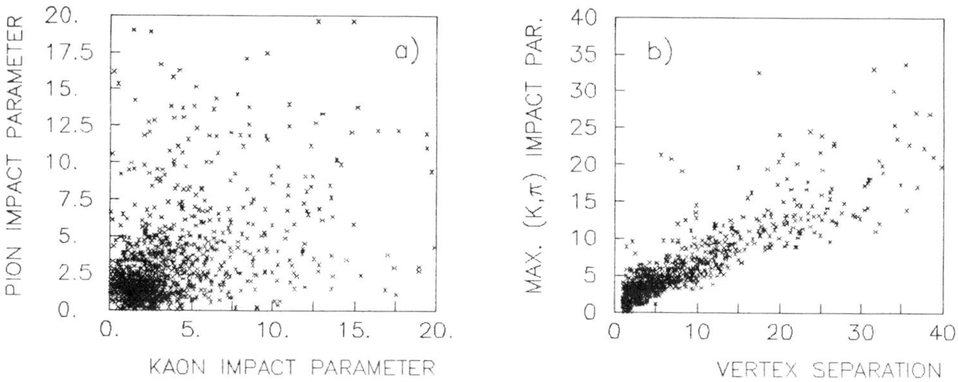

Fig. 12. Impact parameters of $D^0 \rightarrow K^-\pi^+$ decay products. (a) Impact parameters of K^- vs. π^+ ; (b) Vertex separation N_σ vs. $\max(K,\pi)$ impact parameter. All quantities are expressed in terms of numbers of standard deviations.

We briefly address here the question of whether there is any advantage to using a different criterion for selecting the charm events, namely the significance of the impact parameter at the main vertex of the individual charm decay tracks, Δb, rather than the significance of the intervertex separation Δz which we actually used. These two quantities are obviously correlated since a large Δb results from a large decay opening angle, which in turn gives a small vertex error and hence a more significant Δz. First we can check whether the impact parameter of the decay products are correlated amongst themselves. Fig. 12a shows a scatterplot of $\Delta b/\sigma_b$ for the decay kaon track versus the decay pion track from the above sample of D^0 decays; only a weak correlation is evident. Next we choose the larger of the K or π impact parameters as a potential variable to use and plot it in Fig. 12b against the vertex separation, both expressed in numbers of standard deviations. The expected correlation is there, but the impact parameter variable appears to be less powerful in separating charm from background than the intervertex separation variable, which uses the information from both decay tracks. Obviously an impact parameter selection must be used for modes containing only one charged track emanating from the decay vertex, such as the potentially interesting channels $K^0\pi^+, K^0K^+$, and K^0p.

MEASUREMENT OF CHARMED MESON LIFETIMES

It is known that the D^+ lifetime is longer than that of the D^0 and D_s^+. The reason for this is conjectured to be that Pauli interference effects suppress the D^+ decay and/or that non-spectator decays enhance the D^0 and D_s^+ decay rates; however, other considerations such as QCD effects, colour matching and final state effects must also be taken into account. Clearly accurate lifetime measurements are essential to resolve these theoretical issues.

First we estimate the lifetime resolution σ_τ we can hope to achieve. This depends on the proper time resolution of the detector and on the statistics available. The detector proper time resolution σ_t is directly estimated from the spatial resolution of about 500 μm on the vertex separation and from the typical γ factor of 30 (from a mean charm momentum of 60 GeV/c), giving $\sigma_t \sim 0.05$ picoseconds. The effect of the number N of charm decays in the sample is obtained by noting that a single measurement drawn from an exponential distribution of mean lifetime τ gives an estimate of the lifetime with error also equal to τ. One can then combine these factors to obtain an approximate expression for the lifetime resolution :

$$\sigma_\tau \sim \frac{\sqrt{\sigma_t^2 + \tau^2}}{\sqrt{N}}.$$

In our case the detector resolution is much smaller than the lifetimes we are measuring which are of order 1 ps. For the D^0 and D^+, the number of events we have is of order 1000, and so according to this formula we can reach a statistical error level of a few percent. The controlling factor on the measurement will then be our understanding of the systematic errors.

Details of our lifetime analysis have recently been published [10], so the treatment here will be brief.

The aim of the analysis was to make the systematic errors as small as possible by making the corrections required to the pure exponential as small as possible. We obtained clean charm signals by demanding, as discussed above, a minimum decay length $z^{min} = \alpha\sigma_z$, where α is chosen for each channel and σ_z is calculated for each event. The distance from z^{min} to the decay vertex is then used to calculate the lifetime; the validity of this procedure is seen by noting that the exponential form of the lifetime distribution does not depend upon the choice of the point from which the decay is measured, as long as the charmed particle has been created upstream of this point, and as long as the choice is independent of the actual decay point. The advantage of this method is that any confusion surrounding the main vertex position, in particular any shift caused by including tracks from the second charm decay or from secondary interactions, is removed. The fiducial region for decays was defined to end at the first SMD plane.

The time spectrum was fitted over about 4 lifetimes to the form

$$Nf(t)\frac{e^{-t/\tau}}{\tau} + B(t)$$

where the parameters of the fit are N, the number of events in the charm signal and τ, the mean lifetime. The acceptance function $f(t)$ is obtained from Monte-Carlo simulation, and the function $B(t)$ representing the time distribution of the background is obtained from the wings of the distribution. The results are not sensitive to the exact parametrization of $f(t)$. This function corrects for the following effects which deplete the time distribution at small values: occasional vertex misidentifications giving z^{min} upstream of the true primary vertex (and hence of the charm creation point), absorption of the D decay tracks through secondary interactions (important only for the higher multiplicity decays), and a weak dependence of the event sample on the charm vertex position (geometrical acceptance, χ^2 cuts etc.) There is also a small correction, necessary only for the D^+, for a depletion at long lifetimes caused by the fiducial cut. All these effects give a negative correction to the lifetime while the background subtraction gives a positive correction, since the background has a shorter lifetime than the signal. In all cases the difference between

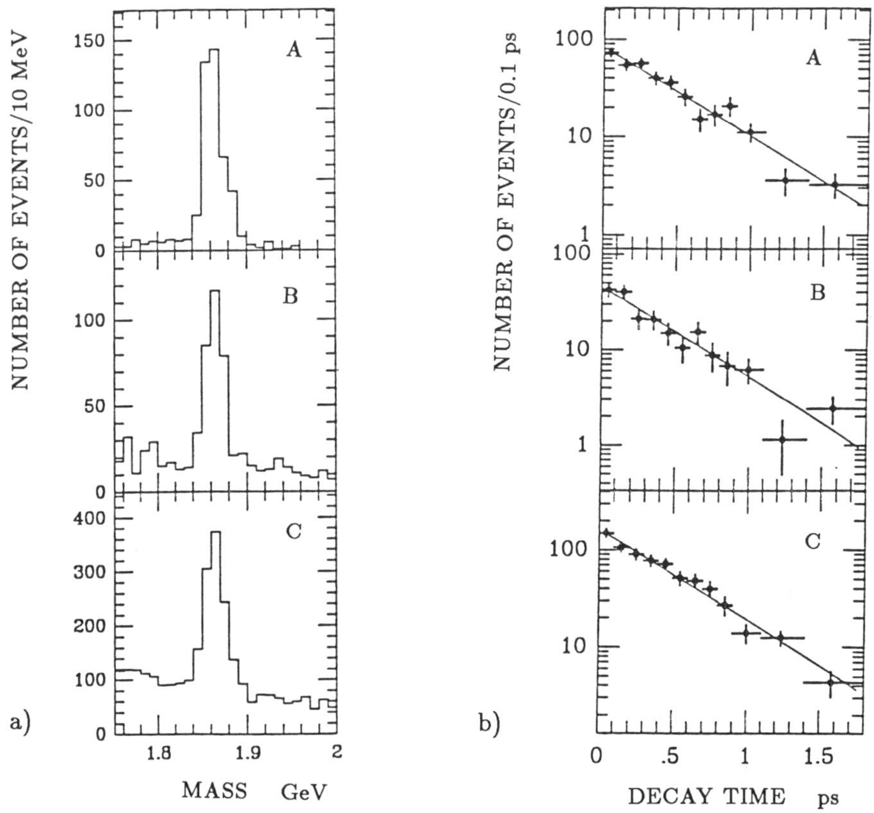

Fig. 13. (a) Mass distributions for the three D^0 channels : A) $D^0 \to K^-\pi^+$ via D^*, B) $D^0 \to K^-\pi^+\pi^+\pi^-$ via D^*, C) $D^0 \to K^-\pi^+$ with no D^*; (b) Lifetime distributions for the three D^0 channels.

the lifetime measured with these corrections and the result which would be obtained from a fit to a pure exponential was less than 10%.

We have measured the lifetime of the D^0,D^+ and D_s^+ (formerly called the F^+) using signals obtained from 30% of the total data [11].

For the D^0 lifetime measurement, we used three independent samples:

(A) $D^{*+} \to \pi^+ D^0, \quad D^0 \to K^-\pi^+$
(B) $D^{*+} \to \pi^+ D^0, \quad D^0 \to K^-\pi^+\pi^+\pi^-$
(C) $D^0 \to K^-\pi^+, \quad (\text{no} \quad D^{*+})$.

For each channel there was a requirement on the minimum Cerenkov joint probability for the appropriate particle identifications. For samples (A) and (B) the mass differences $M(K2\pi) - M(K^-\pi^+)$ and $M(K4\pi) - M(K^-\pi^+\pi^-\pi^+)$ respectively were required to be between 0.1435 and 0.1475 GeV. Events which satisfied the requirement for sample (A) were excluded from sample (C), so the samples are independent. The choice of the parameter α defining the minimum decay length accepted for each mode is given in Table I. The mass distributions for the events

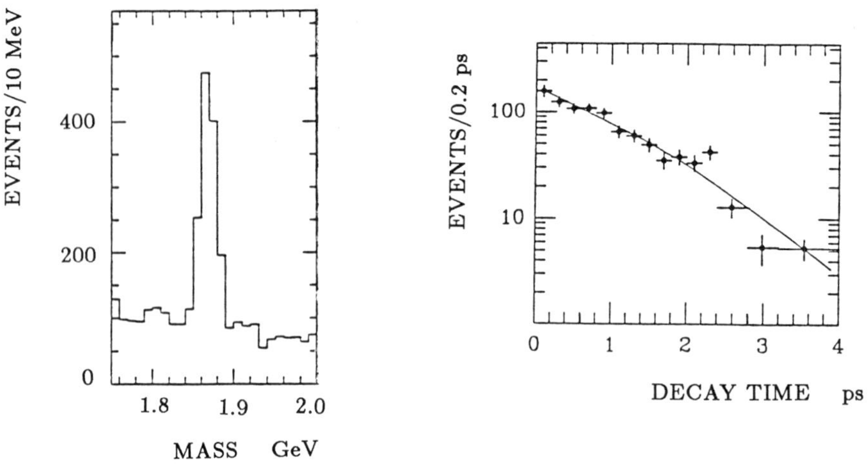

Fig. 14. Mass and lifetime distributions for $D^+ \to K^-\pi^+\pi^+$

so selected are shown in Fig. 13a. There is a clean D^0 peak in all 3 channels, with very small backgrounds for the channels selected using the D^* requirement The background-subtracted proper time distributions for the events in the D^0 mass range 1.839 - 1.889 GeV are shown in Fig. 13b, together with the maximum likelihood fit to the form given above. The measured lifetime in each sample, and the numbers of events involved, are given in Table I. The measurements for the 3 independent samples, which have different corrections and backgrounds, agree within statistical errors. A simultaneous fit to the 3 samples gives $\tau(D^0) = 0.435 \pm 0.015 \pm 0.010$ ps.

For the D^+ lifetime measurement, we used the decay mode $D^+ \to K^-\pi^+\pi^+$. The mass spectrum obtained and the background-subtracted lifetime distribution for events in the D^+ mass range 1.848 - 1.890 GeV are shown in Fig. 14. The selection and fit details are given in Table I. The measured mean lifetime is $\tau(D^+)$ $= 1.06 \pm 0.05 \pm 0.03$ ps. Our measurement of the ratio of the lifetimes $\tau(D^+)/\tau(D^0)$ is $2.44 \pm 0.14 \pm 0.08$.

The D_s^+ decay is of particular interest because the non-spectator diagram involved is an annihilation one. We observed the D_s^+ decay into two $K^+K^-\pi^+$ modes: $D_s^+ \to \phi\pi^+$ and $D_s^+ \to \overline{K}^{*0}K^+$. The mass spectra and background-subtracted lifetime plots for events in the D_s^+ mass range 1.953 - 1.985 GeV are given in Fig. 15. There are clear peaks in both modes from the Cabibbo-suppressed decay of the D^+. The contamination of the D_s^+ from D^+ or Λ_c^+ decays, by a π/K or a p/K misidentification respectively, is reduced to a negligible level because of our good mass resolution and the use of channels involving resonances. Details of the selection, the numbers of events and the lifetime results are given in Table I. A simultaneous fit to both modes gives the result $\tau(D_s^+) = 0.50^{+0.08}_{-0.07} \pm 0.02$ ps. Within the errors the D_s^+ lifetime is consistent with that of the D^0.

Our results are consistent with the 1986 Particle Data Group [12] value of $0.43^{+0.05}_{-0.04}$ ps for the D^0 lifetime averaged from 10 previous experiments, and with the D^+ lifetime of $0.92^{+0.13}_{-0.18}$ ps averaged from 7 previous experiments. Several new measurements and updated results [13] available after the PDG value for the D_s^+ lifetime was calculated now give a world average closer to our measurement.

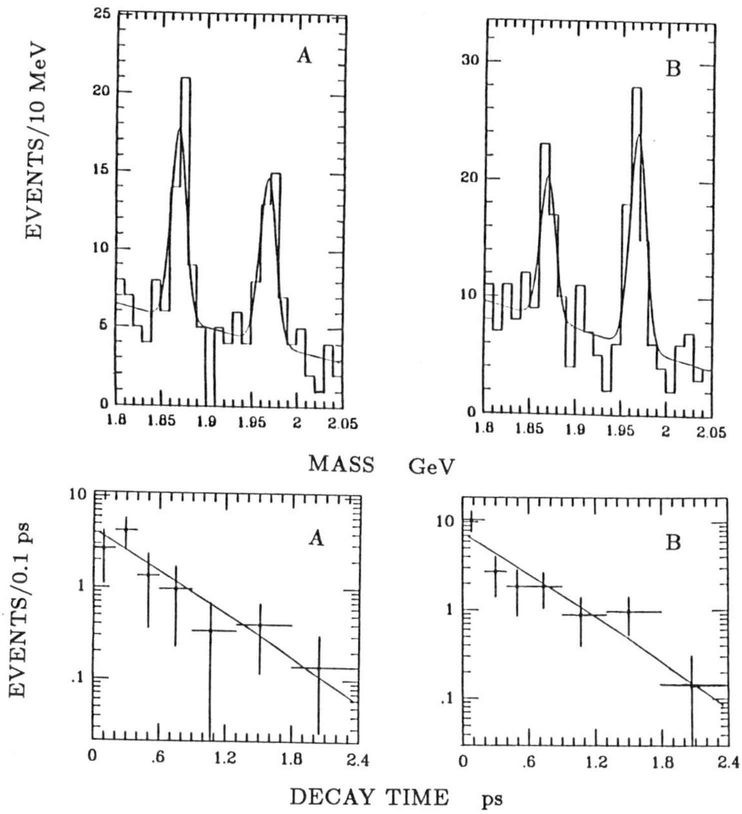

Fig. 15. Mass and lifetime distributions for the two D_s^+ channels : A) $D_s^+ \rightarrow \overline{K}^{*0} K^+$, and B) $D_s^+ \rightarrow \phi\pi^+$.

Table I. Charmed Meson Lifetime Measurements from 30% of the Full Data.

Mode	α	Signal	B/ground	Mean lifetime (ps)
$D^0 \rightarrow K^-\pi^+$ via D^*	5	385	16 ± 2	0.46 ± 0.03
$D^0 \rightarrow K^-\pi^+\pi^+\pi^-$ via D^*	7	228	37 ± 3	0.40 ± 0.03
$D^0 \rightarrow K^-\pi^+$, no D^*	8	746	450 ± 10	0.43 ± 0.02
All D^0		1359	523 ± 11	$0.435 \pm 0.015 \pm 0.010$
$D^+ \rightarrow K^-\pi^+\pi^+$	10	969	383 ± 9	$1.06 \pm 0.05 \pm 0.03$
$D_s^+ \rightarrow \phi\pi^+$	6	45	15 ± 3	$0.49\ ^{+0.11}_{-0.08}$
$D_s^+ \rightarrow \overline{K}^{*0} K^+$	10	24	10 ± 3	$0.52\ ^{+0.15}_{-0.10}$
All D_s^+		69	25 ± 4	$0.50\ ^{+0.08}_{-0.07} \pm 0.02$

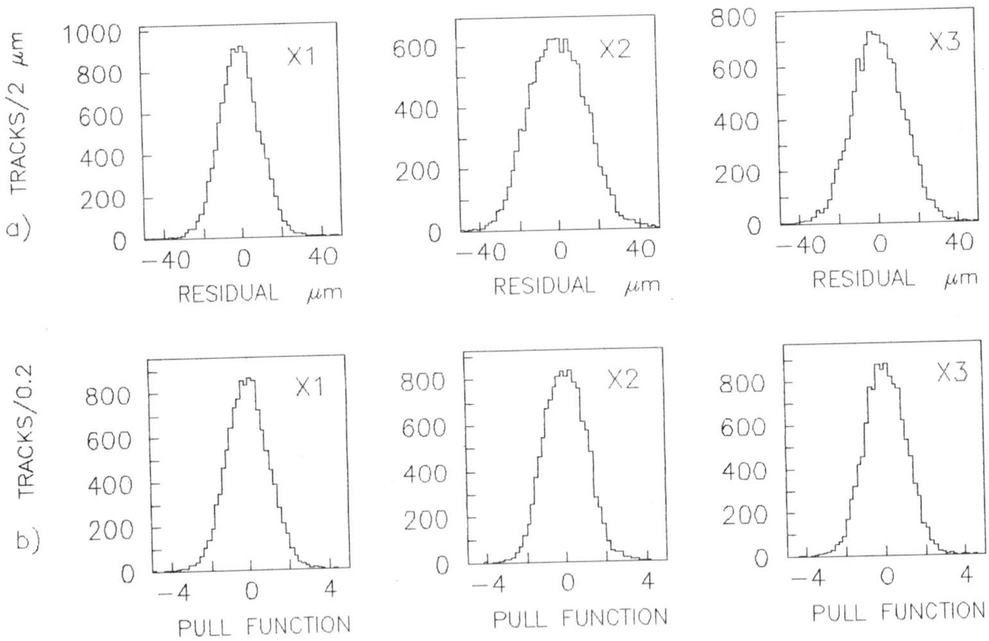

Fig. 16. (a) Residuals $x_{meas} - x_{fit}$; (b) Pulls $(x_{meas} - x_{fit})/\sqrt{\sigma^2_{meas} - \sigma^2_{fit}}$, for the three SMD X planes from track fits to SMD hits.

SMD PERFORMANCE

Detector performance

Overall the SMD performance was very good. The system took about a year to build, including a test beam trial. Once installed at the experiment, it was ready to take data after a week's beam. Setting and maintaining uniform thresholds by hand for the 6840 channels was the most time-consuming part of operating the detector. The only significant feature noted in operation was that the first three Enertec detectors showed a big increase in leakage current after about a week of beam, going from 1 μA to 50 μA, but stayed constant and operated satisfactorily thereafter. It is not known whether the deterioration was halted by a change in operating conditions implemented then – the bias voltage was applied only during the spill. The leakage current of the 6 larger downstream detectors was steady at a few μA. The total dose of the experiment was small by radiation-damage standards – about 5 10^{11} electrons through the detectors.

The detector alignment was such that the nominal resolution of 14.4 μm was actually achieved. Fig 16a shows the residuals $x_{meas} - x_{fit}$ in the three SMD X planes from tracks fitted with this nominal resolution (only tracks with momentum > 5 GeV/c are used here in order to minimize multiple scattering effects). The middle plane has a wider residual than the others, as expected from a least-squares fit. The 'pull' or 'stretch function' $(x_{meas} - x_{fit})/\sqrt{\sigma^2_{meas} - \sigma^2_{fit}}$ plotted in Fig. 16b has an RMS of about 1.0 for all three planes, showing that the nominal resolution is a good description of the data.

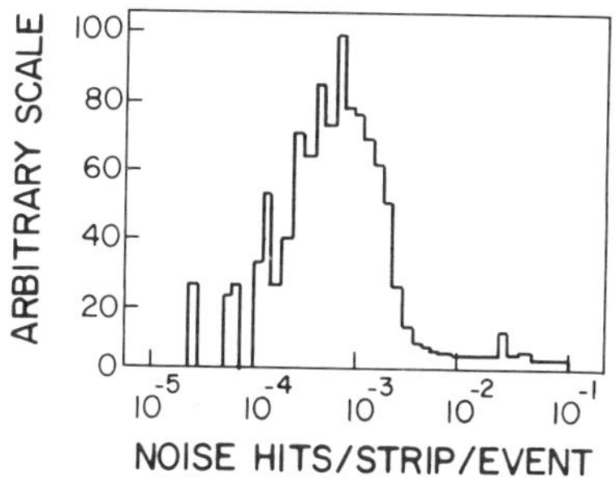

Fig. 17. SMD noise rate.

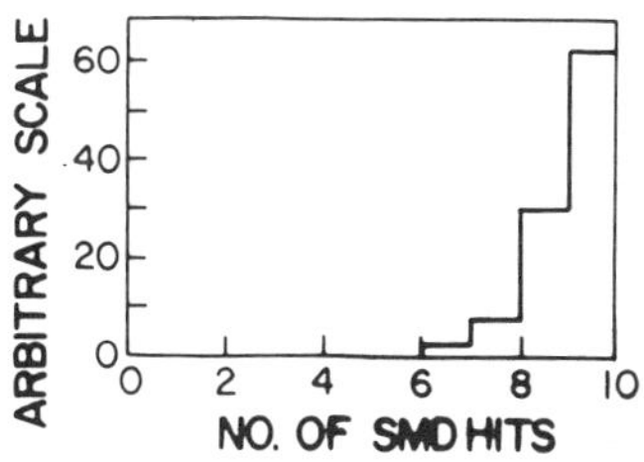

Fig. 18. No. of SMD hits on tracks of momentum > 5 GeV/c which go through all 9 SMD planes.

Electronic noise

We were able to run the SMDs with a gate width of 50 ns, and the 1 or 2 noise hits typically found per plane per event did not affect the pattern recognition. The noise rate, defined as the number of noise hits/strip/trigger is shown in Fig. 17; it has a fairly wide distribution with an average of 10^{-3}. It was not found necessary to ignore in the software the 'hot' channels with noise rates greater than 10^{-2}.

Per-plane efficiency

An estimate of this can easily be found from the number of SMD hits per reconstructed track. Fig. 18 shows this distribution for tracks of momentum > 5 GeV/c which go through all 9 planes; a number of tracks with fewer than 9 hits is evident. From the ratio of 9 to 8 hits we obtain a value of 0.95 for the per-plane efficiency. The loss of efficiency comes from channels which are not working for the following reasons : strips disconnected because of large leakage currents, strips with bad wire bonds, non-working readout channels, bad contact in cable connectors, and bad preamplifier channels, with the last two causes being rare.

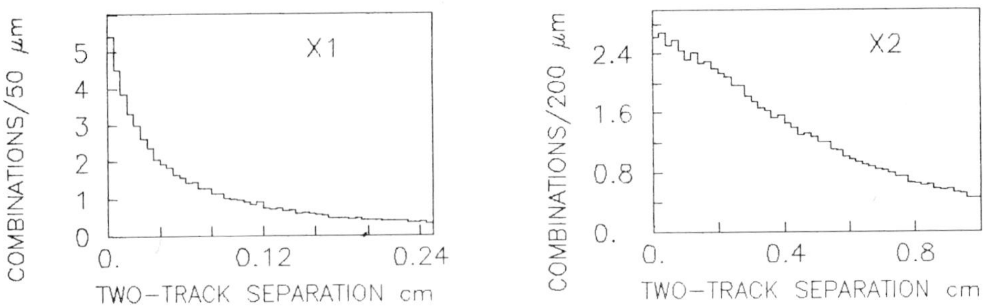

Fig. 19. Two-track separation of all track pairs at SMD planes X1 and X2. The vertical scale is in units of 1000 combinations.

Per-track efficiency

The efficiency for reconstructing a track with 6 or more SMD hits out of 9, given a per-plane efficiency of 95%, is expected to be 99.5 % in a clean single-track situation. Of course, this is not achieved in the actual events for straightforward reasons: multiple scattering causes hits to lie outside the predicted trajectory, and the differing geometrical acceptance of the planes means that some tracks do not go through all planes. However, even allowing for this, there is a loss of efficiency from confusion in the pattern recognition caused by track clustering in the central region. An estimate of the actual track efficiency can be obtained from the charm data itself, since the D^* sample is selected without requiring that the 'slow' pion from the $D^* \to D\pi$ decay be identified in the SMDs. We find that in about 8% of the events in the D^* signal the slow pion is reconstructed as a DC-only track despite having travelled through the SMDs. This estimate of 92% for the SMD+DC reconstruction efficiency is probably conservative since the slow π of mean momentum 5 GeV/c is more affected by multiple scattering than the average track in the data which has about twice this momentum. Monte Carlo events indicate that the efficiency rises with momentum, reaching 0.5 at 2 GeV/c and saturating above 10 GeV/c.

Two-track resolution

The clustering of the tracks is much higher in the first SMD planes than in the downstream ones, as seen in Fig. 19 which shows the two-track separation of all track pairs from the D^* sample in the first and second X planes . Another measure is provided by counting the number of tracks with another track falling within 50 μm of it at each plane. The mean number of such 'overlapped' tracks is (without double counting) 2.27 in X1 but falls to only 0.45 by X2. The high degree of overlap in the early planes only is not a severe problem for the track finding, because some hit-sharing is allowed and because the pair of tracks firing the same or adjacent strips in one view may be separated in another view. However the overlapping contributes to the phantom track problem.

The efficiency for detecting two tracks whose trajectories are very close can be estimated from $\gamma \to e^+e^-$ conversions. The finite opening angle m_e/E_γ for pairs and the multiple scattering they undergo mean that typically the e^+ and e^- are in the same or adjacent strips. The pairs resulting from conversions in the target are detected in one of three classes : both electrons are seen in the SMDs, or one track is seen in the SMDs and the other in the DCs, or both tracks are seen only in the DCs. From the relative size of the γ signal in the e^+e^- mass spectrum for these

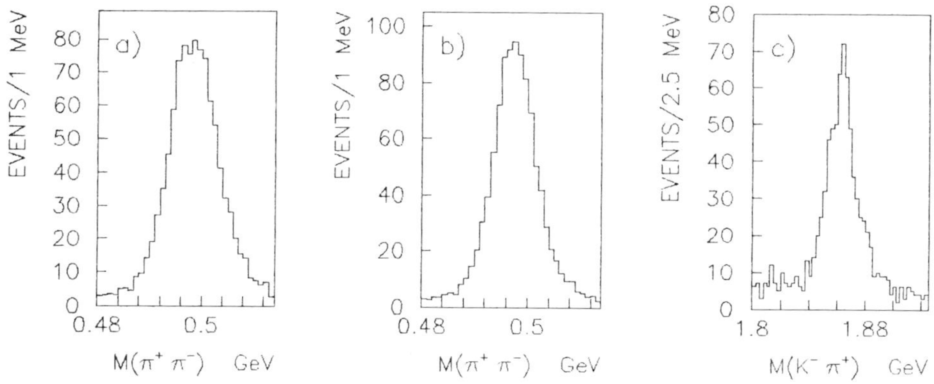

Fig. 20. Mass resolution. (a,b) $\pi^+\pi-$ mass spectra showing K^0 signal, for pion tracks measured in (a) drift chambers only and (b) SMDs and DCs. The plots are a subset of the full K^0 spectrum, selected by requiring that one track goes through both magnets and the other one through one magnet only. Plot (b) was obtained by requiring that the dipion vertex be well separated from the event primary vertex. The normalizations of (a) and (b) are arbitrary; (c) $K^-\pi^+$ mass spectrum showing D^0 signal, from D^* sample.

three classes, one can estimate that the SMD track efficiency for this extreme case is only about 50%.

Mass resolution

The presence of the SMDs improved the momentum resolution of most tracks over the measurement by the DCs alone. This improvement is most noticeable in those tracks which have fewer DC measurements and is reflected in a better mass resolution (which of course depends on the angular resolution as well as the momentum resolution) for particle combinations involving such tracks. An example of this is the $K_s^0 \to \pi^+\pi^-$ decay which results in two classes of events – about 10 % of the K^0's have decay lengths short enough to give tracks with SMD hits, but the bulk of decays occur in or beyond the SMDs giving reconstructed pion tracks with DC hits only. In the cases where the momenta of the pions are such that both tracks have gone through both magnets (and so have up to 35 DC measurements) there is not much difference in mass resolution between the two classes, but in the subset (about half the cases) where one pion is a 2-magnet track and the other is a 1-magnet track (with at most 20 DC points), the SMD hits give a noticeable improvement in mass resolution, as seen in Fig. 20 a) and b) where the mass resolution for the DC-only K^0's is $\sigma = 4.4$ MeV, while for SMD+DC K^0's it is 3.4 MeV.

The mass resolution of the $D^0 \to K^-\pi^+$ (from the D^* sample), shown in Fig. 20c on an expanded scale, is already fairly good at $\sigma \sim 10$ MeV. In principle a further improvement might be expected from using track parameters obtained from a constrained vertex fit. However, this must offset the slight loss of mass resolution caused by starting with track angles derived from the SMD hits alone rather than from the entire track trajectory, and does not appear to be worthwile. Our good knowledge of the track errors also raises the possibility of kinematic fitting of the charm decays, which would also lead to a mass resolution improvement.

FUTURE PLANS AT TPS

A new experiment (E769) is scheduled to run at the TPS in the next (1987) fixed-target running period using a hadron beam of 250 GeV/c. The physics aims are to study charm hadroproduction, to compare charm production (particularly D_s^+ production) by π and K beams, and to study the charm A-dependence.

The raw charm signal to background in this experiment is expected to be about 10 times worse than in photoproduction because of two factors : the charm/total cross section ratio is less favourable in hadroproduction, and the combinatorial background is higher, both because of the intrinsically higher multiplicity in hadroproduction and because of the higher beam energy. To overcome this, one needs better background rejection and more data.

The following changes are being made to the system to achieve these requirements :

To provide increased vertex resolution, 2 more SMD planes (one X and one Y) will be added between the target and the present SMDs. They will be Micron L detectors with 685 channels each, with the central 354 channels having 25 μm pitch and the outer channels 50 μm pitch. Two more SMDs of this type will measure the beam tracks and also help with the vertexing. New preamplifiers (Laben MSP1) on the new planes will provide more gain than before but with even less noise. Two MWPC Y planes will be added between the SMDs and the first DC station to provide additional redundancy in the tracking.

To enable more data to be recorded, the data acquisition will be upgraded by using a cluster of ACP processors in a VME crate and a 'smart' crate controller which will reduce the dead time per event from 3 ms to 1 ms. Together with the larger spill buffer in the ACP which will permit tape-writing between spills, this will result in a data rate to tape of 6000 events/minute, and hence allow up to 600 million triggers to be recorded in the experiment. Alternatively some on-line filtering is possible with the ACP processors to reduce the number of events recorded; the computation time available is the equivalent of 50 VAX ms/event and should allow for an impact-parameter filter.

CONCLUSIONS

Using basically a conventional setup, we have obtained a large sample of charm events. The first results from 30% of the total data give charm lifetime measurements comparable in precision to the previous world average. We hope to obtain much new physics from the full data sample (including detailed studies of the D_s^+ and Λ_c^+, production mechanisms, D^0-$\overline{D}^0$ mixing, excited charm states, exclusive charm semileptonic decay modes, and rare D decays). Clearly, silicon microstrip detectors are very powerful tools for charm physics.

Finally, as the experiment presented here is a representative of the past at this Workshop aimed at the future, it is appropriate to conclude by observing that the success obtained despite the use of now-obsolescent technology augurs well for the future of silicon vertex devices.

ACKNOWLEDGEMENTS

I would like to thank G. Hartner, P. Karchin, J. Raab and M. Witherell for many helpful discussions.

REFERENCES

† The members of the TPS Collaboration are: T.E. Browder, P.E. Karchin, S. McHugh, R.J. Morrison, G. Punkar, J.R. Raab, M.S. Witherell (UCSB), P. Estabrooks, J. Pinfold, J.S. Sidhu (Carleton), J.C. Anjos, A.F.S. Santoro, M.H.G. Souza (CBPF), L.M. Cremaldi, J.R. Elliott, M.C. Gibney, U. Nauenberg (Colorado), J.A. Appel, P.M. Mantsch, T. Nash, M.V. Purohit, K. Sliwa, M.D. Sokoloff, W.J. Spalding, M.E. Streetman (Fermilab), M.J. Losty (NRC), C.O. Escobar (Saõ Paolo), S.B. Bracker, G.F. Hartner, B.R. Kumar, G.J. Luste, J.F. Martin, S.R. Menary, P. Ong, A.B. Stundžia (Toronto).

1. See, for example, K. Sliwa et al., Phys. Rev. D32(1985)1053, and references therein.
2. T. Nash et al., Proceedings of the Conference on Computing in High Energy Physics, Amsterdam, June 1985, (North-Holland). T. Nash et al., 'The ACP multiprocessor system at Fermilab', Fermilab-Conf-86/132, presented at the Berkeley Conference, July 1986.
3. More details on our SMDs and test-beam results are given in P. Karchin et al., IEEE Trans. Nucl. Sci. NS-32(1985)612.
4. J. Kemmer, Nucl. Instr. and Meth. 226(1984)89
5. The solid-state physics aspects of silicon devices, as well as their use as detectors, are well described in C.J.S. Damerell, 'Developments in Solid State Vertex Detectors', SLAC Summer Institute, 1984, and Rutherford Appleton Lab preprint RAL-84-123.
6. P. Jarron and M. Goyot, Nucl. Instr. and Meth. 226(1984)156
7. E.H.M. Heijne et al., Nucl. Instr. and Meth. 226(1984)63
8. The expressions for unequal detector plane spacing are given in R.L. Gluckstern, Nucl. Instr. and Meth. 24(1963)381
9. See for example, M. Regler, 'Multiple Scattering in Least Squares Fitting', in 'Formulae and Methods in Experimental Data Evaluation', European Physical Society, 1984.
10. R.J. Morrison, Proceedings of the Seventh Vanderbilt High Energy Physics Conference, May 1986. P.E. Karchin, Proceedings of the XXIII International Conference on High Energy Physics, Berkeley, July 1986. J.C. Anjos et al., submitted to Phys. Rev. Lett.
11. The results for the D_s^+ given here have been updated from the results for 15% of the data presented at this Workshop.
12. Particle Data Group, Phys. Lett. 170B(1986)1
13. Reviewed by M. Gilchriese at the Berkeley Conference, July 1986.

Discussion B.R. KUMAR

R. NANIA

Do you have preliminary results on D^0 branching ratios?

B.R. KUMAR

So far we have only looked at modes whose branching ratios are relatively well measured. With the full statistics we will be able to investigate branching ratios down to the 10^{-3} level.

J. KIRKBY

One of the most interesting charm measurements which has not yet been done is the ratio $c \to s / c \to u$. The Cabibbo – suppressed hadronic decays have proved very difficult to interpret theoretically and the best measurement is $D^0 \to K^- e^+ \nu / D^0 \to \pi^- e^+ \nu$. Do you have sensitivity to this measurement?

B.R. KUMAR

Yes,we do, we are just beginning to look at this and it looks promising.

G. LUTZ

1) Am I right in understanding that multiple scattering in the MSD arrangement has not been included in track error assignement for vertex finding?
2) Your good S/B ratios are somewhat worse than, e.g., in the NA32 experiment working in a hadron beam with factor 10 lower charm ratio and increased multiplicity. For your hadron experiment I would expect some harder conditions than those you experience now.

B.R. KUMAR

1) Yes. Multiple scattering in the target is always included, but we found that including the detector multiple scattering did not give a significant improvement in the charm signal/background. Remember that the decay products of the D, which has mean momentum ~60 GeV/c, are generally above the range where multiple scattering plays a major role. For the main vertex we can afford to ignore low momentum tracks which suffer significant multiple scattering and so do not give a satisfactory fit to the vertex. There may be some decay modes where we have to include this.
2) I agree. We can in our present data get considerable S/B improvement, in the inclusive $D^0 \Rightarrow K^- \pi^+$ channel you refer to, by cutting harder on the number of sigmas in intervertex separation. We also hope that the 25 μm planes to be added to the microstrip detector will help in this regard.

W. KELLY

What was the reason for the difference in the behaviour of the two types of MSD after irradiation, i.e. why does the leakage increase in the ENERTEC detectors?

B.R. KUMAR

We are not sure. The rates in the detectors and the total dose of the experiment are very small compared to the numbers generally throught to cause radiation damage. In our case the leakage current increase did not have a detrimental effect on the experiment since it levelled off at our acceptable point.

PRESENT AND FUTURE SEMICONDUCTOR TRACKING DETECTORS

G. Lutz[1], P. Holl[1], J. Kemmer[2], U. Prechtel[3],
J. Sedlmeir[1], L. Strüder[1] and W. Welser[3]

Abstract

Semiconductor detector principles, properties and limitations of presently available devices are described. First results from more advanced detectors are given and new detector configurations and principles are proposed. They include CCDs with gating and fast clearing possibility and a integrated detector - amplification structure which could be used for a two dimensional pixel detector with fast random access readout.

---

1) Max-Planck-Institut für Physik und Astrophysik, München
2) Messerschmitt Bölkow Blohm GmbH and Technische Universität, München
3) Messerschmitt Bölkow Blohm GmbH, München

1. Introduction

Since a few years semiconductor detectors have become an appreciated tool in experimental high energy physics. This has been true for silicon strip detectors as well as for photo diodes in calorimetric applications and more recently for charge coupled devices (CCDs) employed as particle detectors. Strip detectors and CCDs have been used already in experiments and interesting physics results concerning the properties of charmed particles have been obtained with them. Their use is forseen in many new experiments presently in the planning or preparation stage. Fairly recently a device operating on a new principle, the semiconductor drift chamber, had been proposed [1]. It has been built and tested [2] and shows very interesting properties which make the use in future experiments very likely. Triggered by the experience with this device many new concepts for further semiconductor detectors have been proposed [3]. Some of those may be extremely well suited for applications in high energy experiments.

It is the purpose of this paper to sketch the working principles of semiconductor detectors, to discuss properties, limitation and further development of present semiconductor devices and to present new detector configurations and principles which might be interesting for application in high energy experiments. In doing so we will limit ourselves to position sensitive devices and concentrate on developments going on in our laboratories in Munich. We will furthermore restrict ourselves to detectors only and leave our closely connected development of readout electronics out of this presentation.

2. Basic Semiconductor Structures

Standard semiconductor detectors are based on the diode and on the MOS (Metal-Oxide-Semiconductor) structure. The former is the basis for strip detector and drift chamber while the latter is used in most CCDs (Charge Coupled Devices) and also in FETs (Field Effect Transistors). More sophisticated detectors will use combinations of both or may even use different structures altogether. We shortly give a simplified description of the two basic structures.

The diode shown in Fig. 1a consists of low doped n-type (conduction by electrons) bulk material sandwiched in between thin layers of highly doped p-type

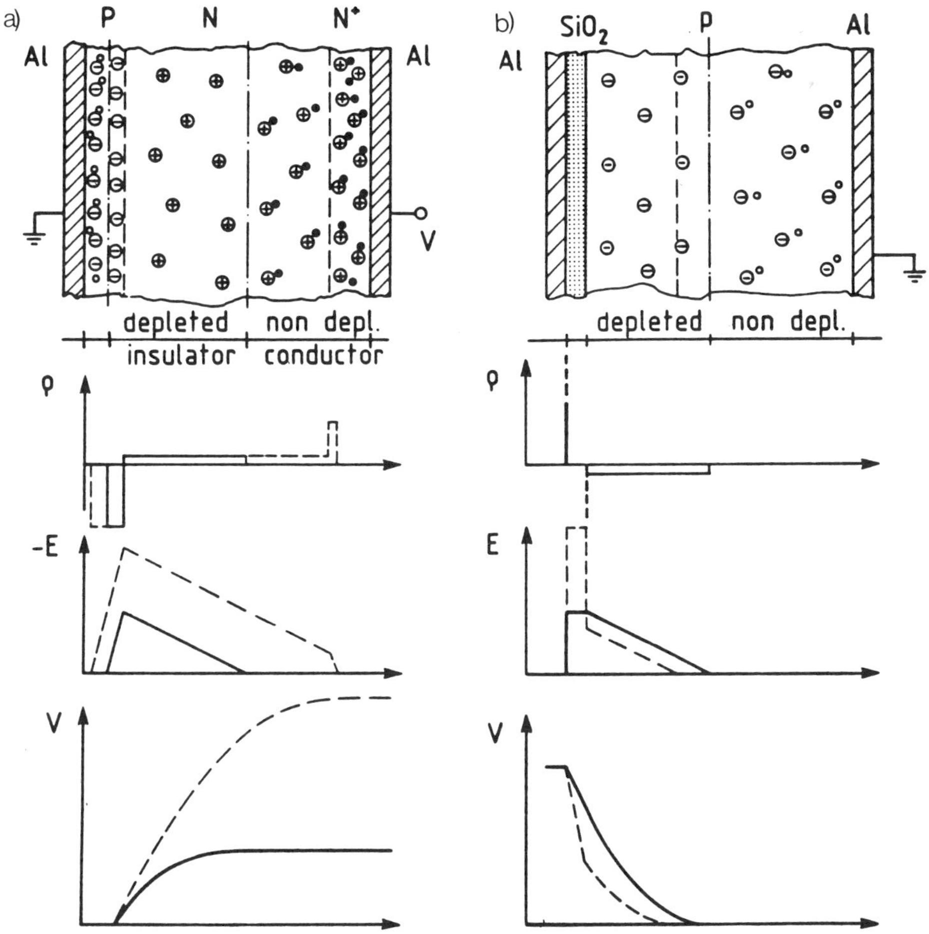

Fig. 1: Schematic drawing of diode (a) and MOS (b) structure including space charge density ρ, electric field E and Potential V. Fixed excess charges are indicated by + and −, the mobile electrons by dots and holes by small open circles.

(conduction by holes) at the left and n-type material at the right. If a positive voltage is applied at the n-side with respect to the p-side (reverse bias) movable charge carriers are pulled away from the pn interface and one obtains an insulating layer in which the now uncompensated fixed excess charges form a space charge region. In Fig. 1a this region extends into the bulk region (partially depleted case). The corresponding charge, electric field and potential are drawn as continuous lines in the lower parts of the figure. The electric field rises linearly, the voltage quadratically with the depth of the space charge region.

The highly doped N^+ region on the right side prevents the space charge region from reaching the metal electrode which would lead to massive charge injection from the metal into the bulk. It thereby allows operation of the detector in fully depleted mode. Charge, field and potential distributions for this case are drawn as dashed lines.

Electron hole pairs may be created in the space charge region either by thermal agitation or by ionizing radiation. These charges are seperated by the electric field and move towards the conducting electrodes thereby generating an electric signal.

The MOS (Metal-Oxide-Semiconductor) structure in Fig. 1b consists of a conducting electrode separated from the p-type semiconductor by a thin insulating layer. Charge, electric field and potential drawn as continuous lines correspond to a situation reached after application of a positive voltage on the metal electrode above the SiO_2 insulating layer. A negative space charge region develops below the oxide.

This situation is not stable however. Electrons created in the space charge region, either by thermal agitation or by ionizing radiation will move towards the oxide where they build up a thin inversion layer, while holes will move into the undepleted bulk. This results in a shrinking of the space charge region and corresponding changes (dashed lines) of charge, field and potential.

3. Semiconductor Properties

Semicondutors posess some striking properties which make them extremely useful for particle detection. We will mention the most important ones and quantitative values will be given for Silicon, presently the most widely used material.

Due to the small band gap (1.1 eV) the average energy needed for creating a electron-hole pair is 3.6 eV compared to 30 eV [4] for gas detectors. This results in a large signal to energy loss ratio.

The large energy loss per unit length (dE/dx = 3.8 MeV/cm for minimum ionizing particles) makes building of very thin detectors, typically 300 μm Silicon, possible. The corresponding most probable signal charge of 24000 e for m.i.p. can easily be measured directly and the problems connected with multiplication in gas detectors are avoided. The same property leads to a small range of δ electrons (few μm). Therefore extremely precise position measurement is possible.

The high mobility of electrons (1380 $\frac{cm^2}{V \cdot sec}$) and holes (480 $\frac{cm^2}{V \cdot sec}$) in the crystal makes fast charge collection, typically 10 ns, possible.

Doping of crystals and application of external voltages results in fixed space charge regions in which electrons and holes can move freely with high mobility. With some ingenuity it is therefore possible to form sophisticated field configurations with striking properties. A first example of such a configuration for which no gas detector equivalent exists is the semiconductor drift chamber. Furthermore integrated detector – signal amplification configurations seem to be feasible. This will be discussed in a later chapter of this paper.

4. Present Semiconductor Detectors

In this chapter we will shortly review working principles and properties of strip detectors, CCDs and drift chambers when used as particle detectors.

Diode Strip Detectors

Based on the diode (Fig. 1a) one can easily design position sensitive devices if one divides the diode into many independent sections. In the most straightforward version these will be narrow strips providing the measurement of one coordinate.

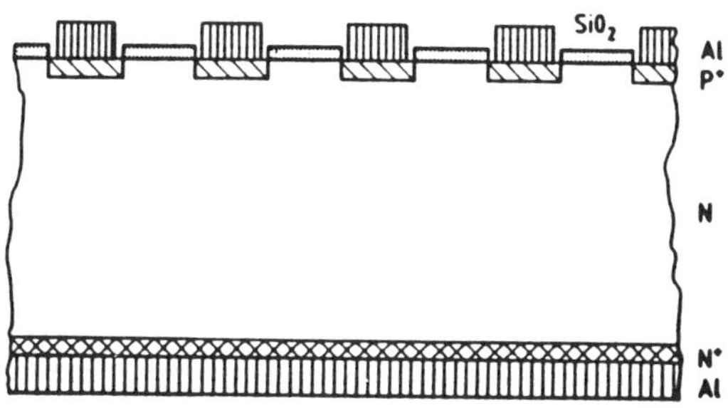

Fig. 2: Diode strip detector

A cross section through such a detector is shown in Fig. 2. With digital readout the expected measurement precision will be given by the pitch (typically 20 to 50 μm) divided by $\sqrt{12}$. Better precision is obtainable with analog readout as one may profit from the distribution of charge over more than one strip due to charge diffusion [5]. With analog readout one may also reduce the number of electronic channels making use of the principle of capacitive charge division [6].

Charge Coupled Devices

A very large number of CCD types has been invented. We restrict ourselves to one of the most simple ones based on the MOS structure.

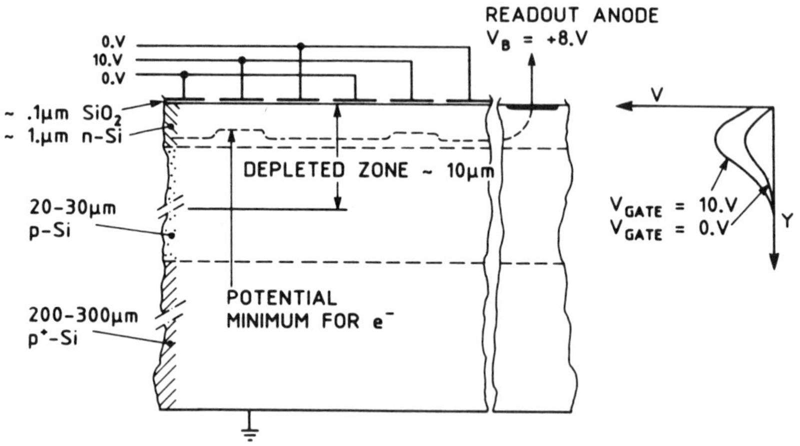

Fig. 3: Schematics of a buried channel CCD

As it was the case for strip detectors one can build position sensitive devices from the MOS structure by dividing the metal electrode into (overlapping) strips. Such a device can then be used to trap the signal electrons in potential maxima below the SiO_2 surface. Potential maxima (minima for negative charges) are generated by applying a periodic potential to the metal strip electrodes.

In the buried channel CCD shown in Fig. 3 these maxima are moved slightly away from the surface into the bulk by the introduction of a shallow n-doped surface layer just below the oxide. This way one prevents signal charges from reaching the oxide – silicon interface where high damages to the crystal lattice

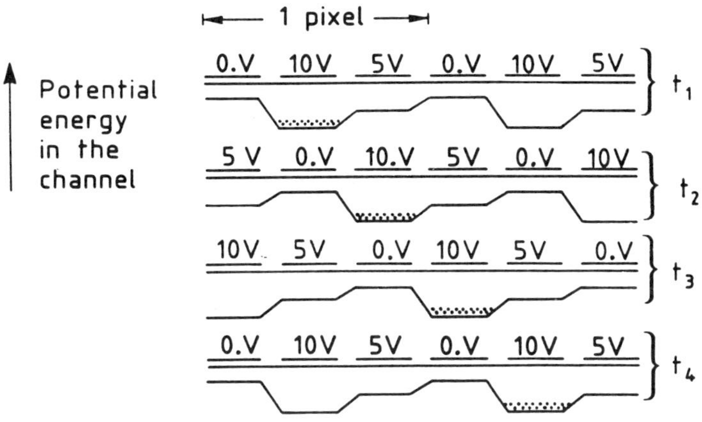

Fig. 4: Charge coupling in a three phase CCD. The position dependent potential is plotted for four successive time sequences

cause a reduction of signal transfer efficiency. The signal charge can be moved below the oxide towards a N^+ output node by suitable variation of the periodic potential at the electrodes (Fig. 4).

Further measures have to be taken in order to prevent spreading of the charge along the direction of the strip electrodes. This can be achieved by the introduction of channel stops, strips with increased p-doping perpendicular to the shift electrodes. This is shown in the schematics of a two-dimensional CCD (Fig. 5) where the signal charges may be moved downwards line by line into a one-dimensional CCD and then cell by cell to the single output node. Usually a first amplification stage is included into such a device in order to take advantage of the small capacitance of the output node.

CCDs have been used successfully for particle detection in one high energy experiment [7]. These were selected commercially available devices used normally for optical applications. The shallow active region and corresponding small signals posed quite severe challenges on the readout system as these small signals had to be read at high speed with sufficiently low noise.

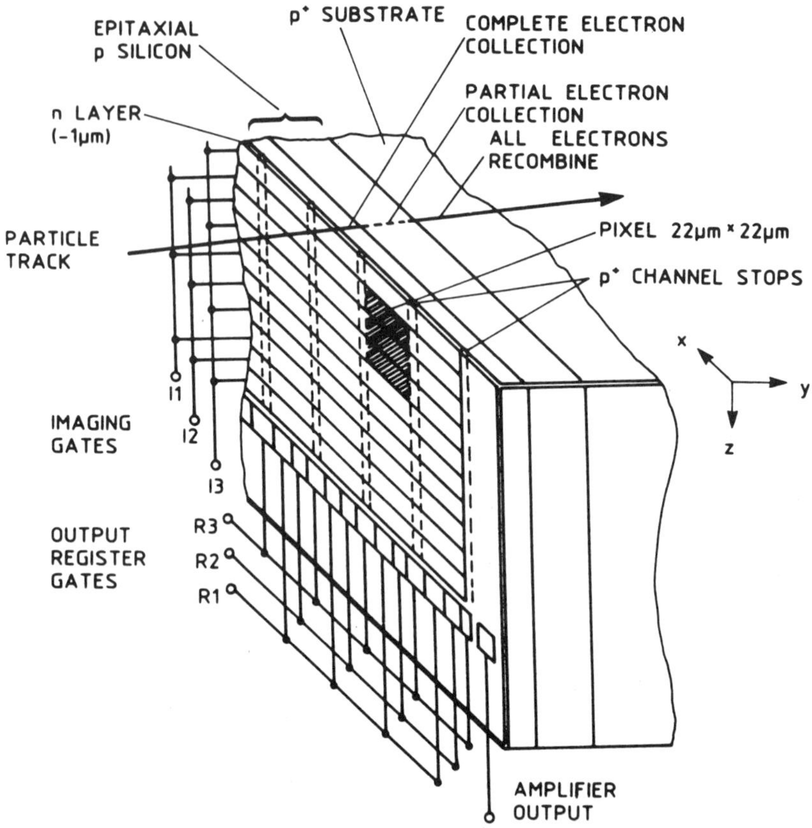

Fig. 5: Perspective view of a schematically drawn two dimensional MOS buried channel CCD

Semiconductor Drift Chambers

This ingenuous device has been invented by E. Gatti and P. Rehak [1]. First satisfactorially working devices were built in a collaborative effort of J. Kemmer at the Technical University Munich, the Max Planck Institute in Munich and the inventors [2]. One and two dimensional devices have been built.

The working principle may be explained by starting from the diode (Fig. 1a and Fig. 6a) if one realizes that the backside N^+ contact does not have to extend over the full area, but can instead be put at any place of the undepleted conducting bulk (Fig. 6b). We have then space to put diodes on both sides of the wafer (Fig. 6c). At

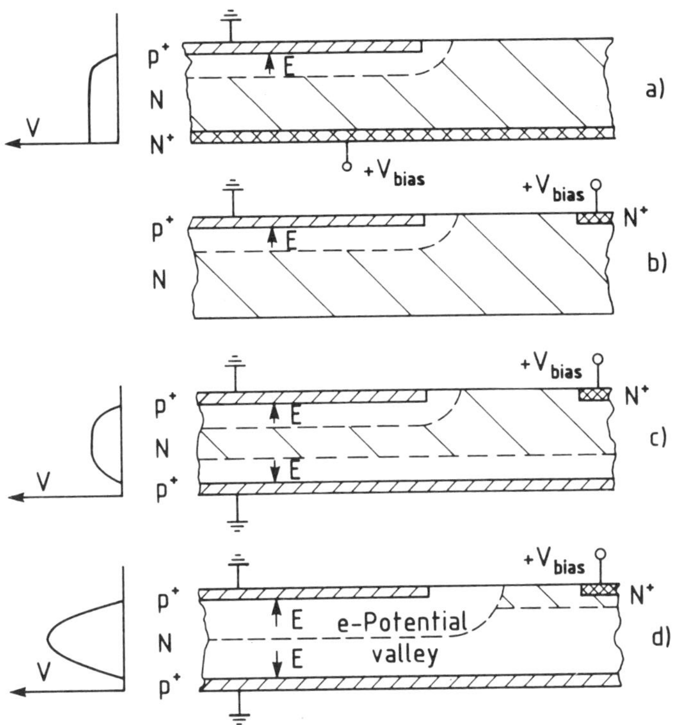

Fig. 6: Basic structures leading towards the drift chamber: diode partially depleted (a); diode with depletion from the side (b); double diode partially depleted (c); double diode completely depleted (d)

small voltages applied to the N^+ electrode we have two detectors separated by the conducting undepleted bulk region (hatched in the figure). At high enough voltages (Fig. 6d) the two space charge regions will touch each other and the conducting bulk region will retract towards the vicinity of the N^+ electrode. Thus one obtains a potential valley for electrons in which thermally or otherwise generated electrons assemble until they (slowly) diffuse towards the N^+ electrode (anode) while holes are drifting (fast) in the electric field towards the P^+ electrodes.

Based on this double diode structure one can easily arrive at the drift chamber if one adds an additional electric field component parallel to the surface of the wafer so as to provide for a drift of electrons in the valley towards the anode. This can

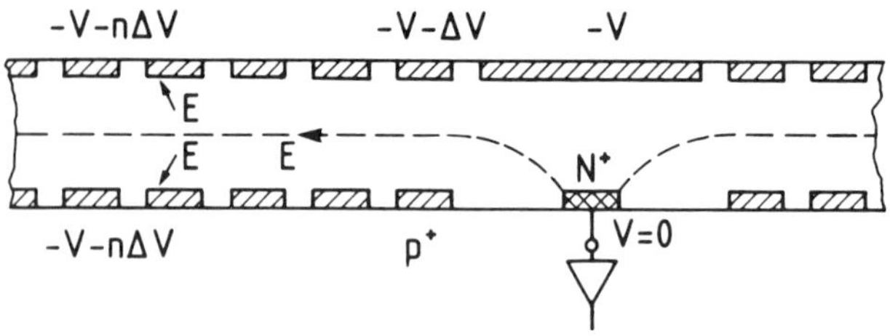

Fig. 7: Semiconductor drift chamber

be accomplished by dividing the diodes into strips and applying a graded potential to these strips on both sides of the wafer (Fig. 7).

Other drift field configurations (e.g. radial drift) can be obtained by suitable shaping of the electrodes. Drift chambers may be used for position and/or energy measurement of ionizing radiation. In the first case position is determined from the drift time. Dividing the N^+ strip anode in Fig. 7 into pads one may furthermore arrive at two dimensional position measurement.

5. Semicondutor Detector Properties

We will now try to compare merits and drawbacks of present semiconductor detectors in order to allow a good choice for specific applications and to give a guide for further development.

Diode strip detectors are fast devices. Charge collection is finished in the order of 10 ns and the speed of the system is usually limited by the properties of the electronics. They are also very precise, a position measurement precision of $2.7\mu m$ has been verified [8]. Disadvantages compared to other detectors are however the restriction to one dimensional measurement and the large number of output channels which lead to bulky electronics with large heat dissipation.

CCDs reach similar position measurement precision as strip detectors, but with the extremely important advantage of true two-dimensional measurement through usually only one output channel. Connected with these excellent properties are however several drawbacks which may in many circumstances prevent or limit their usefulness. Present commercially available devices are small, typically

$1 cm^2$, or exorbitantly expensive. Readout is slow (tens of milliseconds), devices are not clearable and non gatable and therefore sensitive to particles during readout. Furthermore due to the shallowness of the sensitive region signals are small. Extremely low noise electronics and operation at low temperature for suppression of leakage currents was therefore needed in high energy applications [9].

Drift chambers are fairly fast, typical drift times are in the few μsec range. Their precision may approach that of strip detectors and CCDs if good low noise electronics is employed. As CCDs, they are continuously sensitive. One and two dimensional drift chambers have been built. Compared to strip detectors one dimensional drift chambers allow an enormous reduction in the number of electronic channels however two track resolution is limited due to diffusive and repulsive spread of the signal charge cloud during drift time. Compared to CCDs two dimensional drift chambers have the advantage of much faster readout time and one order of magnitude higher signal, however the number of electronic channels is increased to a value close to that of strip detectors.

Further problems in using drift chambers are the dependence of position measurement on variations of the drift field due to inhomogeneities of the detector material and the variation of mobility with temperature. Extensive calibrations of the system for correction of systematic errors will therefore be necessary if something close to the ultimate accuracy is wanted. For strip detectors and CCDs these problems should be minor as position accuracy is determined by the geometry of the electrodes only.

With respect to radiation damage only limited information is available at present. Strip detectors seem to be quite radiation resistant, they have been operated up to a few times 10^{13} m.i.p./ cm^2 with the effect of some increase of dark current and a change in bulk doping concentration which however did not drastically degrade their performance [10]. CCDs seem to be much more sensitive to radiation probably due to damage in the oxide layer and corresponding decrease in charge transfer efficiency [9]. Of these devices drift chambers are expected to be most prone to radiation damage, the reason being the rapid degradation of position measurement precision with increased dark current and the introduction of bulk inhomogenuities in the case of non uniform irradiation.

6. New Developments in Semiconductor Detectors

Considering the previous discussion of detector properties one might wonder on how to combine the good properties of the different types and avoid drawbacks as far as possible. This might be done by modifications of already existing devices or by invention and application of new principles. This ongoing work has already lead to first experimental results which will be presented in this chapter. New concepts will be treated in the following chapter.

Double Sided Readout of Strip Detectors

Double sided readout brings about the obvious advantage of providing twice the information for the same amount of scattering material. With crossed strips on the two detector faces and analog readout it is furthermore possible (to some degree) to correlate signals from the two sides making use of Landau fluctuations. Thus one may obtain a strip detector with quasi two dimensional readout.

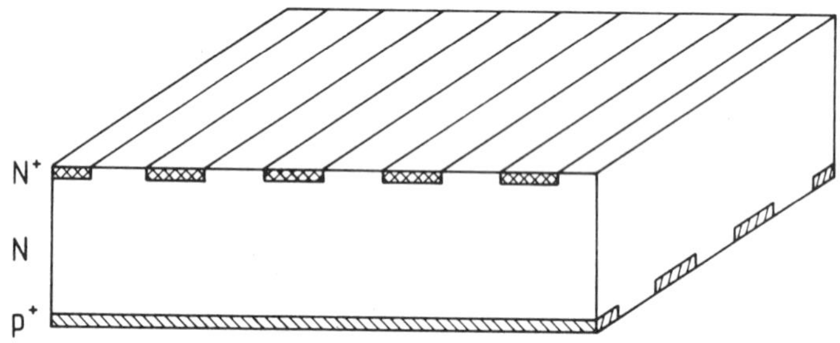

Fig. 8: Double sided readout strip detectors

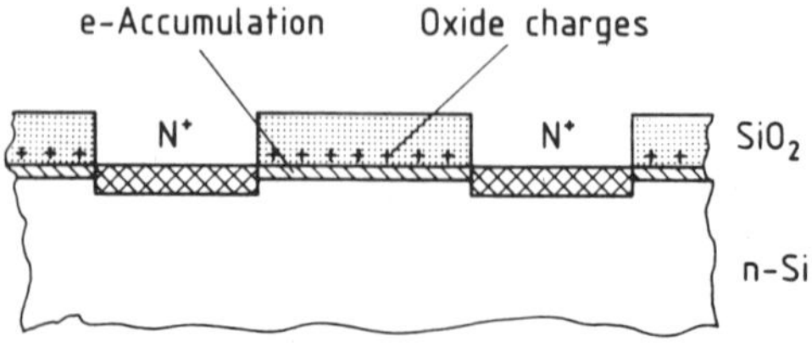

Fig. 9: Insulation problem for N^+ - strips

A problem in producing this kind of detectors is the simultaneous insulation of strips from each other on both sides of the detector. This problem and several solutions for circumventing it have been described in [8]. The naive solution of simply providing highly doped n- and p- strips on the two sides of the detector (Fig. 8) fails due to the buildup of an electron accumulation layer (inversion layer on p- type material) inbetween the N^+ strips below the insulating oxide (Fig. 9).

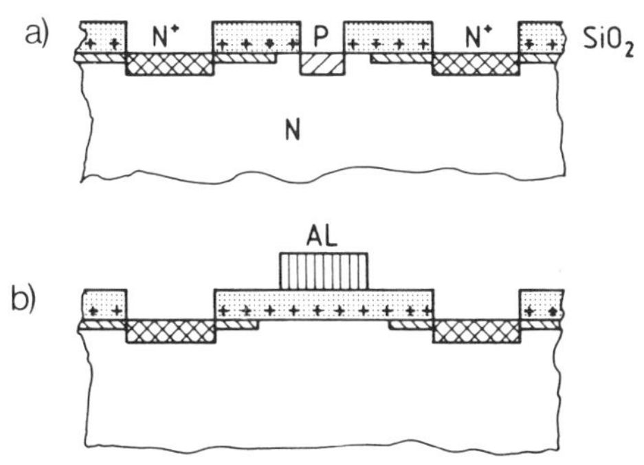

Fig. 10: Insulation by structured P - inplantation (a) or biased MOS structure (b)

This layer of electrons produces an electrical shortening of neighbouring strips. It is caused by the positive charges which are always present at the Silicon - Oxide interface. There are three possibilities for curing the problem:

a) Large area surface doping. In this case the oxide charges are compensated by the negative acceptor ions and the buildup of the electron layer is prevented

b) Disruption of the electron layer by implantation of p-strips (Fig. 10a).

c) Disruption of the electron layer by a suitably (negatively with respect to the N^+ strips) biased MOS structure (Fig. 10b).

Detectors using method b) have been produced and successfully tested at Munich. The prototype detectors had an area of $1cm^2$ and a pitch of $50\mu m$ and of $25\mu m$ respectively in between the 1 cm long N^+ readout strips. As the diode side is not expected to pose any new problems a single large area diode was produced instead of crossed strips for the first prototypes.

Measurements with a $50\mu m$ prototype detector will be presented in the follow-

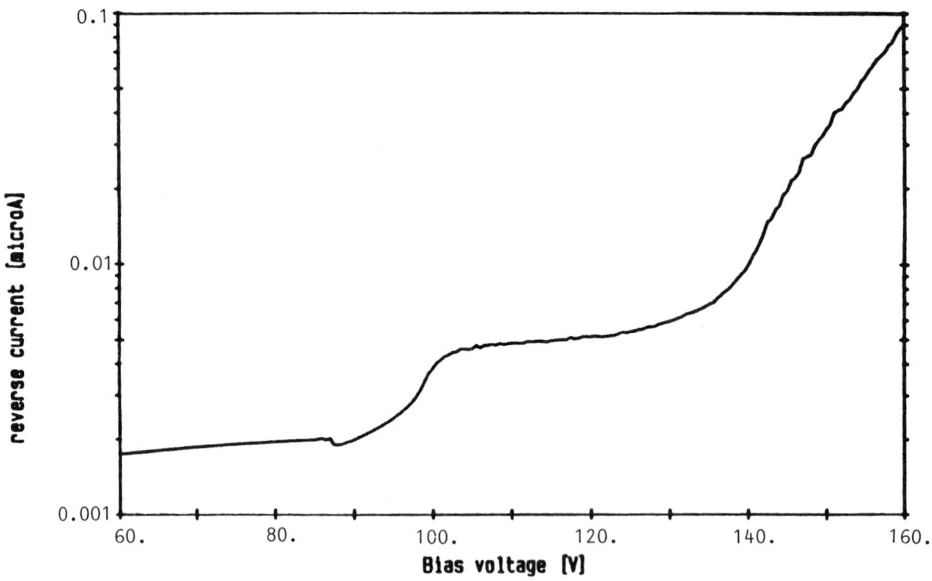

Fig. 11: Reverse bias current of strip detector with electron readout

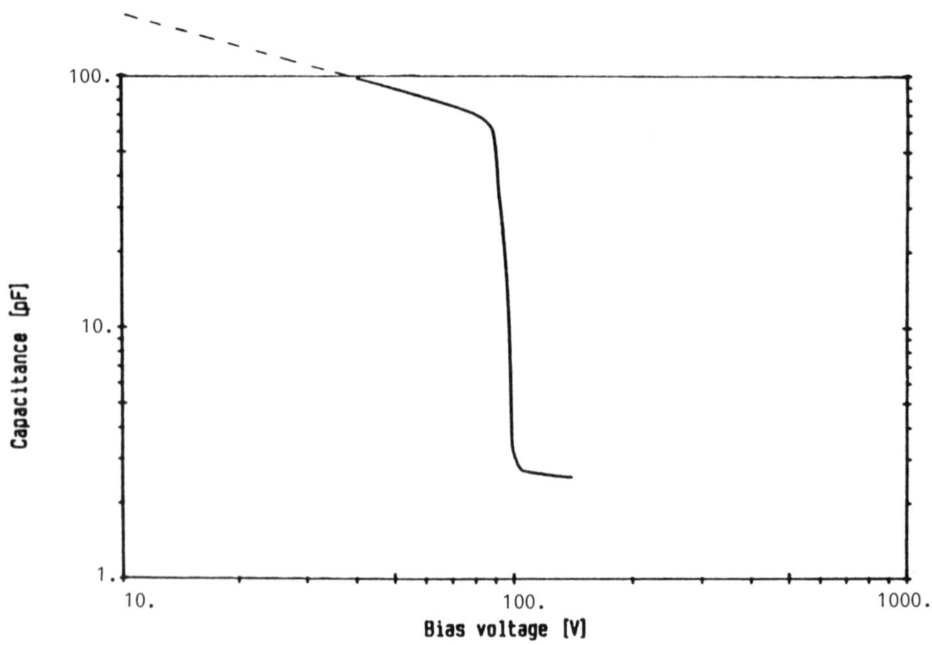

Fig. 12: N^+ strip to ground capacitance as function of bias voltage

ing. Fig. 11 shows the reverse bias current as function of the bias voltage applied between one single strip and the large area diode back side.

Notice the extremely low current of 2 nA which rises only to twice the value once the space charge region reaches the second surface.

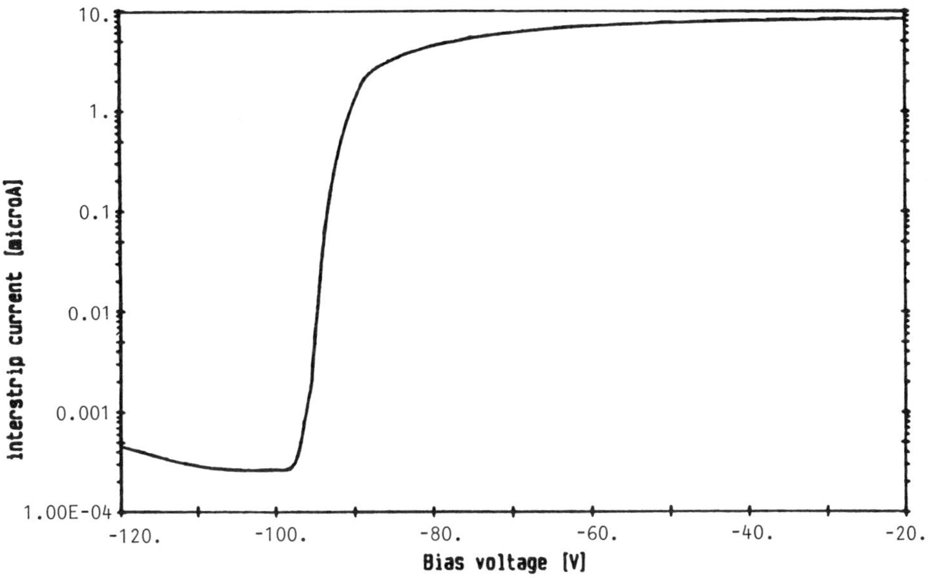

Fig. 13: Current between neighbouring strips as function of bias voltage

Fig. 12 gives the capacitance for the same arrangement in logarithmic scale. The steep drop between 80 and 90 V indicates an insulation of the strip from the rest of the detector caused by the disruption of the connecting path through the bulk when the space charge region from the diode expands to touch the strip side of the detector. This can also be seen directly in Fig. 13 where a small voltage has been applied between neighbouring N- strips and the current has been measured as function of the reverse bias voltage. As may be read from the figure, the interstrip resistance rises by four orders of magnitude by application of the bias. We thus have clearly demonstrated that method b) is working very satisfactorily.

Fully Depleted CCD Detectors

These detectors are based on the semiconductor drift chamber principle. Compared to standard MOS CCDs they are expected to have the advantage of providing an order of magnitude larger signal and of being much less prone to radiation damage.

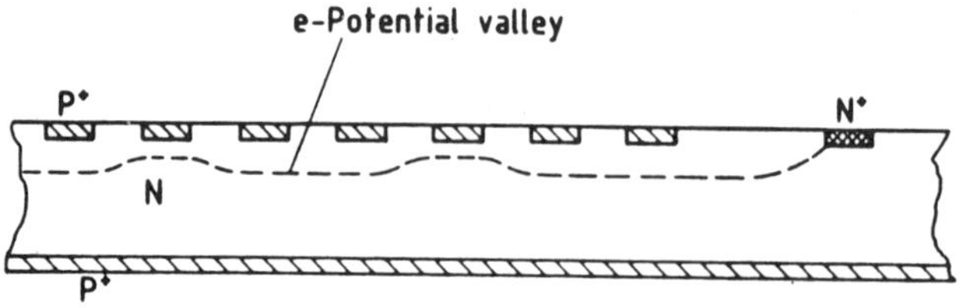

Fig. 14: Schematics of a fully depleted CCD

Starting from the drift chamber one may arrive at the CCD by applying non equal voltages on the top and bottom P^+ electrodes, thereby shifting the electron potential valley close to the top surface (Fig. 14). Application of a periodic potential to the strip electrodes creates potential maxima in which signal electrons will assemble. Shifting to the N^+ output node may be performed in an completely analogous way to that described earlier for normal CCDs.

Although this simple description is correct in principle one has still two problems to solve. The first one concerns injection of holes into the bulk from the top strip electrodes when the electron potential valley is moved close to the surface and the shallow potential maximum is insufficient to provide a barrier for holes. This causes a hole current to flow from top to the bottom surface of the detector. Secondly one has to produce a channel stop in order to prevent signal electrons to spread along the direction of the strips.

Both problems can be solved simulataneously by the introduction of a channel guide, an increase of n-doping near the surface on the strip side as shown in Fig. 15. The increase in doping leads to a stronger bending of the potential close to the surface and thus provides a barrier, preventing holes from crossing to the opposite side of the wafer. Channel stops necessary for preventing a lateral spread of the signal charge are formed by a short interruption of this additional n-doping.

Extensive simulations have been made to determine geometry and technological parameters for a working device. They are described in [11]. For illustration we show here two dimensional simulations of potential, electron and hole distributions

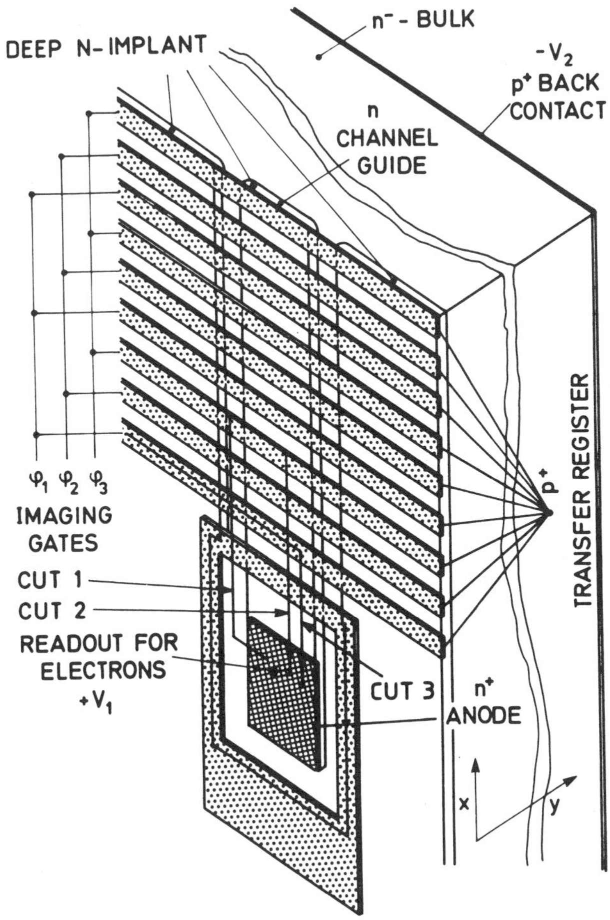

Fig. 15: Perspective view of a schematically drawn fully depletable CCD

(Fig. 16 and Fig. 17) along the three cuts indicated in Fig. 15.

The left side of Fig. 16 corresponds to the situation without channel guide implantation, the right side to that with n-channel guide. We notice that without

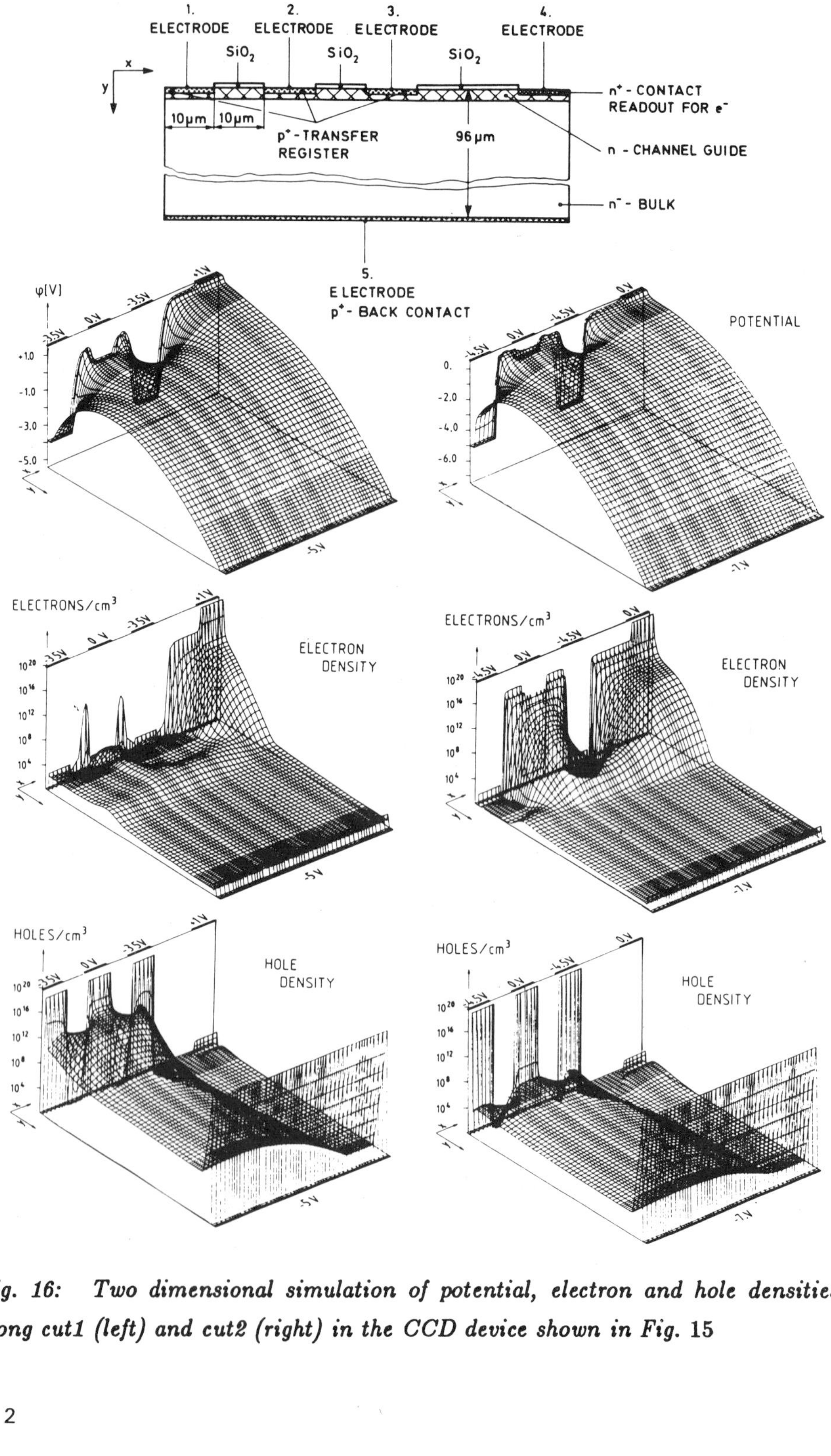

Fig. 16: Two dimensional simulation of potential, electron and hole densities along cut1 (left) and cut2 (right) in the CCD device shown in Fig. 15

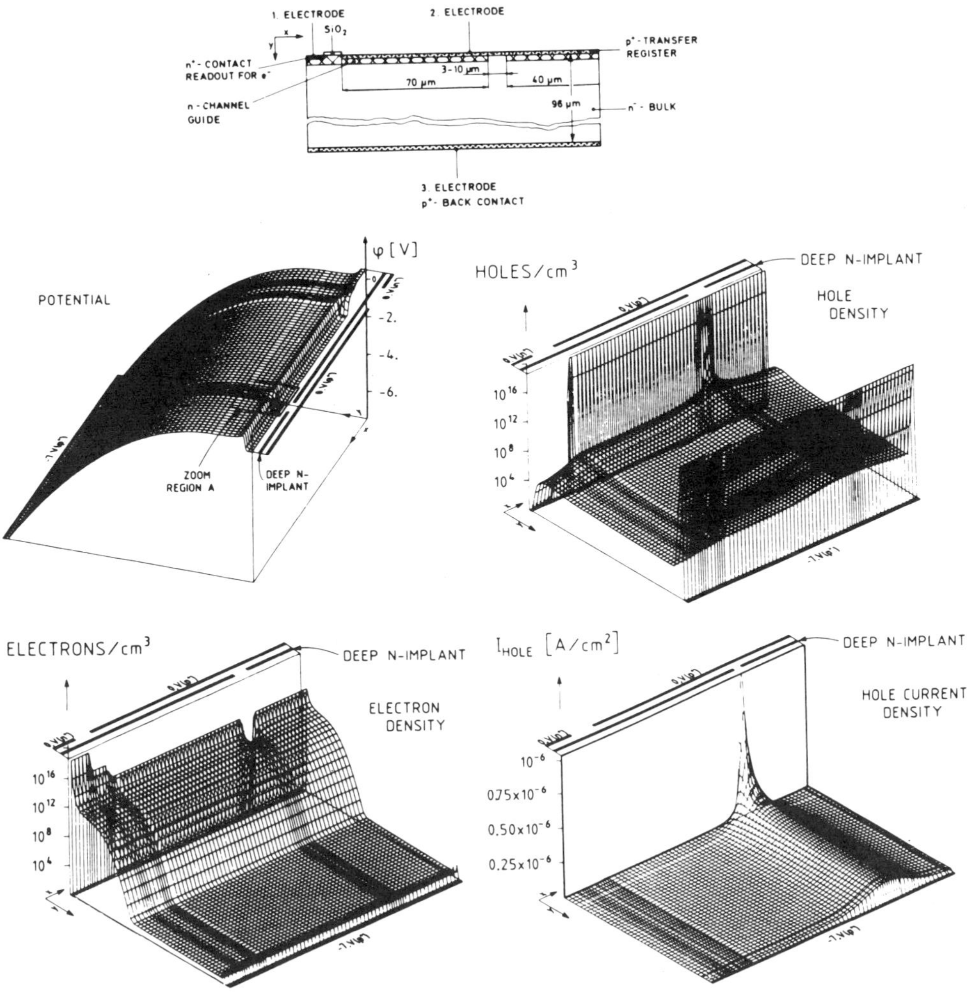

Fig. 17: Two dimensional simulation of potential, electron and hole densities along cut3 across the channel stops in the CCD device shown in Fig. 15

channel guide only very few electrons can be stored below the gate, contrary to the situation with channel guide where we find a high electron concentration in the completely filled pixel below the gate at 0 V potential. Considering the hole concentration on the contrary we find a fairly high value throughout the bulk in the figure at the left side. This leads to a high current from front to back side across the bulk as one readily concludes considering the potential distribution. This current will disappear with narrow channel stops (disruption of channel guide) and proper three dimensional consideration of the situation. The same conclusion may also be drawn from the two dimensional calculation of a cut along the gate across the channel stop (Fig. 17) where we have shown in addition the hole current density. A small hole current originates in the $3\mu m$ wide channel stop.

First prototypes of linear and matrix CCDs have been built and partially tested. The devices perform in static tests as expected showing also a very low leakage current of only a few nA/cm^2. Furthermore it could be demonstrated that charges can be shifted towards the readout electrode although in first, non optimum operating conditions with low transfer efficiency.

7. New Semiconductor Detector Concepts

As already mentioned in chapter 3, semiconductors allow invention of a large variety of new detector configurations which in many cases have no analogon in gas detectors. We will in the following describe a few of them.

Advanced Types of Fully Depleted CCDs

Compared to standard CCDs the fully depleted CCDs described in the previous chapter have the advantage of decreased sensitivity to radiation damage and much larger signals which probably will make possible the operation in an experiment without cooling. Still there remain the problems of gating and clearing and the extremely low readout speed.

These problems may be solved and further advantageous properties may be obtained by a combination of the following measures:

- use of both types of charge carries for signal generation
- other types of gate structures
- shifting of the electron potential valley to either surface of the detector.

In the example shown in Fig. 18 we have replaced the junction gate structure on the top surface by a MOS gate structure. Furthermore there exists a strip structure on the bottom surface consisting of P^+ strips interleaved with N^+ strips.

The electron potential valley divides the fully depleted bulk of the detector into two regions above and below the valley and the position of the valley is controlled

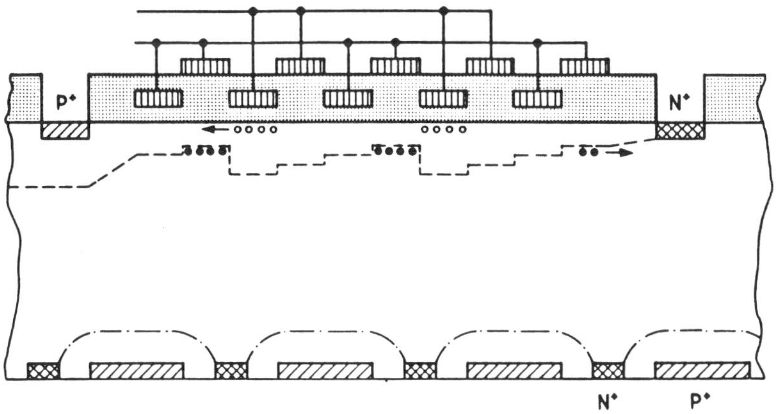

Fig. 18: Fully depleted MOS-CCD – strip detector combination

by the voltage difference between top and bottom surface.

Holes created above the valley will move towards the oxide surface where they may be controlled (shifted) the same way as it is done in a normal MOS CCD. Holes created below the valley will move towards the bottom P^+ strips where they may be used as prompt signal in the same manner as in diode strip detectors. Electrons created anywhere in the bulk will assemble in the potential valley and are then controlled by the top gates if the valley is close enough to the top surface.

Thus one has three detectors in a single device:

- a surface MOS CCD for minority carriers (holes) with a shallow sensitive region extending from the top surface to the potential valley.
- a strip detector with a sensitive region extending from the potential valley to the bottom surface providing a prompt signal
- a majority carrier CCD with a sensitive region extending over the whole bulk.

Combining the fast strip detector information with the slow CCD information may be useful for triggering and for increase of readout speed as the strip detector gives knowledge in advance in which region of the CCD to expect useful information. Readout speed could then be drastically increased by skipping empty rows of pixels. It may also serve to identify in - time and out of time hits in the CCD.

Clearing of electrons can be acomplished by shifting the potential valley to the bottom surface. With suitable relative biasing of N^+ strips with respect to the P^+ strips at the bottom surface the valley splits up in many seperate sections ending at the N^+ strips when it is moved towards the bottom. Electrons are thus collected in the N^+ strips.

Operation of the device during data collection with the valley close to the bottom may also be of considerable interest as it allows to run it as a gatable minority carrier (holes) CCD. Gating is accomplished by shifting the valley close to the top surface thereby reducing the sensitive volume to a very small value. The device may be cleared by moving the potential valley even closer to the surface or by removing it completely with additional lowering of the bottom side potential.

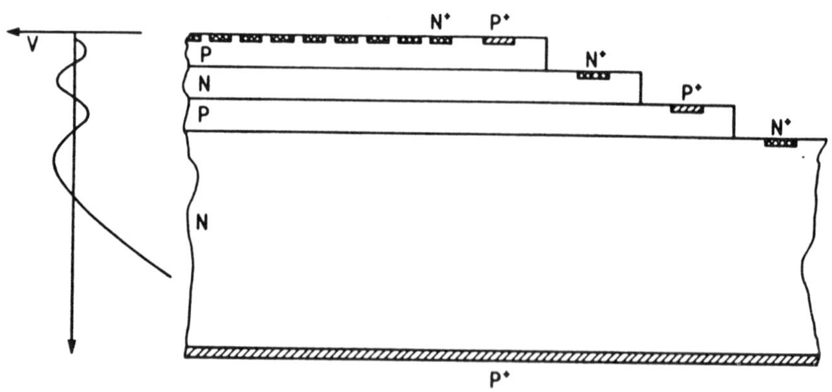

Fig. 19: Multi image CCD

Multi Image CCD

Although the multi image CCD shown in Fig. 19 may not have practical applications for high energy particle tracking but rather for distinguishing radiation with different penetration depth, it is quite interesting as an example for showing the potential available in semiconductor detector structures. Several layers of alternative doping are deposited on the top surface of the wafer thereby creating alternating potential minima for electrons and for holes in the completely depleted detector. The top gate structure controls charge transfer in all layers simultaneously moving electrons and holes either in opposite or in parallel direction, depending on the way how potentials are varied at the gates.

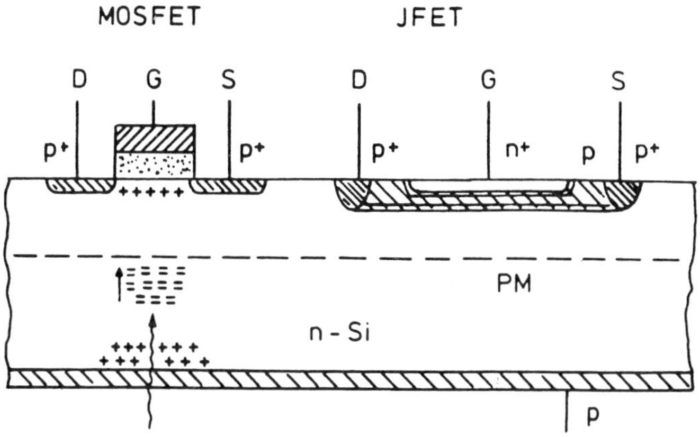

Fig. 20: DEPMOS and DEPFET detector – amplification structures

Non Destructive Readout and Amplification

The new detector - amplifier structure presented in this section promises to be extremely well suited for many applications including high energy particle tracking. It is based on the Field Effect Transistor (FET) and on the double diode structure (Fig. 6) used in chapter 4 to explain the working principle of the drift chamber. Both types of FETs, MOSFETS or Junction FET structures may be used (Fig. 20).

As before an electron potential valley is generated in the fully depleted bulk close to the top surface which carries the field effect transistor structure. A conducting channel parallel to the surface connects source and drain of the transistor. The conductivity of this channel can be modulated by the gate voltage. The structure behaves thus very much the same as a normal transistor, with the exception of the substrate being completely depleted and therefore the unwanted parasitic capacitances to the substrate being reduced to a negligible value.

The very new interesting feature comes from the fact that the conductivity of the transistor channel may also be influenced by the charge created e.g. by a traversing particle and collected in the valley below the gate. The steering mech-

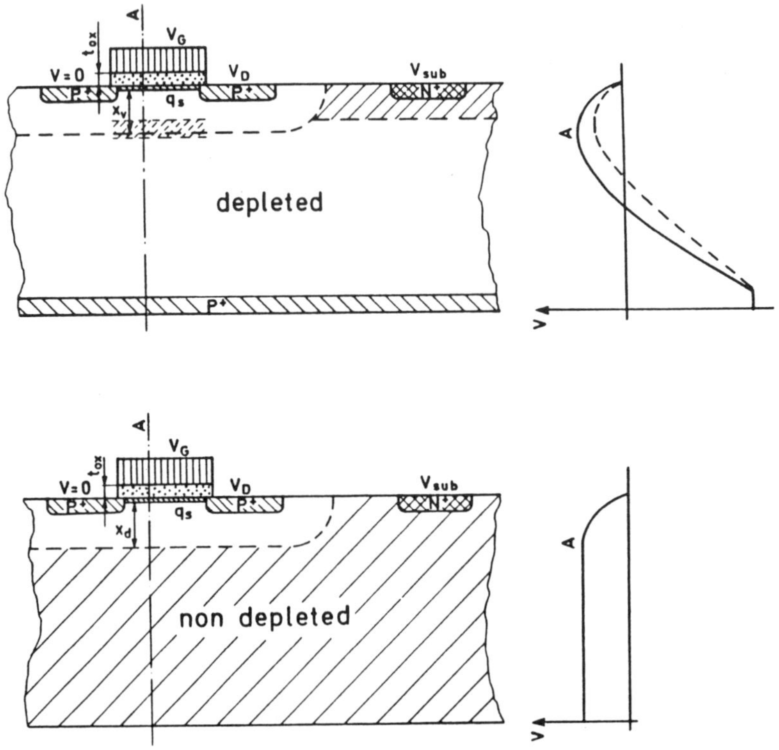

Fig. 21: Comparison of DEPMOS and standard PMOS transistor

anism is somewhat similar to that of a normal FET when changing the substrate voltage, with the important difference, that the transistors are electrically separated from each other.

This is indicated in Fig. 21 where we compare a normal MOSFET to a DEPMOS (DEpleted P channel Metal Oxide Semiconductor) structure. An analysis of such a structure ([3]) leads to the result that modulation of the conducting channel by the signal charge (collected in the e-potential valley below the gate) is almost equal to that obtained if the same charge was brought onto the floating gate of the transistor.

Many applications of these structures are possible. Several examples are given in [3], including proposals on integrating these devices in further electrical circuitry. We will present here one of the most interesting one, a two dimensional pixel device with random access non-destructive readout and optional fast clear

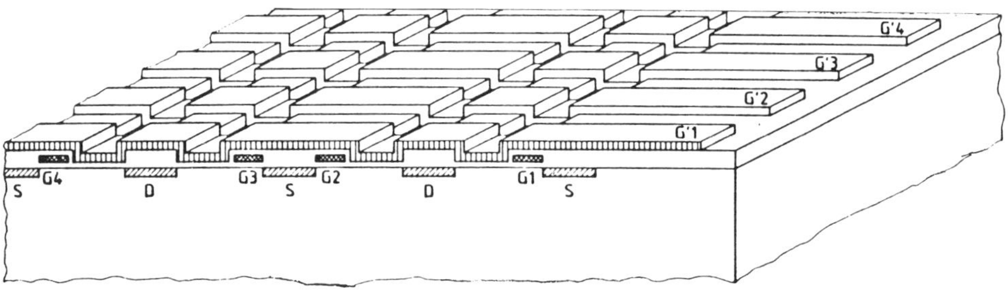

Fig. 22: Random access non-destructive readout pixel device

and gating possibility. This device is shown in Fig. 22. It consists of a two dimensional array of double gate DEPMOS structures. The first gates of each transistor are connected row wise the other gate column wise, as are source and drain of the transistors. A individual transistor can be turned on by putting on negative voltages simultaneously on a row and a column of gates. This way an individual pixel of the detector can be read out. Alternatively one may also turn on simultaneously many transistors, e.g. a whole column, for a fast scan in order to decide where individual readout is appropriate.

Fast clearing of the device could be performed by the insertion of the structure shown in Fig. 23 in between the gates of neighbouring rows in Fig. 22. Proper biasing of the transfer gates and the N^+ clearing electrode would either keep the electrodes below the DEPMOS gate or drain them into the clearing electrode.

This or a similar structure may also be used for gating of the device. Although no exact field calculation have been done it seems very likely that one may by proper biasing of the clearing structures form the electric field in such a way that electrons originating from the depth of the bulk may end up at the N^+ electrodes rather than below the DEPMOS gate. This may be accomplished by putting a positive voltage at a sufficiently large N^+ electrode while the gates still have to provide a barrier against electrons moving from below the DEPMOS gates towards the clearing electrode.

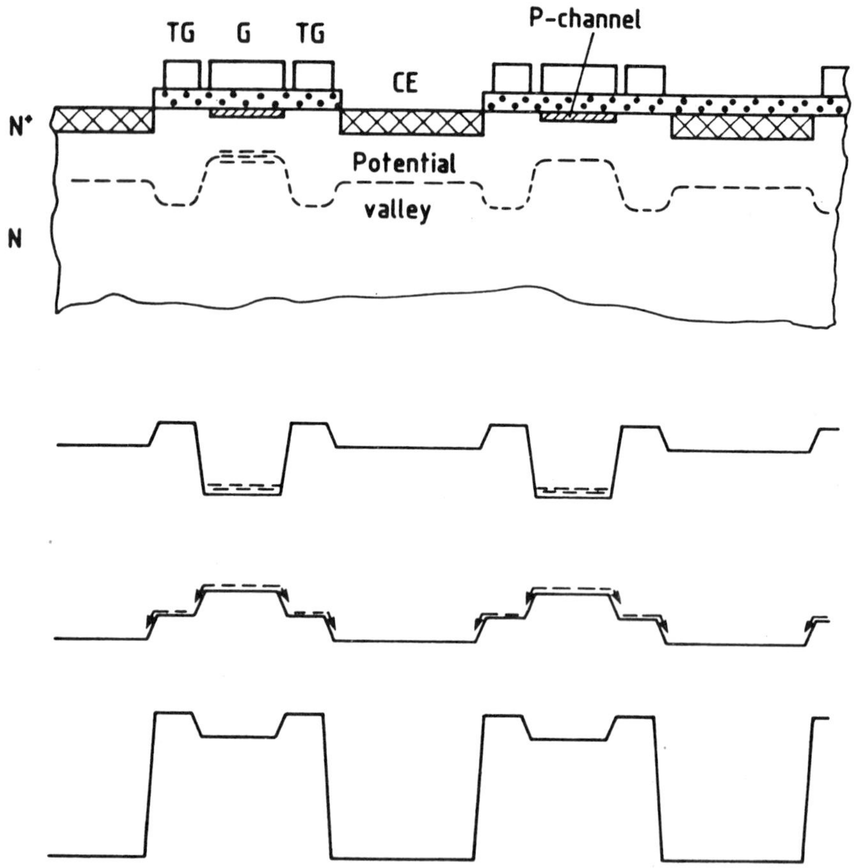

Fig. 23: Fast clear and gating structure for the pixel device shown in Fig. 22. Cut through the transistor gates G perpendicular to the channel. TG are the transfer gates for clearing, CE the clearing electrodes. The electron potential along the bottom of the valley is shown below for (top to bottom) data collection mode (gate on), clearing mode and gate off mode.

8. Summary

The use of semiconductor tracking detectors has become wide spread in High Energy Physics experiments and proposals due to their excellent properties. So far this has been limited to diode strip detectors and CCDs. Work on further perfection of these detectors is going on in several laboratories.

The special properties of semiconductors however make possible in addition a large variety of novel detector structures which may be very well tailored to High Energy Physics needs. The first example of such a structure, the semicondutor drift chamber has already been built and tested successfully and the development of fully depletable CCDs is in progress. We have furthermore proposed novel detector principles and configurations.

One of the most interesting one is a random access pixel device with fast clear and gating possibilities. Work on realizing such a structure should start in Munich very soon.

Acknowledgements

We are indebted to P. Solč and W. Ziemann for their excellent work in fabrication of the detectors.

REFERENCES

1. E. Gatti and P. Rehak, Nucl. Instr. and Meth. 225 (1984) 608; Semiconductor Drift Chamber - An Application of a Novel Charge Transport Scheme

2. E. Gatti, A. Longoni, P.Rehak, J. Kemmer, P. Holl, R. Klanner, G. Lutz and A. Wylie, Nucl. Instr. and Meth. A235 (1985) 224; Semiconductor Drift Chambers for Position and Energy Measurement

3. J. Kemmer and G. Lutz, Nucl. Instr. and Meth. A253 (1987) 365; New Detector Concepts

4. A.H. Walenta, Nucl. Instr. and Meth. A253 (1987) 558; Principles and New Developments of Semiconductor Radiation Detectors

5. E. Belau, R. Klanner, G. Lutz, E. Neugebauer, H.J. Seebrunner and A. Wylie, T. Böhringer, L. Hubbeling, P. Weilhammer, J. Kemmer, U. Kötz and M. Riebesell, Nucl. Instr. and Meth. 214 (1983) 253; Charge Collection in Silicon Strip Detectors

6. B. Hyams, U. Kötz, E. Belau, R. Klanner, G. Lutz, E. Neugebauer, A. Wylie and J. Kemmer, Nucl. Instr. and Meth. 205 (1983) 99; A Silicon Counter Telescope to Study Short Lived Particles in High Energy Hadronic Interactions

7. S. Barlag et. al., CERN - EP 86 - 173, submitted to Phys. Lett. B; Measurement of the Life Time of the Charmed Baryon Λ_c

8. J. Sedlmeir, Untersuchungen über einseitig und zweiseitig auslesbare Siliziumstreifendetektoren, Diplomarbeit, Technische Universität München, 15. August 1985

9. R. Bailey, C.J.S. Damerell, R.L. English, A.R. Gillman, A.L. Lintern, S.J. Watts and F.J. Wickens, Nucl. Instr. and Meth. 213 (1983) 201; First Measurements of Efficiency and Precision of CCD Detectors for High Energy Physics
C.J.S. Damerell, R.L. English, A.R. Gillman, A.L. Lintern and F.J. Wickens, Nucl. Instr. and Meth. A253 (1987) 478; CCDs for Vertex Detection in High Energy Physics

10. H. Dietl, T. Gooch, D. Kelsey, R. Klanner, A. Löffler, M. Pepé and F. Wickens, Nucl. Instr. and Meth. A253 (1987) 460; Radiation Damage in Silicon Strip Detectors

11. L. Strüder, P. Holl, G. Lutz and J. Kemmer, Nucl. Instr. and Meth. A253(1987)386; Device Modelling of Fully Depletable CCDs
MPI-PAE/Exp. EL. 166, submitted to Nucl. Instr. and Meth. ; Development of Fully Depletable CCDs for High Energy Physics Applications

Discussion G. LUTZ

A. H. WALENTA:

1) How do you plan to produce the multilayer CCD/drift chamber and what will be the minimum tickness of the layer? What will be the purity needed?
2) The advanced CCD with random access to simple cells looks, with respect to its organisation, similar to dynamic RAM's. The development costs of these devices may have reached 1B$. Why would your devices be cheaper to develop?

G. LUTZ

1) A possibility for producing the required alternate doped multilayer is epitaxial growth. Reasonable ticknesses for the layer would be 5 to 10 μm with doping densities in the range of 10 Ω cm. In order for the device to work, the total doping in each layer has to be rather closely the same.
2) Just for clarification this is not an advanced CCD but a new detector concept based on field effect transistors built on top of a fully depleted crystal. CCD type structured components may be added for shifting the stored signal charges from transistor to transistor or into a dump for fast cleaning of the device. Compared to RAM's the structures could be much coarser, so we expect higher yields as we would not exploit the limits of technology. Furthermore we could tolerate a certain fraction of non working cells contrary to requirements in computer memories. Profiting in addition from the enormous investiment of industry already done one should expect several orders of magnitude lower development costs than given in your example.

P. WEILHAMMER

What is the requirement on the bulk material for cheap depleted CCD's in particular resistivity?

G. LUTZ

The requirements are the same as for standard detectors, however additional doping close to the surface is required in the process.

TODAY AND TOMORROW FOR SCINTILLATING FIBRE (SCIFI) DETECTORS

Jasper Kirkby

CERN
Geneva, Switzerland

ABSTRACT

We review the current status of SCIFI detectors in experimental particle physics and their prospects in tomorrow's supercollider detectors. Our main focus concerns tracking devices although we also discuss the important aspects of SCIFI calorimetry. We could say, in summary, that SCIFI detectors will have a bright future.

SECTION I: TODAY

1. PRINCIPLE OF OPERATION

In the late 1950's an intensive effort was devoted to the observation of the tracks of charged particles in, initially, solid scintillators such as NaI (Tℓ) and then in targets made from coherent (parallel) bundles of plastic or glass scintillating fibres[1]. The fibre approach results in an excellent effective depth-of-field since a certain fraction of the scintillation light is 'trapped' inside the fibres by total internal reflection and consequently the optical image is preserved by the coherent geometry and transported to an output face (Fig. 1). The early work achieved considerable success but came to a halt in the early 1960's when bubble and spark chambers emerged as the clear leaders for particle tracking.

However, since that time there have been considerable advances in the technologies associated with SCIFI detectors, namely scintillators, fibre optics, image intensifiers, multi-anode phototubes, CCD cameras and data acquisition systems. As a result, and in view of their attractive properties, a renewed development of SCIFI detectors is taking place.

The modern wave of SCIFI development started with the work on plastic fibres by Borenstein et al.[2] and then by Saclay[3]. The contemporary work on glass SCIFI was initiated by Potter[4] and Ruchti et al.[5]. Additional groups are now involved in the development of tracking detectors based on scintillating fibres of plastic[6–8], glass[9,10] and liquid in glass capillaries[11] and in the

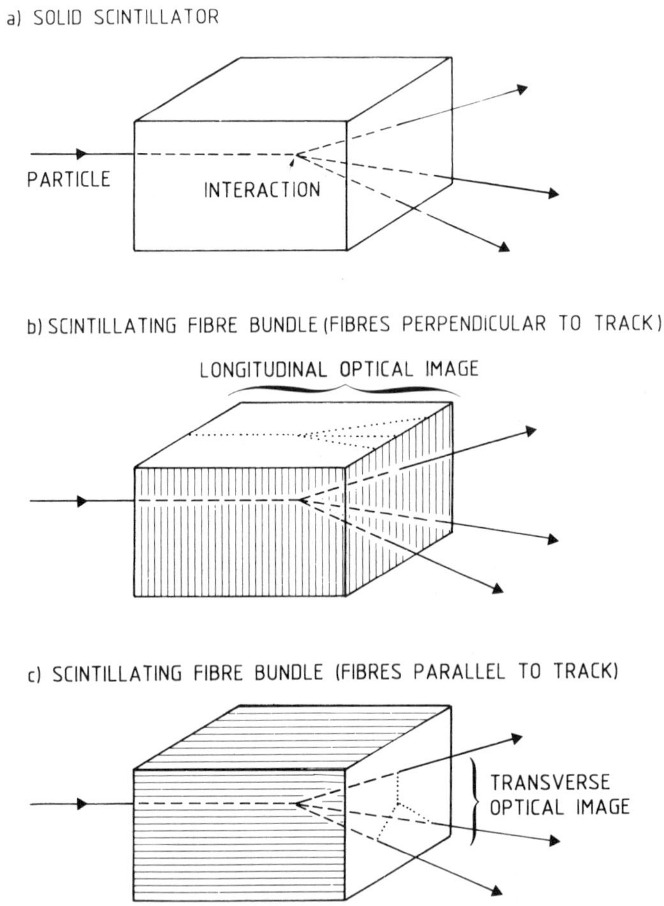

Fig. 1. Principle of the scintillating fibre (SCIFI) target. The solid scintillator a) is replaced by scintillating fibre bundles which transport either a longitudinal b) or transverse c) optical image of the interaction to an output face.

development of calorimeters based on plastic scintillating fibres[12-14], plastic wave-shifting fibres[15,16] and liquid scintillator in teflon tubes[17].

The operation of an individual scintillating fibre is illustrated by the example shown in Fig. 2. All types of scintillating optical fibre possess a scintillating core material surrounded by a non-scintillating optical cladding of a lower refractive index (producing a 'step index' fibre). The cladding provides a highly efficient reflective surface for transporting along the fibre the light which is contained inside a narrow cone of half angle, $90°-\phi_{critical}$ (see Fig.2). In order to perform efficiently, the interface between the core and cladding must be optically smooth and, since the electromagnetic wave extends (with an exponentially-decreasing amplitude) into the cladding, the cladding material must be transparent and of depth several wavelengths (≥ 5 in these applications). The cladding serves, in addition, to isolate the reflective surface from mechanical or chemical damage; fibres can, in principle, be more easily handled than sheets of scintillator where the reflective surface is exposed.

The fraction of the total scintillation light which is trapped in each direction is approximately 0.5 [1-n(cladding)/n (core)], where n is the refractive index. (This expression is approximate because it does not account for non-meridianal rays; the effect is to underestimate the trapping fraction by a factor of 15-30%, depending on fibre geometry.) The trapping efficiencies of glass, plastic and liquid SCIFI are seen (Table 1) to be small, between 2.5-4% in each direction. This can be compared with 10-15% for the optical collection efficiency of typical sheets of plastic scintillator. In order to absorb the large (92-95%) fraction of untrapped scintillation light, each fibre may include an optional outer layer of a thin opaque material (extra mural absorber, EMA) as indicated in Fig. 2.

Table 1

Approximate SCIFI optical trapping efficiencies

Scintillator medium	Core material	n (core)	Cladding material	n (clad.)	Trapping efficiency ε (%) *)
Glass	GS1	1.59	N51A	1.51	2.5
Plastic	Polystyrene +PBD+POPOP	1.59	Polyvinyl acetate	1.46	4.1
Liquid	Methyl-naphthalene +bis-MSB	1.62	N51A	1.51	3.4

*) The trapping efficiency for scintillation light, in each direction, $\varepsilon \sim 0.5$ [1-n(cladding)/n(core)].

2. OVERVIEW

2.1 SCIFI Characteristics

The key merits of SCIFI detectors are as follows:

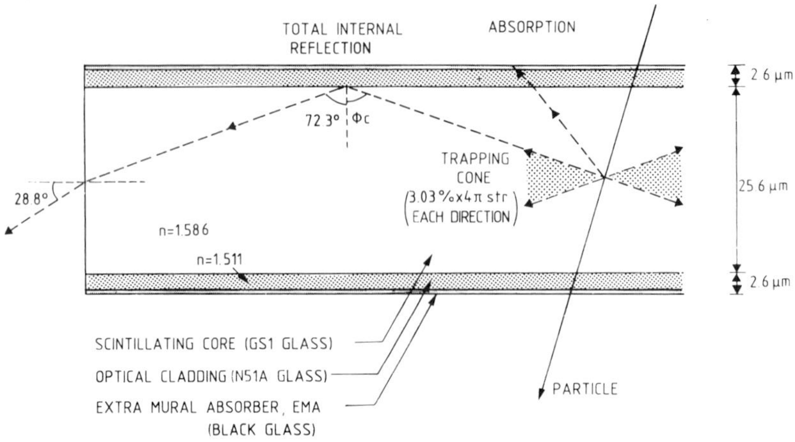

Fig. 2. Geometrical and optical parameters of an individual scintillating glass fibre (target 2 in Ref. 10 and Figs. 17 and 18).

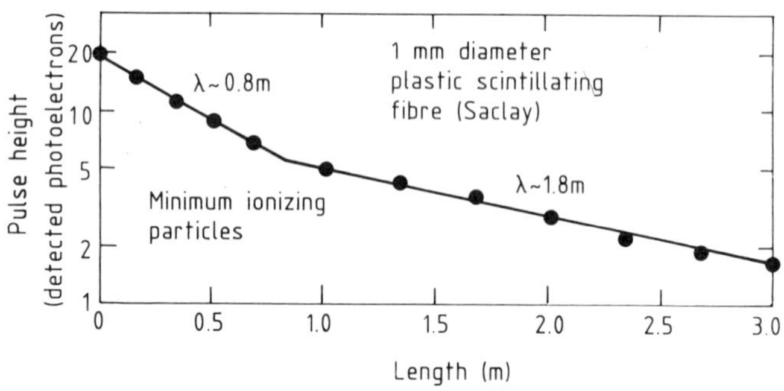

Fig. 3. Pulse height vs. length observed in plastic SCIFI using a bialkali phototube (Ref. 3).

1. Compactness/high density of hits per unit path length, e.g. minimum ionizing particles generate 5 photoelectrons (or equivalently 'hits') per mm in plastic SCIFI at a length of 1 m.

2. Fine spatial resolution, e.g. $\sigma \sim 20$ μm for glass SCIFI.

3. Fine two-track resolution, e.g. 50 μm for glass SCIFI, and absence of signal saturation up to extremely high particle densities.

4. Very high granularity, up to 10^5 cells cm^{-2}.

5. Radiation hardness. Both glass and plastic SCIFI can withstand ≤ 10^6 rads with tolerable losses in light output.

6. Flexible/can follow curved surfaces.

7. Intrinsically fast response of the detector itself, e.g. in the case of plastic SCIFI, the pulse time can be determined to less than 1 ns and the pulse integration time is ≤ 20 ns.

8. Insensitivity to magnetic fields.

9. In addition, for calorimeters:

   a) High density, e.g. with a packing fraction of 50%, a Pb/SCIFI calorimeter has a radiation length of 1.1 cm.

   b) Good hermeticity and uniformity of response with respect to transverse, longitudinal and angular impact position. 'Cracks' due to wavebars or mechanical boundaries are essentially absent.

   c) Good energy resolution for electromagnetic showers, due to the fine sampling. The measured values are $\sigma_E/E \leq 10\%/\sqrt{E(GeV)}$, extending down to very low energies (~ 20 MeV) in the case of phototube readout.

   d) An imaging capability which can provide, for example, a precise location of the conversion point or shower core.

   e) The capability to construct a 'compensating' hadron calorimeter (relative electron/hadron response = 1) from non-exotic absorber materials such as Pb.

2.2 Applications

These characteristics point to the following applications of SCIFI detectors:

1. Live targets either for detecting short-lived heavy-flavoured particles[18] or for rare decay experiments involving stopped particles[19].

2. Tracking in the environment of very high multiplicity and particle density such as high-energy ^{16}O interactions[20].

3. Precise, compact tracking in collider detectors[21].

4. Charge-flow tracking, i.e. the measurement of charged particle densities in jets, rather than to measurement of individual particles.

5. z-vertex measurement in supercollider detectors.

6. Electromagnetic calorimetry[22] with an imaging capability.

7. Hadron calorimetry which is hermetic and compensated.

In the remainder of Section I we first examine the choice of scintillators and their performance in fibres. We will point out the limitations of present scintillating fibres and indicate possible solutions. This is followed by a description of the readout techniques and a review of the experimental status of SCIFI tracking detectors. In Section II we consider the influence of SCIFI on the design and performance of tomorrow's supercollider detector. This section also discusses the readout of SCIFI detectors at ultra-high rates.

3. CHOICE OF SCINTILLATOR

3.1 Overview

We compare in Table 2 the main characteristics of glass and plastic SCIFI. Plastic scintillators are seen to be generally well-suited to SCIFI applications: good light yield, long attenuation length, fast, low mass, reasonably radiation-hard and relatively easy (i.e. low temperature) fibre production. The major limitation concerns the lack of efficient narrow-diameter fibres; present scintillator/waveshifter combinations are unsuitable below fibre diameters of ~ 500 μm. Glass scintillators, however, do not suffer from this problem and can be drawn into extremely fine fibres without loss of efficiency. With the exception of this property and also of radiation hardness, glass SCIFI is generally inferior in comparison with plastic SCIFI. The most serious limitation is the very short attenuation length.

In view of these characteristics the development of glass SCIFI has centred on precise small-volume active targets whereas plastic SCIFI has largely been used in compact tracking ofmoderate resolution or in calorimetry. We will now discuss the characteristics of these scintillators in more detail.

3.2 Plastic scintillator

3.2.1 Scintillation mechanism

In order to understand the performance of plastic SCIFI we must first consider the mechanism of light production by organic scintillators and wave-shifters.

Table 2

Comparison of the key properties of glass and plastic SCIFI

Property	Glass ($GS1-Ce^{3+}$)	Plastic (Polystyrene +PBD+POPOP)	
1. Photon yield of bulk material [NaI(Tℓ) = 40]	1.5 ± 0.5	8-10	photons keV^{-1}
2. Detected photoelectron yield of fibres (@ specified fibre length)	2(@ 1 cm)	5(@ 100 cm)	pe mm^{-1}
3. Attenuation length, λ	2-20	80-200	cm
4. Scintillation decay time constant, τ	55	3	ns
5. Radiation length, X_o	9.8	41.3	cm
6. Radiation hardness*	10^6	$\leq 10^6$	rads
7. Minimum diameter of efficient fibres	~ 10	~ 500	μm
8. Fibre manufacture	Difficult	Easy	

* We define radiation hardness as the exposure which causes an induced attenuation of ~ 8.7 dB m^{-1} (equivalent to λ = 50 cm) in the region of the scintillation spectrum.

Light emission in organic materials[23] is an inherent molecular property, occurring not only in the crystalline state but also in vapour and dilute solution in a liquid or plastic medium. Organic phosphors are primarily pure or substituted aromatic hydrocarbons since the benzene ring constitutes a stable system with a variety of (π electron) vibrational levels. This structure is readily excited into a high vibrational state by photons or ionizing particles and then promptly (few ns) de-excites with the emission of radiation. Before radiative decay there is a rapid (10^{-12} s) vibrational relaxation in which excess energy is lost in collisions with neighbouring molecules. This causes a (Stokes) shift of the bands of fluorescent light to somewhat longer wavelengths than those corresponding to the original transition from the ground state vibrational levels. However, in most scintillators there is a significant overlap of these two spectra -- the absorption (or excitation) and emission spectra -- which results in a ('self') absorption length of the primary fluorescent light of only a few mm. Therefore, in order to allow light to be transmitted through appreciable thicknesses of the material, a further fluorescent component is added which preferentially absorbs the scintillation light and re-emits at longer wavelengths. The second component (the 'waveshifter') is typically present in small concentrations (0.01-1% by weight) since it has a large absorption cross-section. Further waveshifters may be added until the final light has a sufficiently long wavelength for good transmission (λ = 1-2 m) through the bulk material. The properties[24] of some scintillator and waveshifter materials in common use are summarized in Table 3.

3.2.2 Overview of properties

We compare the three common classes of plastic scintillators in Table 4. They are characterized by their polymer base material as follows:

1. PVT (polyvinyl toluene or polymethyl styrene). This is the material typically used by industrial manufacturers such as Nuclear Enterprises[25] (NE designation) or Bicron[26] (BC designation).

2. PMMA (polymethyl methacrylate or acrylic). This requires a scintillating material, such as naphthalene, to be added to the non-scintillating polymer base. It was developed[27-28] as an inexpensive and mechanically robust alternative to the PVT scintillators, albeit with somewhat reduced light output. This category is found to be the most susceptible to radiation damage.

3. Polystyrene. This was developed[28-29] as an inexpensive scintillator which could be simply fabricated in large quantities by extrusion, thus avoiding the normal process of molding between two plates of glass.

3.2.3 Light yield and attenuation length

Experimentally we observe, in NE102A (PVT scintillator), 60 photoelectrons (pe) mm^{-1} for minimum ionizing particles, with an optical path length of 5 mm NE102A, using a RCA 8850 quantacon phototube (bialkali photocathode). The calculated light collection efficiency is 21%. In a typical large counter the observed yield reduces to 10-15 pe mm^{-1}.

Table 3

Organic scintillators and wavelength shifters

Material	Emission λ_{max} (nm)	Decay time, τ (ns)	Quantum efficiency	Photon yield (γ keV^{-1})
1. <u>Scintillators</u>				
Naphthalene	322	96	0.23	4.8
Anthracene	400	4.9	0.36	17
p-Terphenyl	340	0.95	0.93	10
Polyvinyl toluene	312	14	0.07	
Polystyrene	332	23	0.03	
2. <u>Waveshifters</u>				
PBD [1.]	360	1.1	0.92	
POPOP [2.]	410	1.4	0.93	
bis-MSB [3.]	420	1.3	0.94	
BBQ [4.]	480	18	-	
K27 [5.]	500	< 10	0.86	
3HF [6.]	510	< 1	0.36	

1. PBD: 2-phenyl-5(4-biphenylyl)-1,3,4-oxadiazole.

2. POPOP: p-bis [2-(5-phenyloxazolyl)]benzene.

3. bis-MSB: p-bis (o-methylstyryl)benzene.

4. BBQ: benzimidazo-benzisochinolin-7-on.

5. K27: xanthene acid derivative (supplied by Hoechst AG, Frankfurt, FRG).

6. 3HF: 3-hydroxyflavone.

Table 4

Properties of plastic scintillators

	PVT	Acrylic	Polystyrene
Commercial names	NE - BC -	PLEXIPOP ALTUSTIPE	KSTI
Typical composition (by weight):			
Polymer base	Polyvinyl toluene	Polymethyl methacrylate (PMMA)	Polystyrene
Additional scintillator	-	10-15% naphthalene	-
Scintillator/waveshifter 1.	2% p-terphenyl (NE102A)	1% butyl PBD	1% butyl PBD
Waveshifter 2.	0.1% POPOP (NE102A)	0.01% POPOP	0.01% POPOP
Density	1.03	1.18	1.06 g cm^{-3}
Mean refractive index, n (emission spectrum)	1.60	1.49	1.59
Softening point	75	180	98°C
Radiation length, X_o	42.4	34.4	41.3 cm
Hadronic interaction length, λ	79.5	70.8	77.4 cm
dE/dx (minimum ionizing particles)	2.0	2.3	2.0 MeV cm^{-1}
Photon yield (minimum ionizing particles)	10	3-9	8-10 photons keV^{-1}
Decay time constant, τ	3	3-7	3 ns
Wavelength of peak emission, λ_{max}	430	430	430 nm
Typical attenuation length, λ_{atten}	2	1-1.5	1-2 m
Radiation hardness (see Section 3.2.4)	$\leq 10^6$	$\leq 10^5$	$\leq 10^6$ rads
Mechanical handling properties	Poor	Good	Fair

Plastic SCIFI, such as produced by Saclay[3], involves a polystyrene - PBD (1%) -- POPOP (0.01%) core of diameter 0.5-1.5 mm surrounded by a polyvinyl acetate cladding of thickness 7-22 μm. This results in a trapping efficiency in each direction of approximately 4% of the scintillation light (Table 1), as compared with ~ 10% in standard (slab) scintillation counters. The observed light yield in plastic SCIFI is therefore expected to be ~ 40% of the yield of a standard counter of equal thickness. This is confirmed, as indicated in Fig. 3, where a yield of 5 pe mm^{-1} is measured[3] at 1 m distance.

Figure 3 also indicates (1-2 m) attenuation lengths for plastic SCIFI which are typical of the bulk material. This confirms the absence of substantial reflective losses at the core-cladding interface. This observation is reinforced by a measurement[3] of the attenuation in a clear (undoped) polystyrene fibre, shown in Fig. 4. Here we see excellent attenuation lengths, $\lambda > 20$ m, in the green-red spectral region. [Reference 15 found $\lambda = 11$ m for a polystyrene fibre (diameter 1.5 mm and length > 4 m) doped with the waveshifter K27 ($\lambda_{max} \sim 500$ nm).]

3.2.4 Radiation hardness

The radiation hardness of detectors was barely relevant to experimental particle physics a few years ago but, starting with the ISR and then the pp colliders, it has now become a crucial issue. Furthermore, as we shall discuss in Section II, it will become increasingly important at the supercolliders.

The experimental data[27-34] are incomplete and in some cases contradictory. This is probably due to the fact that there are many parameters which influence the amount of radiation damage, amongst which are:

1. Total dose.

2. Dose rate.

3. Type of radiation: photons, electrons, hadrons, neutrons or heavily ionizing particles.

4. Surrounding chemical environment, especially the presence of oxygen.

5. Composition of the material, in particular the presence or not of a UV-generating component. In the case of fibres, the composition of both the core and the cladding is relevant.

6. Presence in the material of small concentrations of impurities.

7. Time between irradiation and measurement (recovery), and the surrounding chemical environment during this interval.

8. Heat cycling (annealing).

We can find the following qualitative consensus among the experimental measurements:

1. All data agree that acrylic scintillators are more readily damaged by radiation than polystyrene or PVT scintillators. Equivalent damage (in the region of the scintillation spectra) occurs in the latter scintillators at about an order-of-magnitude higher doses. This is seen (Fig. 5) in transmission measurements[30] made on undoped acrylic and polystyrene fibres. The vertical axis is expressed as the attenuation,

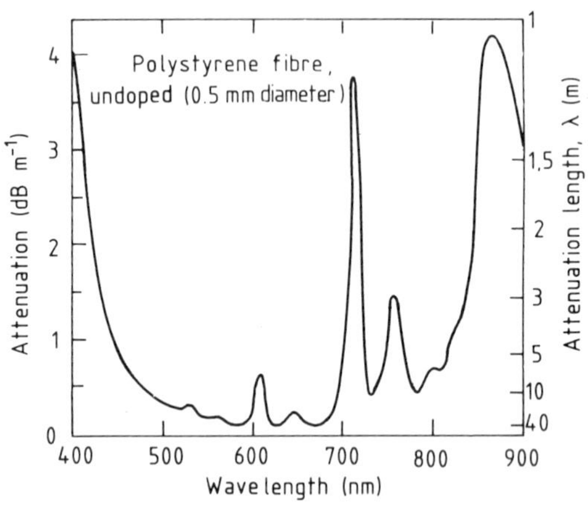

Fig. 4. Attenuation in undoped polystyrene fibres vs. wavelength (Ref. 3).

$$A(\mathrm{dB\ m^{-1}}) = -10 \log_{10} (I_{out}/I_{in})/L \ ,$$

where I_{in} and I_{out} are the input and output light intensities (at a certain wavelength) and L(m) is the fibre length. This quantity is related to the more familiar attenuation length λ (m) by the expression λ (m) = 4.343/A (dB m^{-1}). Figure 5 demonstrates the preferential deterioration in transmission at shorter wavelengths ('yellowing' of the material). Furthermore, polystyrene exhibits a threshold in $\log_{10}$ (radiation dose) below which the induced attenuation is tolerable. This threshold, which is seen in other measurements[35], occurs between 10^5 to 10^6 rads.

2. The amount of damage does not seem to depend on the type of ionizing particle which deposits a given total energy.

3. After periods of a few days in air following exposure, substantial recoveries in transmission are observed.

The data are in less clear agreement over the following issues:

1. The influence of an oxygen atmosphere. Reference 32 finds a strong deterioration of acrylic scintillators after only 10^4 rads, providing the dose rates are small (10 rads hr^{-1}) and air is present. The damage is reduced by a factor of 10 in a N_2 atmosphere. This leads to the explanation that UV light, produced by the butyl PBD, creates free oxygen radicals from the atmospheric O_2 molecules which slowly diffuse into the plastic. Oxygen radicals are highly reactive and result in chemical damage to the material. Reference 34 similarly finds a large oxygen effect in clear (undoped) acrylic fibres, but of opposite sign. This reference also finds that the presence of oxygen leads to an accelerated (~ 4 days) and almost complete recovery of damaged acrylic fibres. A further experiment[30], involving undoped fibres of both acrylic and polystyrene, finds no effect due to dose rate in an air environment (and therefore, equivalently, no effect due to oxygen). These discrepancies may be related to the presence or otherwise of scintillators and waveshifters in the material under test.

2. The relative contribution to the observed loss of light from a deterioration of the scintillation efficiency and from a decrease in the attenuation length.

In summary, it seems that polystyrene scintillators and their fibres will operate with tolerable losses up to several 10^5 rads, and perhaps up to 10^6 rads. This is just at the level where they become viable for supercollider detectors. What is now clearly needed is a new, comprehensive and systematic study to be made of radiation damage to plastic scintillators and plastic scintillating fibres.

3.2.4 Limitations and cures

The main limitation of current plastic SCIFI is the absence of efficient narrow-diameter fibres. This is illustrated in Fig. 6 where we see (Fig. 6a) the primary scintillation light propagating a short distance before absorption and re-emission by the waveshifter. The attenuation of the primary scintillation light is given by the expression,

$$I = I_o \, 10^{-d\varepsilon c} = I_o \, e^{-d\varepsilon c(\ell n \, 10)} ,$$

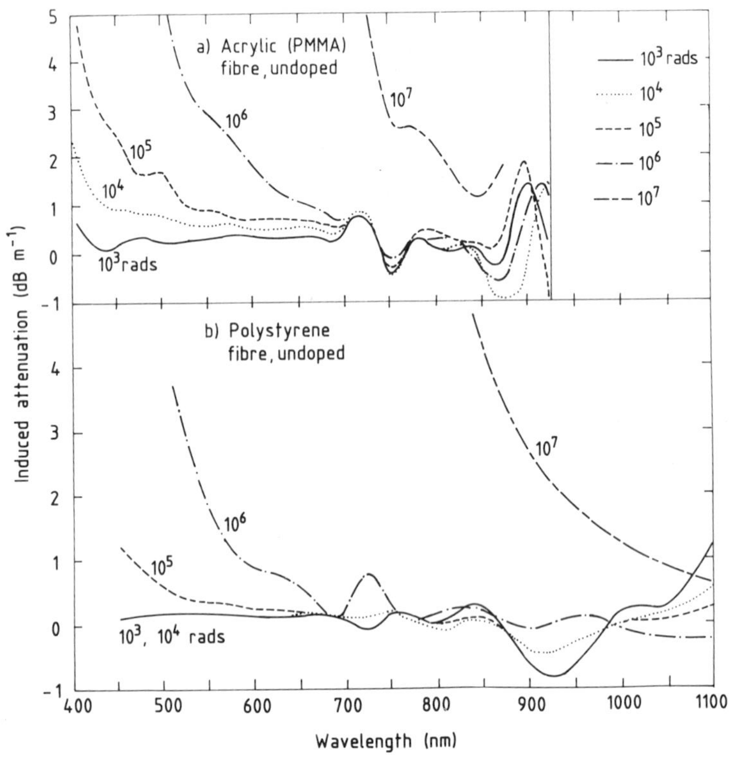

Fig. 5. Radiation-induced attenuation in a) undoped acrylic and b) undoped polystyrene fibres (Ref. 30). The curves show the changes in attenuation relative to the zero radiation values.

where I is the light intensity, d(cm) is the distance of light propagation, $\varepsilon(M^{-1}\ cm^{-1})$ is the molar absorptivity (or molar extinction coefficient) and c (M) is the concentration of the waveshifter in moles $litre^{-1}$ (M). Typical values (for POPOP) are $\varepsilon = 4 \times 10^4\ M^{-1}\ cm^{-1}$ and $c = 0.1\ (g\ litre^{-1})/364$ (molecular weight) $= 2.7\ 10^{-4}$ M. This indicates an attenuation by a factor 1/e in a distance $d = [4 \times 10^4 \times 2.7 \times 10^{-4} \times 2.3]^{-1}$ cm, i.e. 400 μm. Therefore, providing the fibre diameter is more than ~ 500 μm most of the primary scintillation light will be shifted before reaching the core-cladding interface. This shifted light will then be collected with the nominal efficiency ε (~ 4%).

As the diameter of the fibres is reduced, we see (Fig. 6b) that cross-talk will occur, i.e. light is trapped in fibres which are not intercepted by the particle. Furthermore, if one tries to correct this with EMA (Fig. 6c) the efficiency (light output) of the fibres reduces to an unacceptably low value ε^2 (~ 0.16%), since both the primary and the shifted light must be trapped.

A simple solution to this problem would be to increase the waveshifter concentrations. The drawback is this leads to a decreased bulk attenuation length, since most waveshifters in common use suffer some self absorption. It is likely, however, that some optimization can be done of the concentrations of the components in present plastic scintillators to suit specific experiments and fibre lengths.

An alternative solution is to provide each fibre with an external aluminized coating (Fig. 7). This serves to contain the primary scintillation light with a few reflections, before the waveshifting takes place. The waveshifted light is then transported in the standard fashion by reflections at the core-cladding interface. The obvious drawback here is that single fibres as opposed to multifibres (Section 4) must be handled. [Aluminization is in fact utilized for the 1 mm diameter plastic SCIFI of the UA2 track detector (Section 6.1), to provide an EMA.]

The most attractive solution is to find alternative waveshifters with better absorption and transmission properties. The waveshifter POPOP is seen (Fig. 8) to have a small Stokes shift and a region of overlap between the absorption and emission spectra. It is this overlap which leads to self absorption of the emission spectrum, especially at the shorter wavelengths. A promising new candidate to replace POPOP is 3HF[36] (3-hydroxyflavone) which displays (Fig. 9) a strikingly clean Stokes shift and a strong molar absorptivity. Furthermore it is found[37] to have an acceptable quantum efficiency (0.36, compared with 0.93 for POPOP) and a very fast decay time (< 1 ns). Of particular note is the green emission spectrumwhich should result in substantial improvements in both the attenuation length ($\lambda \sim 20$ m from Fig. 4) and, more importantly, in the radiation hardness (see Fig. 5). Studies[37,38] are presently underway to investigate plastic scintillators containing 3HF waveshifter.

Finally we mention that a suitable plastic EMA is not yet available for plastic SCIFI. As we shall see, EMA is essential in tracking devices. This situation reflects the relatively recent development of plastic SCIFI rather than any fundamental obstacle.

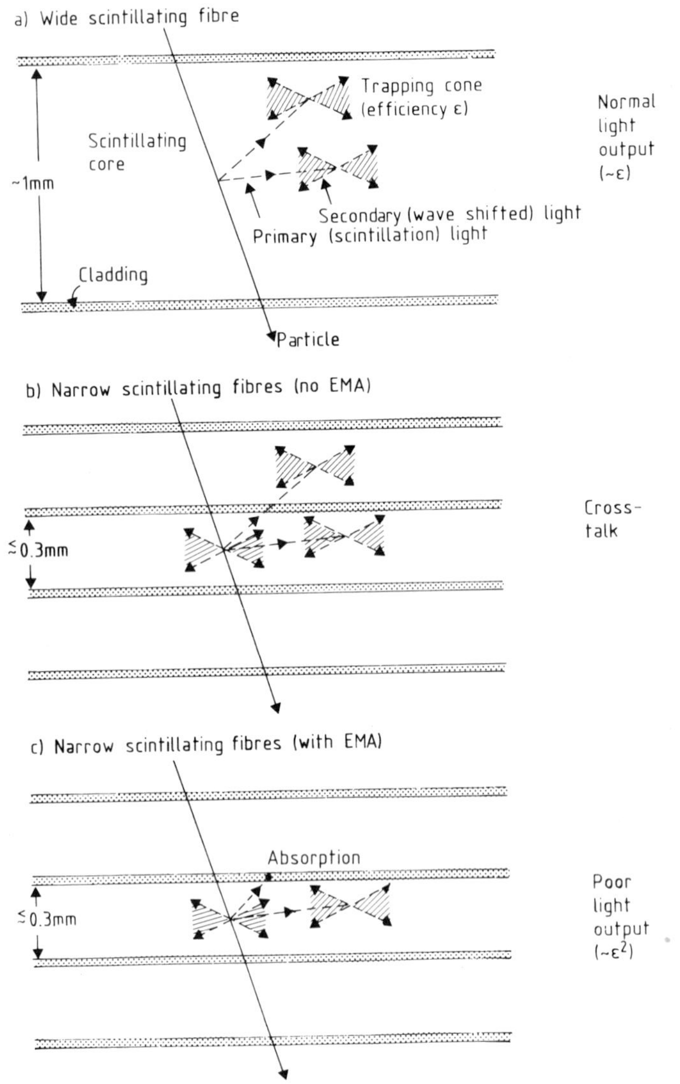

Fig. 6. Schematic representation of the operation of plastic SCIFI with a) wide and b), c) narrow diameters. The narrow diameter fibres function poorly due to either b) cross-talk or c) low efficiency.

3.3 Glass scintillator

3.3.1 Scintillation mechanism

The luminescence of inorganic phosphors differs fundamentally from that of organic phosphors. Here the process is a lattice effect, intimately connected with the crystalline nature of the host material. An inorganic scintillator involves a crystalline or near-crystalline host insulator which is doped with a small concentration of impurity atoms, known as 'activators'. Examples are NaI activated with Tℓ or a glass matrix activated with Ce^{3+} ions. (In certain cases activator centres may also be caused simply by lattice imperfections in a pure crystalline material.) Ionizing radiation causes electrons in the host matrix to be transferred from the valence band to the conduction band. The resulting electron-hole pair (known as an 'exciton') then rapidly moves through the lattice under the influence of the electromagnetic field between the excited ion and a neighbour in the ground state. This mobility results from the crystalline nature of the host. In a pure crystal the lifetime of the exciton, which terminates when the electron and hole recombine, is estimated to be ~ 10 ns, during which time it has travelled a mean square displacement of ~ 1 μm from its origin. However, well before this time it may encounter an activator centre, which is chosen to have energy levels inside the forbidden gap of the crystalline host. This effectively locks the exciton on the activator centre, which can then decay in its characteristic time with the emission of a photon.

There are two other ways in which the activator centre may be excited. One is simply direct absorption of energy from the ionizing particle or direct absorption of photons of the appropriate frequency (for which the host may be transparent if the photons are not energetic enough to cause transitions from the valence to conduction bands). The other can involve the motion of free electrons and holes, evidently in association with conductivity.

After excitation of the activator centre the neighbouring lattice undergoes a rapid relaxation. This occurs in a time of the order of 10^{-13}–10^{-12} s, characteristic of lattice vibrations and the emission of phonons. The result is a Stokes shift of the emitted light to longer wavelengths than those of the excitation bands and, as with organic scintillators, the phosphor becomes largely transparent to its own scintillation light.

In addition to activator centres there may be (both deliberate and accidental) dopants which compete for excitons and which have non-radiative decays. These are known as 'quenchers' or 'killers' and influence both the amount and time constant of the emitted light. This quenching can also be caused by mutual interactions between activator centres when the latter are present in high concentrations. As an example we find quenching effects start to appear in GS1 glass when the Ce_2O_3 fraction by weight exceeds 2%.

A further impurity or lattice defect may give rise to metastable levels in the forbidden gaps. These centres, known as electron 'traps' or 'colour centres', have the property of holding an exciton essentially indefinitely. They may be caused by an inpurity in the crystal or by the transformation or dislocation of atoms from the lattice due to the effects of ionizing radiation, in particular from highly-ionizing particles. They adversely affect the performance of the scintillator by reducing the energy transfer efficiency to the activator centres, by absorbing scintillation photons and thereby reducing the transparency of the material, and by generating phosphorescent (long-time constant) light. Electrons trapped in these centres can be liberated by appropriate UV irradiation or heating (annealing) but, depending on the nature of the trapping centre, this may only offer a temporary cure. Annealing at a sufficiently high temperature can provide enough thermal energy for interstitial atoms to migrate through the crystal and recombine in vacant sites, thus returning the crystal to its undamaged state.

In summary we see that inorganic scintillators have the ability to transfer the ionization energy efficiently through the lattice to an activator centre where light emission takes place. In fact inorganic scintillators have the highest conversion efficiency of ionization energy into light, e.g. NaI (Tℓ) produces 40 γ keV^{-1}, or an efficiency of 12%, as compared with 3% for good plastic scintillators. Finally, the light is emitted $\lesssim$ 1 μm from the particle trajectory, i.e. it is local, in contrast with the situation for plastic scintillator/waveshifters.

3.3.2 Overview of properties

The first challenge facing inorganic SCIFI was to find materials which could be drawn into clad optical fibres. The work has focussed on scintillating glasses, originally[5] containing Tb^{3+} but now on glasses based on the Ce^{3+} activator (Refs. 5,9,10,39) since it has the shortest known decay time constant ($\tau \simeq 55$ ns). Table 5 presents a summary[10] of the properties of the cerium glass GS1, manufactured by Levy Hill Laboratories[40]. We mention that this and similar highly-crystalline glasses are difficult to manufacture and difficult to reprocess into fibres since the glass devitrifies readily, i.e. transforms to an opaque white crystalline ceramic.

3.3.3 Light yield and attenuation length

We observe, in GS1, 40 pe mm^{-1} for minimum ionizing particles, with an optical path length of 5 mm GS1, using a RCA 8850 quantacon phototube (bialkali photocathode). The calculated geometrical light collection efficiency is 28%.

We have also compared the responses of GS1 and a PVT plastic scintillator, NE102A, to highly ionizing particles. Using an ^{241}Am source (5.5 MeV α) we find, for GS1, a light yield of 40% relative to the yield of a minimum ionizing particle which deposits 5.5 MeV. Plastic scintillators show a far more serious saturation; our equivalent measurement with NE102A is 7%. This difference in response dramatically demonstrates the different scintillation mechanisms: the glass matrix efficiently conducts excitons away from the highly ionized region whereas the organic scintillator cannot and saturation occurs. This saturation is formalized by Birks law[23] which expresses the fluorescent light output per path length dx in a scintillator,

$$dI/dx = S\,(dE/dx)/[1+kB(dE/dx)]\ ,$$

where S is the scintillation efficiency, dE/dx is the specific energy loss and kB is the (Birks) constant, characteristic of the material, which represents the probability of quenching (by molecular interaction) relative to photon emission. Typical values for organic scintillators are kB = 0.01–0.015 gm cm^{-2} MeV^{-1}. For fast electrons one ion pair is produced every ~ 200 nm, i.e. ~ 200 molecular distances apart, and so quenching is negligible and dI/dx = S dE/dx. However, for 5.5 MeV α particles one ion pair is produced every ~ 1 nm, i.e.

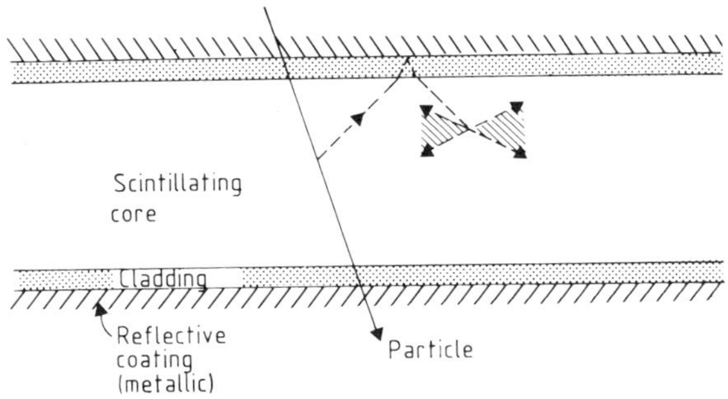

Fig. 7. Operation of narrow-diameter plastic SCIFI with an outer metallic reflective coating.

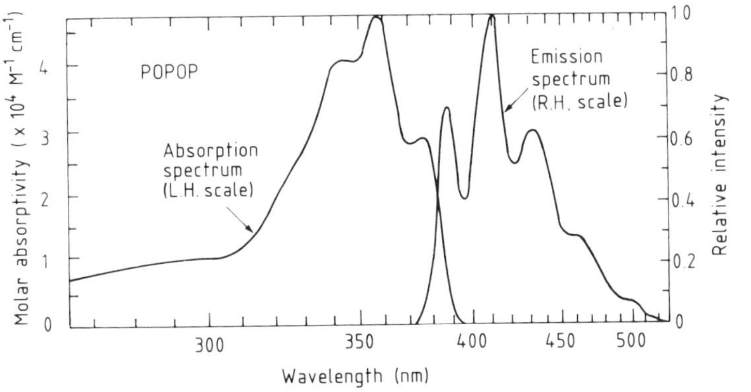

Fig. 8. Absorption and emission spectra of the aromatic waveshifter POPOP (Ref. 24).

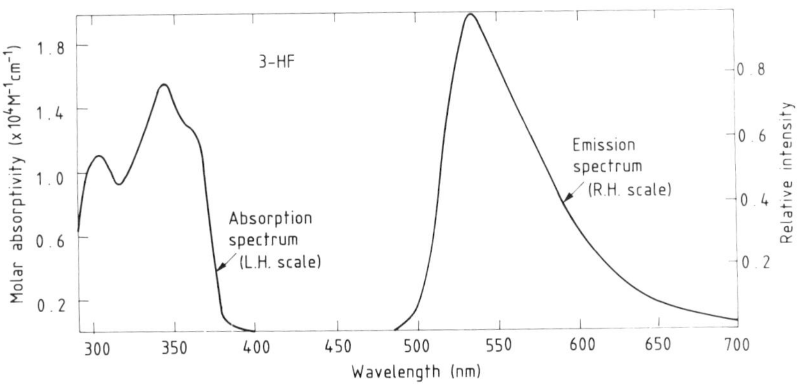

Fig. 9. Absorption and emission spectra of the aromatic waveshifter 3-hydroxyflavone, 3HF (Ref. 36).

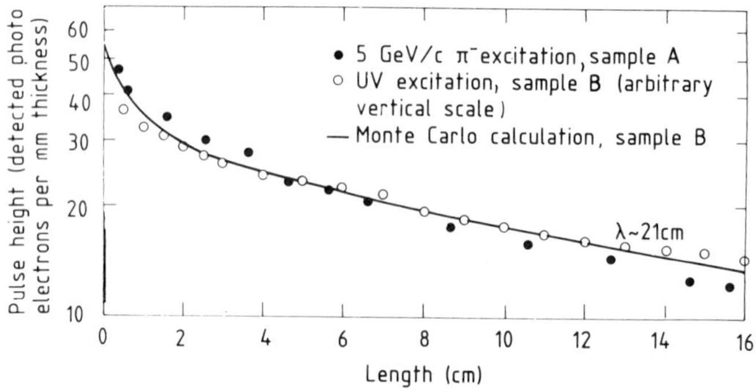

Fig. 10. Light yield, measured with a bialkali phototube, of polished GS1 scintillating glass blocks of length ~ 17 cm (Ref. 10). The curve is a Monte Carlo calculation of sample B based on measurements of the scintillation emission spectrum (Fig. 11) and transmission curve (Fig. 12). The vertical scale shows the observed photoelectron yield for sample A (for which the calculated photon efficiency is 28%).

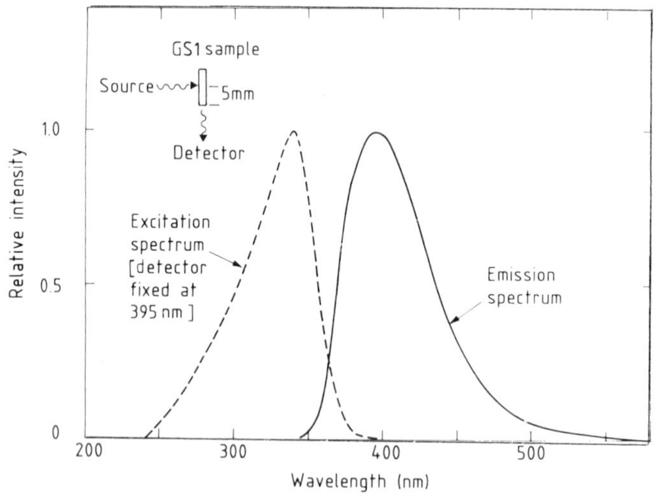

Fig. 11. Excitation and emission spectra of GS1 scintillating glass (Ref. 10). The excitation curve shows the relative emission intensity (with the detector fixed at 395 nm) when GS1 is excited by a UV source at the indicated wavelength and with a fixed intensity.

Table 5

Properties of GS1 scintillating glass

Composition, by weight[1]:	SiO_2	55%
	MgO	24%
	Al_2O_3	11%
	Li_2O	6%
	Ce_2O_3	4%
Density, ρ	2.64 ± 0.1	g cm^{-2}
Mean refractive index (GS1 spectrum), n	1.586 ± 0.005	
Transition temperature	620	°C
Linear thermal expansion coefficient, α	75 x 10^{-7}	$°C^{-1}$
Radiation length, X_o	9.77	cm
Hadronic interaction length, λ	37.4	cm
dE/dx (minimum ionizing particles)	4.6	MeV cm^{-1}
Photon yield (minimum ionizing particles)	1.5 ± 0.5	photons keV^{-1}
Decay time constant, τ	55	ns
Wavelength of peak emission, λ_{max}	395	nm
Attenuation length	2-20	cm
Radiation hardness (see Section 3.3.4)	10^6	rads
Approximate cost	8	\$ US cm^{-3}

1. Levy Hill Laboratories, Ref. 39.

about 1 per molecule, resulting in strong quenching of the photon emission. Not only does GS1 have a good light yield for highly ionizing particles but it also has a high cross-section for slow neutron detection (via ^{7}Li capture). This and similar glasses are therefore used extensively in neutron experimentation[39].

The observed light yield vs. length in bulk GS1 is shown in Fig. 10. We see a sharp attenuation ($\lambda \sim 2.5$ cm) in the first 2 cm followed by a flattening to $\lambda \sim 20$ cm at the furthest distances. This behaviour is easily understood from the observed overlap of the excitation (absorption) and emission spectra (Fig. 11). We discuss in Section 6.4 the light yield and attenuation length of fibres of this glass.

In order to understand the cause of the poor attenuation, several samples of GS1 were prepared with different fractions of Ce^{3+}. The results of transmission measurements for these samples (Fig. 12) show the absorption is not due to an impurity in the host glass but is due to the Ce itself. However the poor quantitative agreement between Ce^{3+} concentration and transmission has led to the hypothesis that the Ce^{3+} may be partially oxidizing to Ce^{4+}, which is known to have a very strong optical absorption in this region. Quantitatively the poor transmission of GS1 would result if only 2% of the original Ce^{3+} was oxidized to Ce^{4+}. These considerations give hope that GS1 or an equivalent glass can soon be produced with a good attenuation length.

3.3.4 Radiation hardness

The transmission of GS1 after exposure to secondary radiation from external targets at CERN is shown in Fig. 13. Up to levels of $\sim 10^6$ rads we see relatively mild deterioration in transmission. At higher levels the glass becomes increasingly opaque to its scintillation light, although it remains fairly clear to visible light even at the highest exposure of 10^8 rads. The good resistance of GS1 to radiation is due to the strong electron affinity of cerium, which inhibits the formation of colour centres by ionizing radiation. In fact cerium glasses are widely used in high radiation environments such as reactors. In the case of SCIFI detectors, it is necessary for both the core and cladding glasses to be resistant to radiation. We therefore anticipate that radiation-resistant SCIFI detectors will require a low refractive index cladding glass containing a small cerium admixture.

The temperature annealing of GS1 has also been investigated. The results (Fig. 14) show a mild recovery at 200°C and a substantial improvement at 400°C. This indicates there are prospects for recovering radiation-damaged SCIFI detectors providing they are constructed in such a way to allow disassembly and baking at high temperature.

3.3.5 Limitations and cures

The primary limitation of the Ce scintillating glasses is the very short attenuation length, as described above. It is now necessary to demonstrate that pure Ce^{3+} glass can be prepared in a suitably reducing environment and to measure the improvement in the attenuation length.

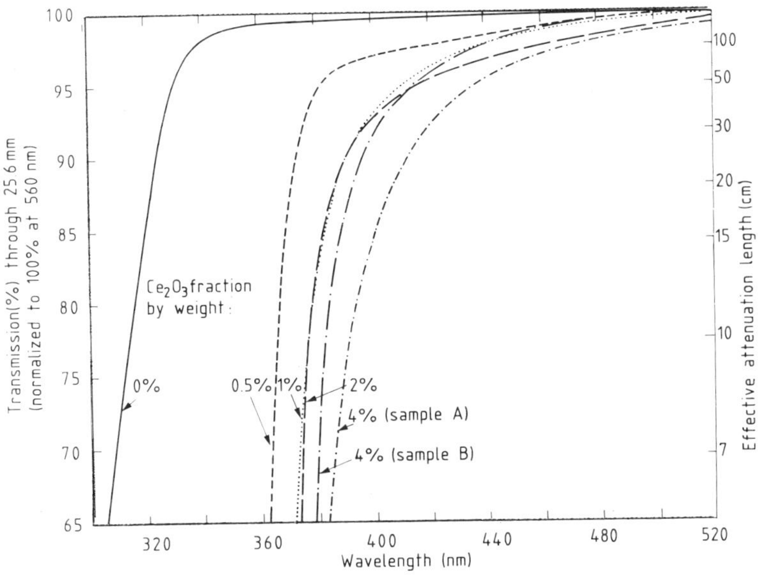

Fig. 12. Transmission vs. wavelength for samples of GS1 scintillating glass involving different fractions of Ce_2O_3 (Ref. 10). Two different samples (A and B) of nominal GS1 (4% Ce_2O_3) are displayed. The curves are normalized to 100% transmission at 560 nm. The measured transmissions at 560 nm range between 80–88%, corresponding largely to reflective losses at the entrance and exit faces and at scattering centres, such as small bubbles, inside the glass.

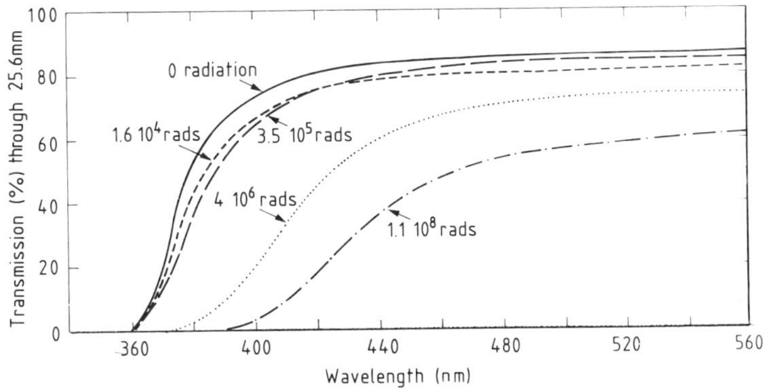

Fig. 13. Transmission vs. wavelength for samples of GS1 scintillating glass after exposure to secondary radiation from external targets at CERN (Ref. 10). The curves are not corrected for reflective losses.

There are two other potential drawbacks with glass SCIFI. The first is the slow scintillation time constant of 55 ns. In the case of readout using image intensifiers and phosphors this actually presents no limitation since the fastest phosphors (which also, in fact, contain Ce) have a similar time response. The second is the short radiation length of 9.8 cm, which may lead to an unacceptable amount of material in certain applications.

3.4 Liquid scintillator

There are two separate developments involving fibres filled with liquid scintillator:

1. Glass capillaries. Small (6 x 6 x 12 mm^3) active targets have been made from fine glass capillaries fused into a coherent bundle[11]. The fibres have 60 μm pores and provide an open volume of 50% which may then be filled with liquid scintillator (see Table 1)

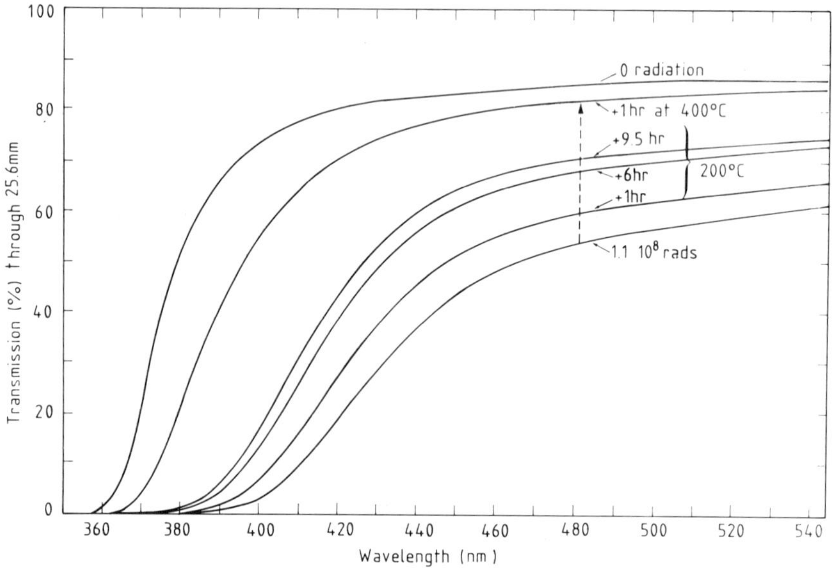

Fig. 14. Temperature annealing of a sample of GS1 scintillating glass after exposure to 1.1 x 10^8 rads (Ref. 10). The indicated annealing times are cumulative (e.g. the total annealing time at 200°C is 16.5 hours). The curves are not corrected for reflective losses.

by capillary action. A yield of 4 hits (pe) mm^{-1} for minimum ionizing particles is observed using an image intensifier. The attenuation length is very short (~ 2.5 cm).

2. Teflon tubes. Extruded teflon FEP tubes, of outer diameter 5 mm and wall thickness 0.4 mm, have been used to read out electromagnetic calorimeters[17]. They were filled with NE235 scintillator[25] of refractive index 1.47. In association with teflon (n = 1.38) this gives an approximate trapping efficiency in each direction of 3.1%. The light yield after a path length of 2 m was found to be (20 ± 4) γ per MeV deposited in the scintillator (equivalent to 1 pe mm^{-1}), with a good attenuation length λ = 140 cm. The light yield showed no significant deterioration after irradiation up to 10^5 rads (a drop of 20% in the attenuation length was observed at 10^5 rads) but at the highest doses (4×10^6 rads) the teflon had become brittle and the transmission was very poor. Further development of this technique is taking place[37] using alternative materials in place of teflon.

4. SCIFI MANUFACTURE

The production of SCIFI from bulk plastic or glass scintillator is illustrated in Fig. 15. (The dimensions shown on this figure refer to the glass SCIFI targets discussed in Section 6.4.). The whole process must take place under clean conditions since dust particles render fibres useless.

The first step (Fig. 15a) is to prepare a preform comprising of a polished scintillator core around which is placed a cladding. The cladding can either be an extruded material, with inner dimensions closely matched to the core, or, in the case of the Saclay plastic fibres, a coating of polyvinyl acetate which is applied directly to the polished surface of the core scintillator. The ratio of the physical dimensions of the core and cladding are preserved throughout the drawing processes and must be properly chosen at the outset. The preform may be completed by a thin EMA coating which, for glass SCIFI, is an opaque black glass. All preform materials must have proper matching of physical properties such as softening points, expansion coefficients, etc. The final preform weighs from 0.1 kg up to several kg, depending on the material and capacity of the pulling station.

The preform is slowly inserted through the top of an oven (Fig. 16) where it is heated to the appropriate softening temperature. The lower tip of the preform is then caught and pulled downwards to form fibres. These fibres may be cut into lengths of several metres or, in the case of plastic SCIFI, stored on a drum. (Saclay stores on each drum a single 1 mm diameter fibre of length 0.6 km from each preform weighing ~ 0.5 kg.)

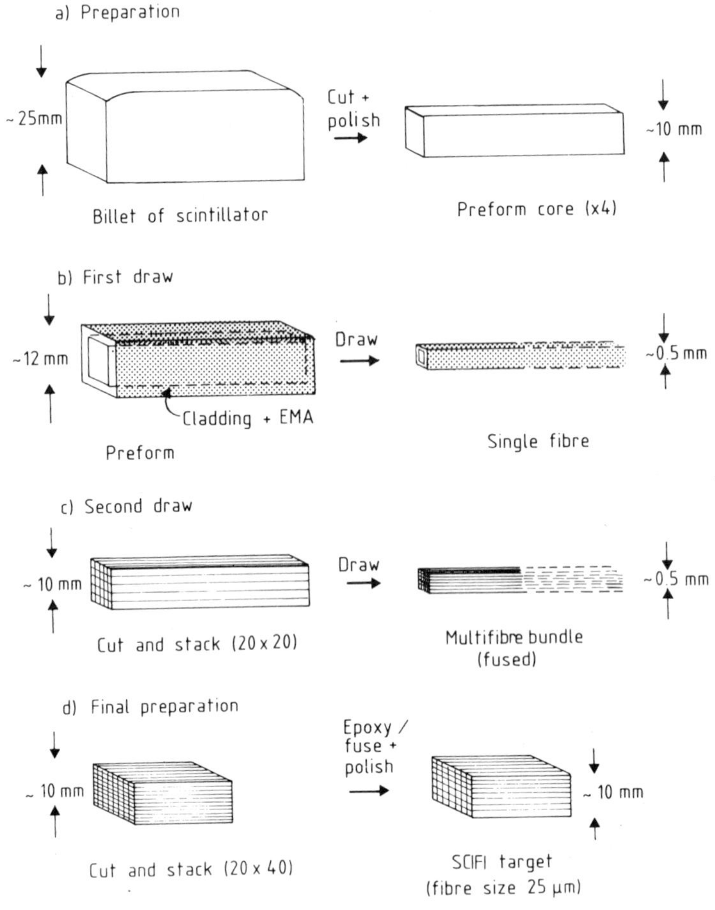

Fig. 15. Steps involved in the manufacture of coherent SCIFI bundles from bulk scintillator. The dimensions refer to the glass SCIFI targets of Ref. 10 and Figs. 17 and 18.

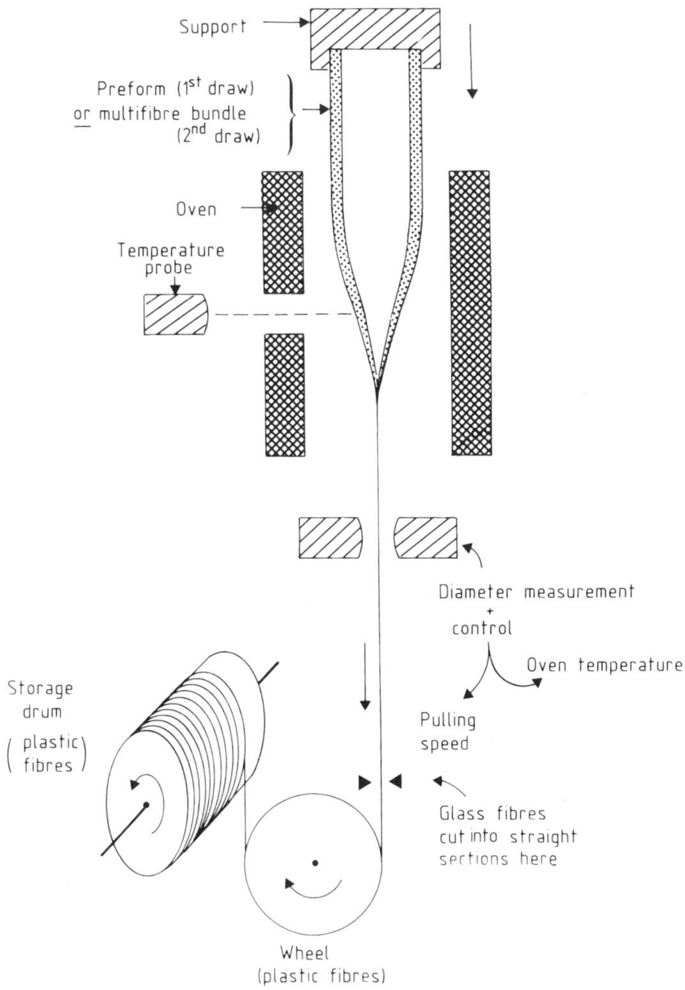

Fig. 16. Schematic representation of a fibre pulling station.

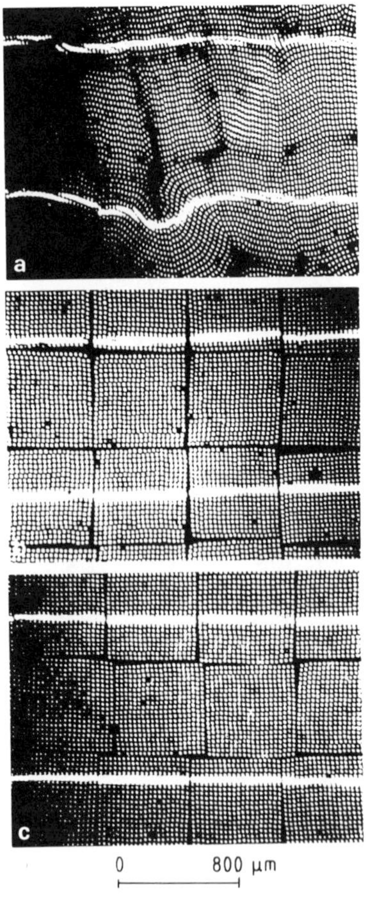

Fig. 17. Microphotographs of glass SCIFI targets (Ref. 10): a) target 1 (no EMA, fusion bonded multifibres), b) target 2 (EMA, epoxy bonded), and c) target 3 (no EMA, epoxy bonded). Photographs b) and c) show clearly the multifibre bundles, of transverse dimension 0.6 mm, which are composed of matrices of 20 x 20 individual fibres. The bright lines are the images of a pair of illuminated parallel straight slits of width 30 μm and separation 1 mm. The slits are placed in contact with the entrance face of each SCIFI target and viewed through 25 mm fibre length at the exit face.

An automatic measurement of the fibre diameter provides feedback control of the pulling speed and oven temperature. Good pulling stations produce fibres with diameter variations of ~ 1%, whereas the best stations can achieve ~ 0.1%.

The first draw (Fig. 15b) produces single fibres of diameter ~ 1 mm which can be used directly for calorimetry or moderate-resolution tracking. In order to make finer fibres a second draw (Fig. 15c) is necessary. This involves the manual assembly of a set of single fibres (typically 100-200) into a mechanical fixture which holds them in a precise array. After the second draw a multifibre bundle is obtained, with individual fibres of the required small diameter (~ 25 μm) which are fused together by their cladding/EMA. The last process (Fig. 15d) is to assemble the multifibre bundles into a device of the required dimensions and to bond them with epoxy or by fusion (heat). The final preparation involves optical polishing of the output faces.

Microphotographs of glass SCIFI targets manufactured[41] by this process are shown in Figs. 17 and 18. (These targets are used by Ref. 10 as described in Section 6.4.) The later targets (Figs. 17b and c) are seen to have reasonably good coherency (parallelism to < 1 fibre diameter over the length of the fibres). However we can see by comparing Fig. 18 with Fig. 19, which shows the near-perfect regularity of a fibre-optic plate used in image intensifiers, that there is room for further improvement.

5. READOUT

5.1 Overview

In order to be sensitive to the single-photon signal levels which are characteristic of SCIFI, the readout system must have a photon gain of ~ 10^5 in addition to an imaging capability. This is achieved by image intensifiers of typically three stages where at least one involves a micro-channel plate (MCP) to provide both high gain and a fast gating capability. After intensification the image is generally stored on a charge-coupled device (CCD) from where it may be either read out in ~ 10 ms or fast-cleared in ~ 1 μs. Fibre-optic coupling is used between all the optical stages to ensure the highest efficiency and simplicity of assembly.

This approach has certain advantages, such as:

1. High degree of multiplexing. One image intensifier/CCD channel handles up to 10^5 fibres.

2. Good physical match between narrow diameter (25 μm) SCIFI, image intensifier resolution (50 μm spot diameter) and typical CCD pixel size (20 × 20 μm^2). (SCIFI of larger diameter may be read out with a demagnifying image intensifier.)

3. Low geometrical distortions.

However, there are also some drawbacks associated with image intensifiers, such as:

a) Rate limitation. The fastest phosphor (P47, Ref. 42) has a time constant τ = 35 ns. In a three-stage tube this implies a minimum integration time of 100-150 ns.

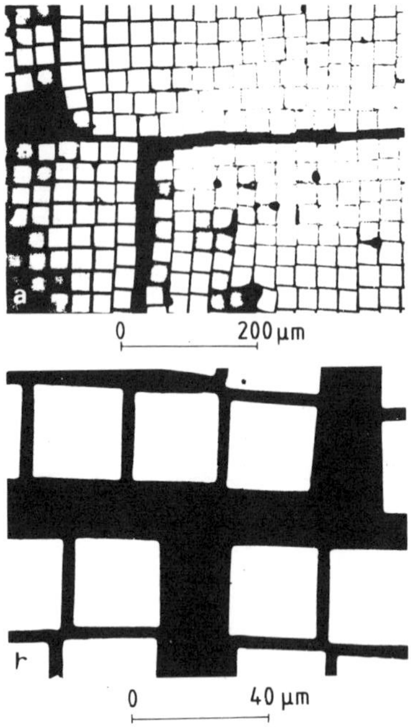

Fig. 18. High magnification views of individual glass fibres from target 2 (Ref. 10). The transverse dimension of a core (bright square) is 26 µm and the centre-to-centre spacing between fibres is typically 31 µm.

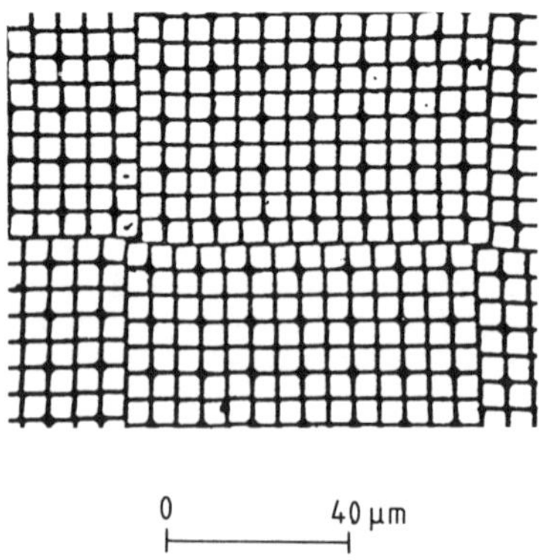

Fig. 19. Microphotograph of a fibre optic plate used in image intensifiers. The centre-to-centre fibre spacing is 5.5 µm.

b) Finite lifetime. Both the MCP and the photocathode deteriorate with time, due to ion bombardment. The gain lifetime (1/e) is typically limited to a charge of $\leq$ 0.05 Coulomb cm^{-2} on the phosphor. (An ion barrier film over the MCP input face can increase this value by an order of magnitude.) These charge limitations can readily be exceeded over the duration of an experiment unless careful attention is paid to gating and minimization of system gain.

c) Cost. The minimum cost of a single image intensifier system suitable for single photon detection is ~ 30 KSF.

d) A certain 'touchiness' in operation. This is a subjective but nevertheless valid comment. It is not unrelated to item c).

In addition to the standard image intensifier/CCD readout chain there are a great variety of electronic imaging systems which could be adapted to SCIFI readout. The most useful is probably the SIT (silicon-intensifier target) tube preceded by suitable image intensification, known as the ISIT. The SIT tube accelerates electrons from a photocathode to about 10 keV before they strike a silicon wafer target. In this way the tube has a charge gain of ~ 1000. The silicon target is a diode array of 14 x 14 μm^2 pixels which can be read out in a few msec by means of an electron beam scanning the back face.

A recent development of multi-anode phototubes by Hamamatsu[43] (R2594-95 and R2486-89) and by Philips[44] (XP4702) provides a potentially attractive future alternative to the readout of moderate diameter ($\geq$ 1 mm) SCIFI. As an example, the Philips tube has an array of 64 (8 x 8) anode pads of spacing 2.54 mm. The dynode structure is formed by a set of thin parallel electrodes spaced by 0.2 mm and containing aligned holes of 300 µm diameter with a centre-to-centre spacing of 770 µm. The device is very compact: 35 mm deep x 50 mm diameter. Initial measurements[45] show a gain of 10^7 and low (< 5%) cross-talk with the nearest neighbour. After further development this class of device, with suitable VLSI electronics, offers the future possibility of a robust, high-speed readout system for SCIFI imaging calorimetry or moderate resolution SCIFI tracking. Finally, for readout of coarse SCIFI calorimetry, the traditional phototubes or Si diodes can be used.

We will return to the readout devices in Section 10 where we look at their operation in a supercollider detector. However we will now describe two present readout systems which have been used to obtain some of the experimental SCIFI results of the following section.

5.2 Examples of SCIFI readout systems

The UA2 SCIFI detector[6] readout is shown in Fig. 20. It involves a three-stage image intensifier[46] as follows:

1. II1: electrostatic focus, 80 mm input diameter, 18 mm output diameter, S20 photocathode, P47 phosphor.

2. II2: gated proximity focus MCP, 18 mm diameter, S20 photocathode, P47 phosphor.

3. II3: electrostatic focus, 18 mm input diameter, 7 mm output diameter, S20 photocathode, P46 phosphor.

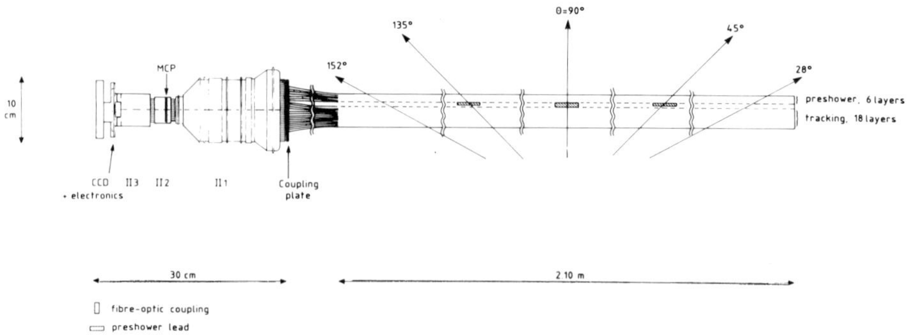

Fig. 20. Geometry and readout system of the UA2 plastic SCIFI detector, SFD (Ref. 6).

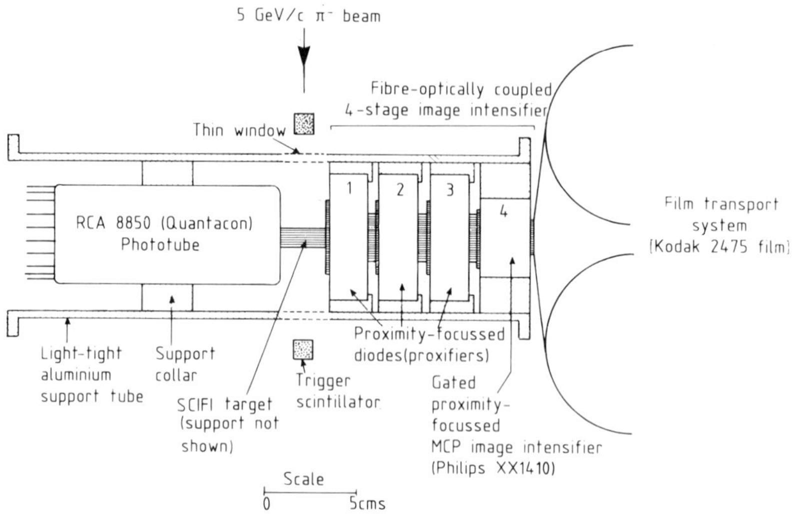

Fig. 21. Geometry and readout system of a glass SCIFI target (Ref. 10). The original film recording medium (as shown) is replaced by a CCD in later tests.

The overall linear demagnification is 0.08 and the operational photon gain is 0.4 10^5. The gain lifetime of tube II2 was measured to be an integrated charge of 5.5 x 10^{-2} Coulombs cm^{-2} at the phosphor. The decrease in gain was found to be caused by equal losses in the photocathode quantum efficiency and in the gain of the MCP. The CCD is a Thomson TH 7852 [47] (active area 4.3 x 5.8 mm^2 and pixel size 30 x 19 μm^2). This project has included the important development[48] of the fast clear (1 μs) of the CCD using the antiblooming diodes adjacent to each pixel.

The CERN-MPI-RAL[10] SCIFI readout is shown in Fig. 21. This involves a four-stage (for film recording) or three-stage (CCD recording) image intensifier as follows:

1. Stages 1-2-(3): proximity focussed diodes[49], 25 mm diameter, bialkali or S20 photocathodes, P20 or P11 phosphors.

2. Final stage: gated proximity focus MCP, 18 mm diameter, S20 photocathode, P20 phosphor.

The system has unit magnification and an overall photon gain of $\sim 10^5$ (three stage). The CCD is a Thomson TH7861 (active area 6.6×8.8 mm^2 and pixel size 23×23 μm^2).

The CCD data acquisition system is shown schematically in Fig. 22. A local preamplifier and control unit clocks analogue data out of each pixel sequentially. This data is fed to a 'frame grabber' or 'frame storer' which digitizes the data in a FADC, suppresses low channels and stores the pulse heights and addresses of the remaining channels into RAM. Typically a clock frequency of 10 MHz is used and the readout of a single frame takes $\sim$ 10 ms.

6. EXPERIMENTAL STATUS OF SCIFI TRACKING DETECTORS

6.1 Plastic SCIFI: Cambridge-CERN-Orsay-Saclay-UA2

A large SCIFI detector (the 'SFD')[6] is under construction as part of the UA2 upgrade programme for the SPS Collider with ACOL. A SCIFI detector was chosen since it provides compact tracking and pre-shower information while allowing sufficient room for a transition radiation detector in the restricted space of the UA2 inner detector. The device involves plastic fibres[3] of diameter 1 mm and length $\sim$ 235 cm arranged into a cylindrical detector of length 210 cm and mean radius 41 cm. The EMA for each fibre is provided by a sputtered aluminium layer of thickness 100 nm. The fibres are oriented into stereo triplets (with angles 0, $\pm\alpha$ relative to the beam axis); 6 triplets (with α = 275 mr) precede a 1.5 X_0 Pb converter sheet and 2 triplets (with α = 367 mr) follow, to provide a total of 24 layers of fibres. The operational mean detection efficiency per single fibre layer is 91% for minimum ionizing particles which cross at 90°. This corresponds to 3.8 pe mm^{-1} at the photocathode of the first image intensifier stage. The SFD involves in total 60 K fibres (140 km) which are read out with 32 image intensifier/CCD units. At the (80 mm diameter) entrance face of the image intensifier chain (Section 5.2) each fibre is inserted into a drilled plate with holes of 1 mm diameter spaced by 1.3 mm between centres. The system will be triggered at 2.6×10^5 Hz and can be read out at up to 200 Hz.

The SFD prototype is producing some very nice results as we shall now see. Some typical (40 GeV) electron events are shown in Fig. 23.

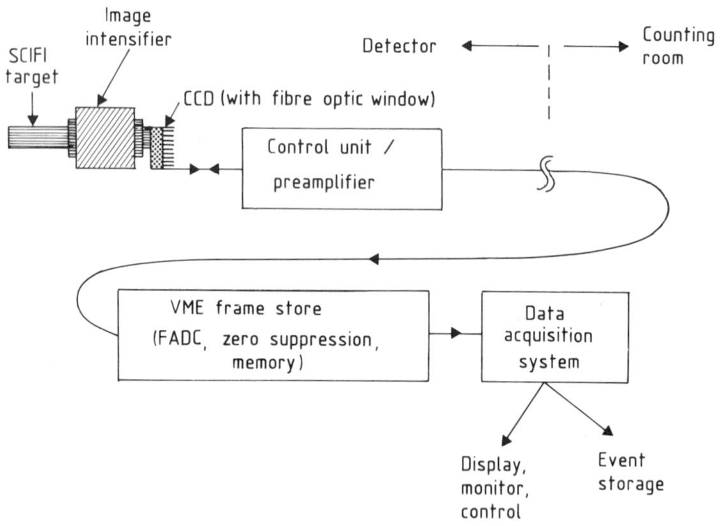

Fig. 22. CCD readout system.

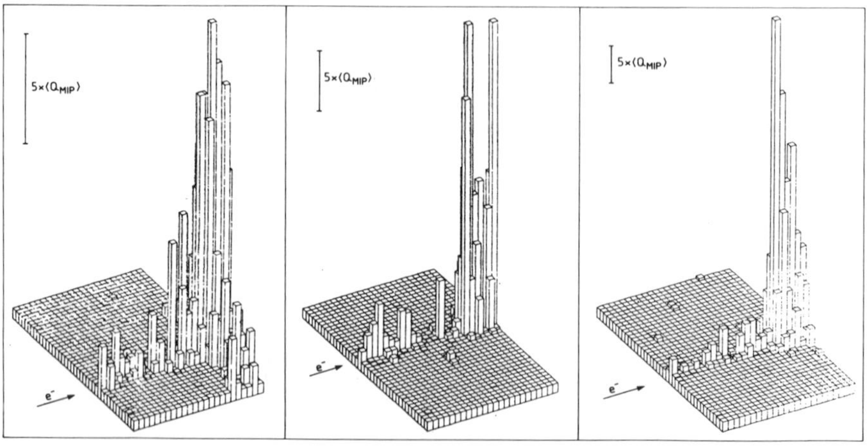

Fig. 23. Typical single events of 40 GeV/c electrons in the UA2 SFD prototype (Ref. 6). Each pixel represents 1 x 1 mm², equivalent to a single fibre. A 1.5 X_0 Pb converter (not represented) separates the first 18 ('tracking') and last 6 ('preconverter') fibre layers.

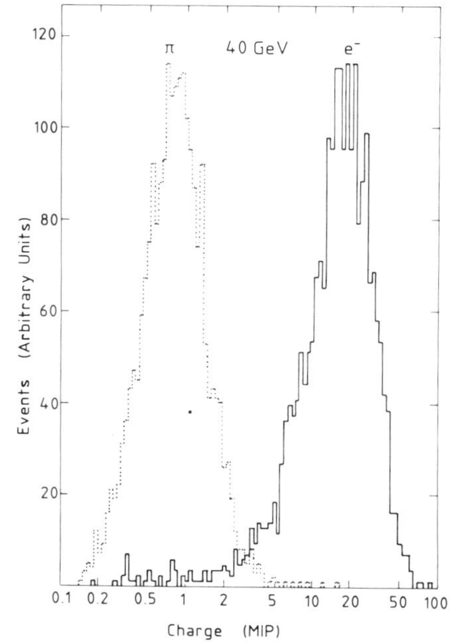

Fig. 24. The total pulse height in the 6 preconverter fibre layers of the SFD for 40 GeV π and 40 GeV e (Ref. 6). The electron data have a small π contamination of ~ 1%.

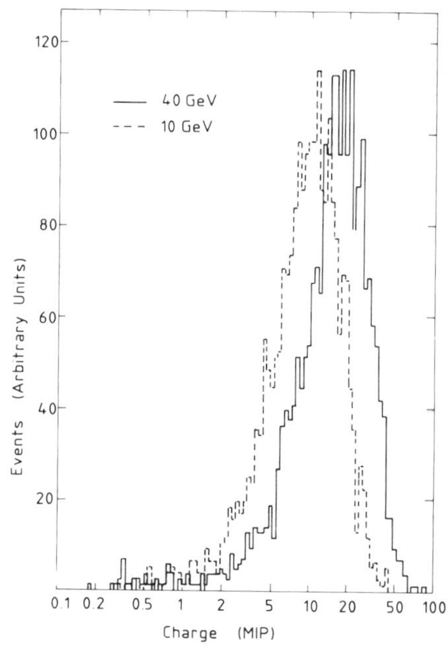

Fig. 25. The same as Fig. 24 but showing 10 and 40 GeV e (Ref. 6).

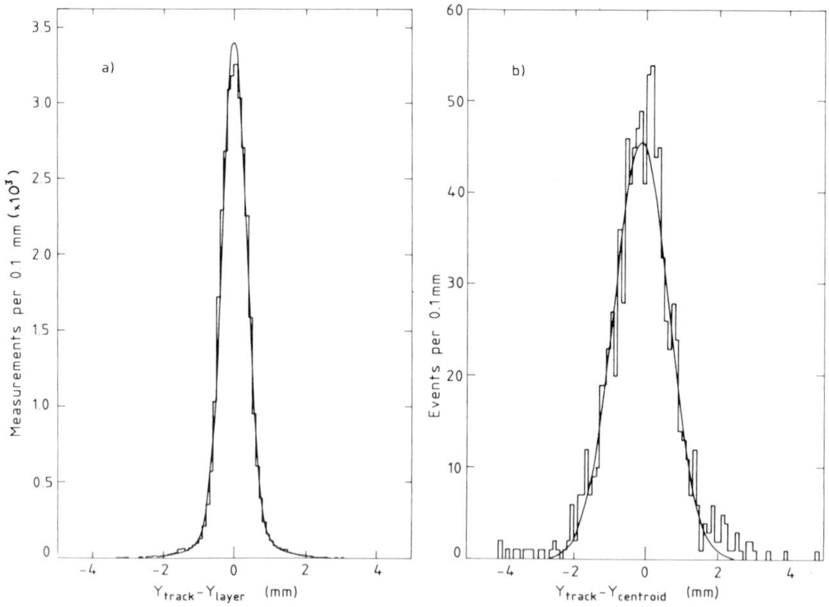

Fig. 26. a) The track residuals in the first 18 tracking layers of the UA2 SFD prototype (Ref. 6). The distance between each spot centroid and the fitted track is plotted. The data are fitted with a central Gaussian, of σ = 0.35 mm (94%) and two satellite Gaussians of σ = 0.58 mm (2 x 3%).
b) The shower location precision for 40 GeV e. The horizontal axis is the difference between the impact point in the Pb converter, as determined by the first 18 layers, and the centroid of the shower pulse height distribution seen after the Pb converter, in the final 6 layers. The fitted curve is a Gaussian of σ = 0.76 mm.

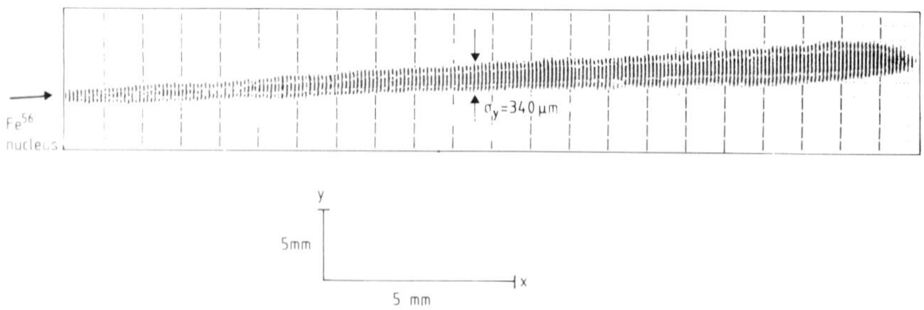

Fig. 27. A stopping ^{56}Fe nucleus in the plastic SCIFI detector of Ref. 7. The non-EMA fibres are square with a centre-to-centre spacing of 100 μm. The observed broadening of the track is largely due to UV cross-talk (see Fig. 6b).

These events clearly demonstrate the suitability of SCIFI to imaging calorimetry. The sensitivity can be easily understood. A high energy electron radiates $1 - e^{-1.5} = 78\%$ of its initial energy in crossing a 1.5 X_0 converter. This appears in a characteristic soft (1/E) spectrum of photons which, in turn, generate electrons by Compton scattering, pair production, etc. A SCIFI array has an excellent sensitivity to this soft cascade since the energy threshold is very low (1 pe per ~ 35 keV deposited energy).

The clear separation between π and e responses in the preconverter fibres is shown in Fig. 24. After correcting for beam contamination, the πe rejection at 40 GeV is measured to be 7.3×10^{-2} at an electron detection efficiency of 98%. The performance is essentially unchanged at 10 GeV (Fig. 25): a πe rejection of 7.3×10^{-2} at an electron detection efficiency of 96%.

The measurement precision of the SFD is illustrated in Fig. 26a. The track residuals in the first 18 layers have $\sigma = 0.35$ mm. The fitted curve also includes two satellite Gaussians ($\sigma = 0.58$ mm) which reflect a $2 \times 3\%$ probability of cross-talk to a neighbour in the readout system. In order to obtain these results the positive (pincushion) distortions in the electrostatically-focussed image intensifier system had first to be corrected. The maximum magnitude of the distortion is 13%: 1 mm in space (i.e. 1 fibre diameter) is equivalent to 70 μm at the centre of the CCD and to 80 μm at the edge of the CCD. The distortions are effectively corrected using a magnification proportional to a polynomial in radius (R mm) at the input to the image intensifier of the form: $r = 0.070\ R\ (1 + 2.0 \times 10^{-6}\ R^3)$, where r (mm) is the radius measured at the output by the CCD. This function is determined by means of a matrix of illuminated fiducial fibres.

Finally the shower location precision is indicated in Fig. 26b. This shows a transverse matching precision $\sigma = 0.76$ mm between a candidate electron and a shower. This precision is necessary to reject false electron signals due to the overlap of a π and γ.

6.2 Plastic SCIFI: Washington University, St. Louis

A SCIFI range detector[7] for heavy nuclei has been developed jointly by Washington University and Fibre Optic Development Systems, Inc.[50]. The detector is made from $100 \times 100\ \mu m^2$ square fibres of polystyrene scintillator (KST1-415), with an acrylic cladding of thickness 7 μm and no EMA. These fibres are arranged into coherent bundles of cross-section $2.7 \times 2.7\ cm^2$ and length 30 cm. The readout chain is as follows: fibre optic reducer (40 mm to 18 mm diameter), dual MCP image intensifier (photon gain up to 10^6), fibre optic reducer (18 mm to 11 mm diameter) and finally a CID readout sensor (pixel dimensions $23 \times 27\ \mu m^2$).

A typical event is shown in Fig. 27. Near to the end of the nuclear range the optical FWHM of the track is approximately 8 fibres, i.e. 800 μm, whereas the physical region of energy deposition is less than 1 fibre width. This is an experimental demonstration of the broadening process shown in Fig. 6b and a measurement of the absorption length of the waveshifter in this plastic scintillator. Although the fibres were found to have an attenuation length of only 11 cm, recent improvements of the procedure and of the scintillator have increased this to 60 cm.

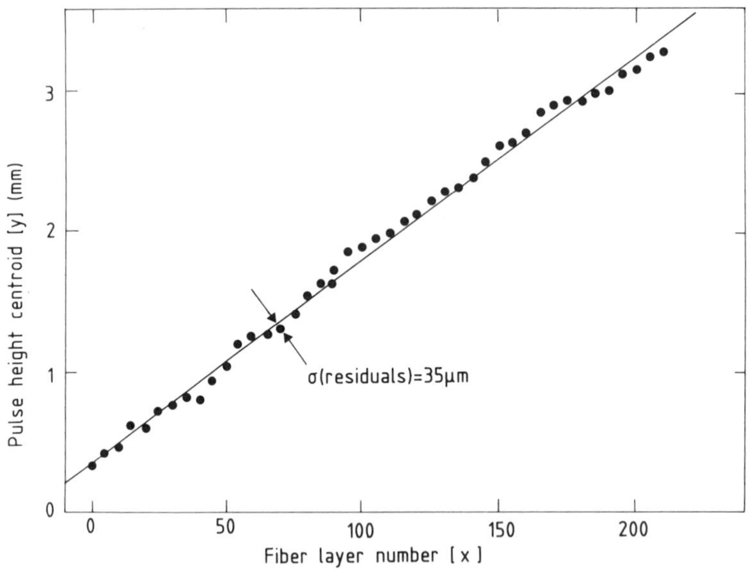

Fig. 28. Track residuals for a single ^{56}Fe event (Ref. 7). The transverse pulse height centroid (y) is plotted vs. fibre layer number (x). Each fibre layer represents 0.1 mm. The residual error is σ = 35 μm.

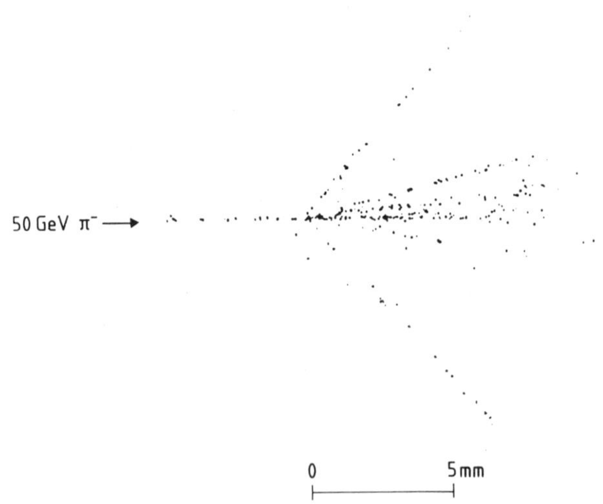

Fig. 29. Photograph of a 50 GeV π^- interaction in a glass SCIFI target (Ref. 5).

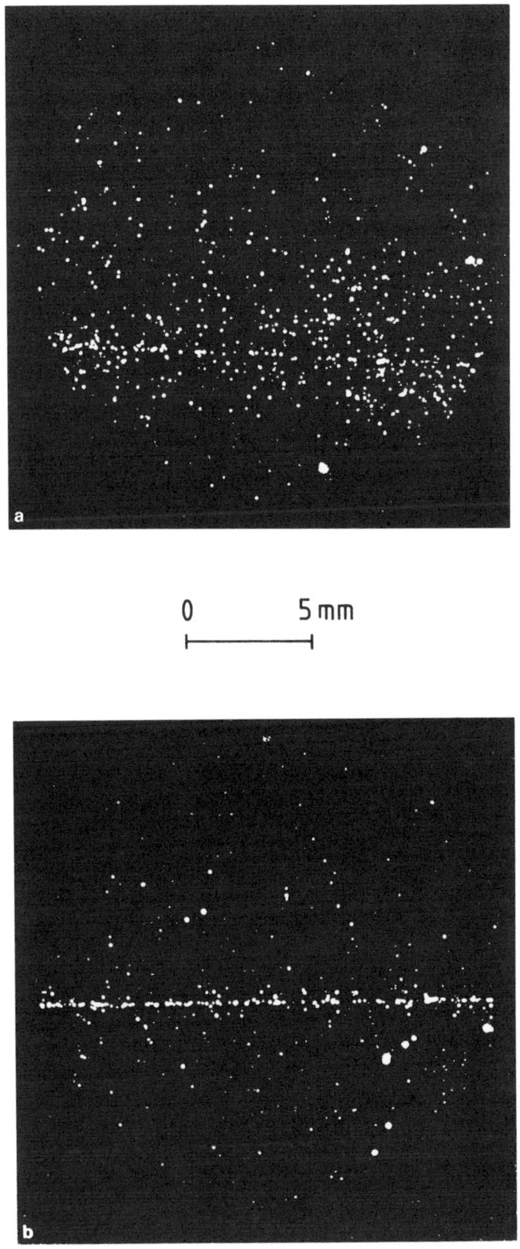

Fig. 30. Minimum ionizing tracks recorded on film using the glass SCIFI targets shown in Figs. 17 and 18 (Ref. 10): a) target 1 (no EMA) and b) target 2 (with EMA).

The track residuals of a single ^{56}Fe event are shown in Fig. 28. The small r.m.s. residual error (σ = 35 μm) is a consequence of the large energy deposition. There may be evidence in Fig. 28 for a further potential improvement if the longitudinal correlations are due to fibre incoherency and not a smoothing due to fibre cross-talk.

6.3 Glass SCIFI: Notre Dame-Fermilab

A SCIFI active target has been developed[5] for Fermilab experiment E687 to observe the photoproduction and decay of heavy-flavoured particles. The target is made[41] from 25 x 25 μm^2 square fibres of GS1 glass[40] with N51A cladding and an EMA. The readout involves a two-stage electrostatic focus image intensifier followed by a gated proximity focussed MCP stage. The overall photon gain is 2 x 10^5. The output is fibre-optically coupled to a Fairchild CCD 3000F CCD camera.

An interaction recorded on photographic film (Fig. 29) shows the potential for the high resolution SCIFI detector to become an 'electronic' bubble chamber. The observed spot frequency for minimum ionizing particles is 4 mm^{-1} in a 1 cm deep target with the opposite face polished and aluminized. The track residuals show a r.m.s. deviation of 28 μm.

6.4 Glass SCIFI: CERN-MPI(Munich)-RAL

This group has been developing[10] the application of glass SCIFI to active targets and compact, precise tracking in collider detectors. The targets[40,41] and readout are already described in Sections 4 and 5.2. The essential features are the fibre size, 25 x 25 μm^2, and length, 25 mm.

Examples of some events recorded with photographic film (Kodak 2475) are shown in Figs. 30 and 31. These photographs dramatically demonstrate the importance of an EMA to absorb the untrapped scintillation light.

Measurements of the spot size using photographic film (Fig. 32) and CCD (Fig. 33) indicate a FWHM of 52 ± 4 μm. The spot size is found to be only weakly dependent on its total brightness (Fig. 34). This behaviour is desirable since it implies that the two track resolution of 52 μm will not be strongly affected by increases in the gain of the readout system.

An example of the track residuals distribution is shown in Fig. 35. The residuals distribution has been investigated at various depths onto the target. The results (Fig. 36) show a constant r.m.s. residual, σ = 22.3 μm. This indicates that the fibres are parallel to better than one fibre diameter over the length of the target, in agreement with optical measurements (Figs. 17 and 18).

The light yield and attenuation length has been measured using the gap length distribution (Fig. 37) familiar in bubble chamber analyses. The results (Fig. 38) show a yield of 2 pe mm^{-1} at zero length and an attenuation length of 29 mm. This initial short attenuation length is confirmed by UV measurements made on long individual multi-fibre bundles (Fig. 39). At longer distances the fibres display an attenuation length of ~ 25 cm. This figure sets a lower limit on the attenuation length caused by scattering losses in the fibres, as opposed to absorption losses.

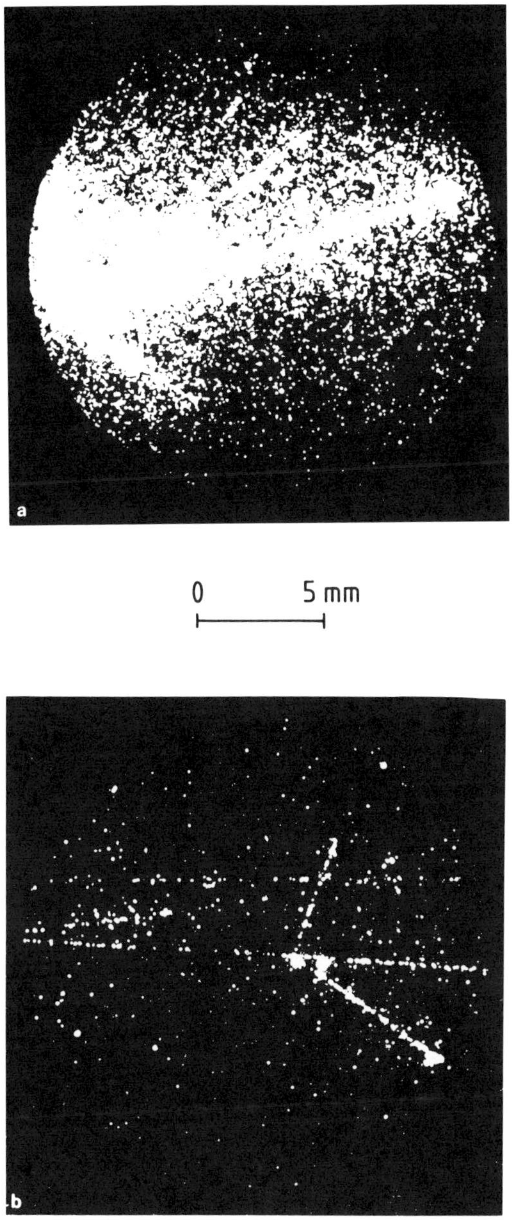

Fig. 31. Interactions of 5 GeV/c π^- recorded on film using the glass SCIFI targets shown in Figs. 17 and 18 (Ref. 10): a) target 1 (no EMA) and b) target 2 (with EMA).

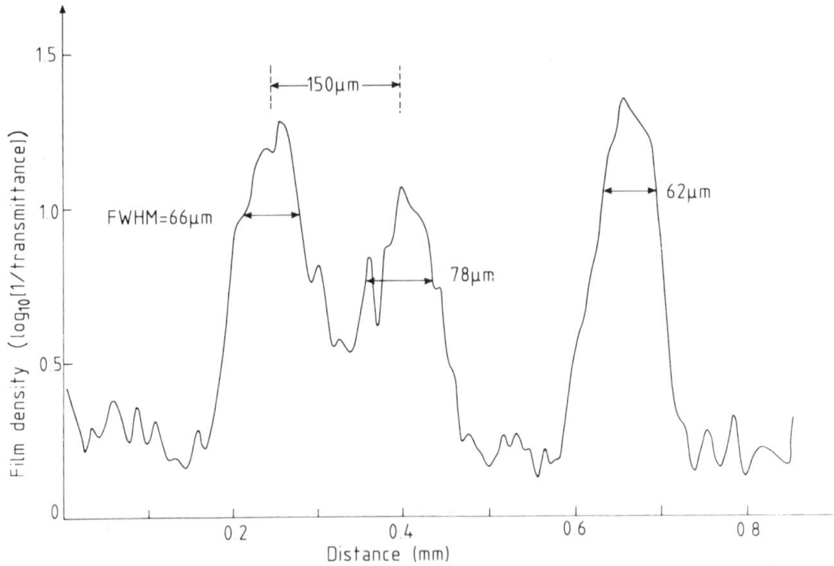

Fig. 32. Densitometer measurement through the centre of three spots along a track recorded on film in glass SCIFI (Ref. 10). The slit width of the densitometer (along the track direction) is 10 µm.

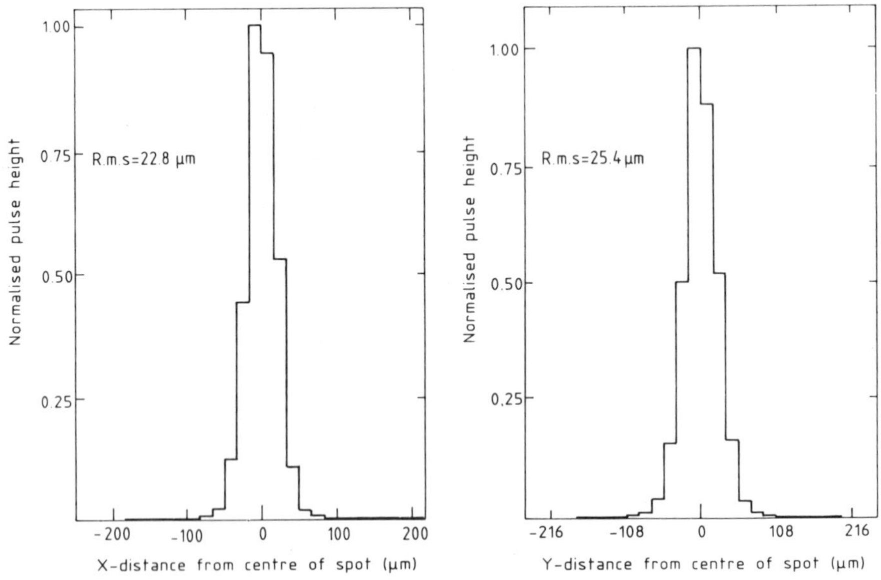

Fig. 33. Spot profiles obtained in glass SCIFI (Ref. 10). Figures 33–38 are results obtained in a 5 GeV π^- beam using the glass SCIFI target 2 (Figs. 17 and 18) with CCD readout.

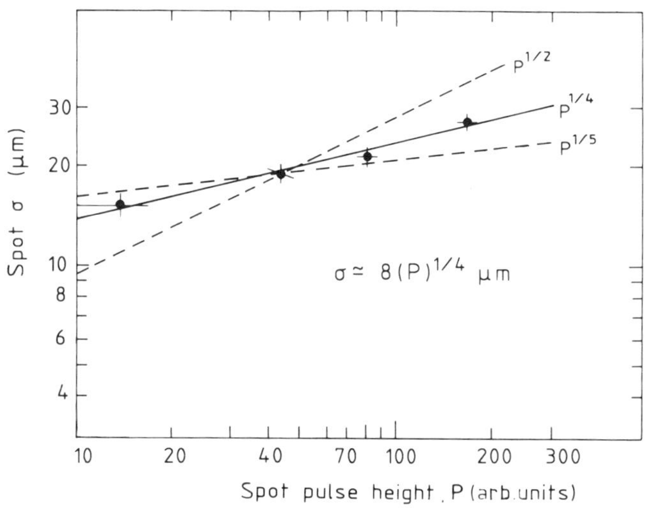

Fig. 34. Spot size vs. pulse height in glass SCIFI (Ref. 10).

SECTION II: TOMORROW .....THE SUPERCOLLIDER

7. THE CHALLENGE OF THE SUPERCOLLIDER ENVIRONMENT

The next threshold in our understanding of the laws of nature is widely expected to lie in the energy range below 2 TeV. This requires an e^+e^- machine with $\sqrt{s}$ = 2 TeV or a hadron collider with $\sqrt{s} \geq$ 12 TeV and, for both machines, a luminosity, L = 10^{33} cm^{-2} s^{-1}. Proton-proton colliders apparently can be built to these specifications using present technologies. The question is: Can the experimenter work in such a harsh environment?

The hadron supercollider environment (Table 6) is characterized by:

1) High energy
- Multi-TeV individual particles.
- Multi-TeV jets, with a high multiplicity and high density near the jet core (Fig. 40).

2) High luminosity
- > 10^8 Hz event rate.
- Multiple interactions per crossing.
- High particle fluxes (cm^{-2} s^{-1}).
- High radiation doses to the detector.

Almost all current detectors would suffer serious limitations in this environment. For this reason we must not only extend the performance of existing detectors but also explore alternatives and new technologies in search of a breakthrough. It is the aim of this section to explore the possibilities for SCIFI in tomorrow's supercollider detector.

8. THE SCIFI MICROTRACKER CONCEPT

8.1 Is tracking necessary?

It is worthwhile first to examine the need for tracking in a supercollider detector. It is easy to see that early experiments or those at ultra-high luminosities (10^{34} cm^{-2} s^{-1}) will include 'beam-dump' styles of detectors, which include only calorimetry or a passive absorber followed by a muon spectrometer. It is equally easy to see that, without an inner tracking device, the overall performance of a detector will be substantially reduced, and a great deal of physics will be inaccessible or ambiguous. Tracking (in many cases associated with a magnetic field) provides the following:

1. Electron identification and measurement (charge, p/E match and elimination of backgrounds such as $\pi\gamma$ overlaps or e^+e^- pairs).

2. Improved muon identification and measurement (track vector before the absorber and p_{inner}-p_{outer} match to reject hadron decays and punch-throughs).

3. Measurement of jet charged multiplicity. This is important, for example, in helping to tag the presence of a W [52]. A 1 TeV W still has $\langle n_{ch} \rangle \sim 17$ but a qg di-jet of 1 TeV energy has $\langle n_{ch} \rangle \sim 75$.

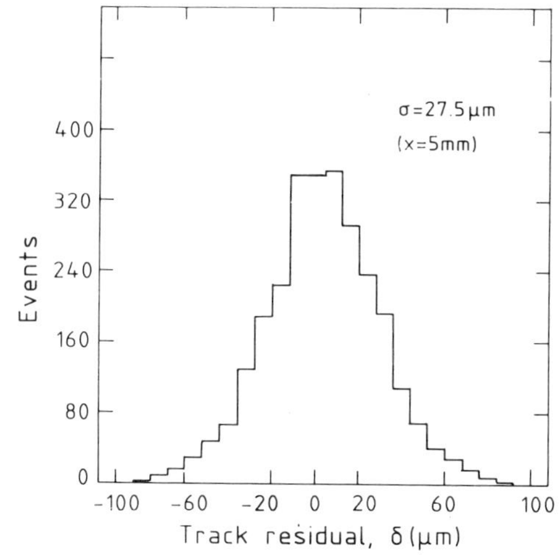

Fig. 35. Track residual distribution in glass SCIFI (Ref. 10).

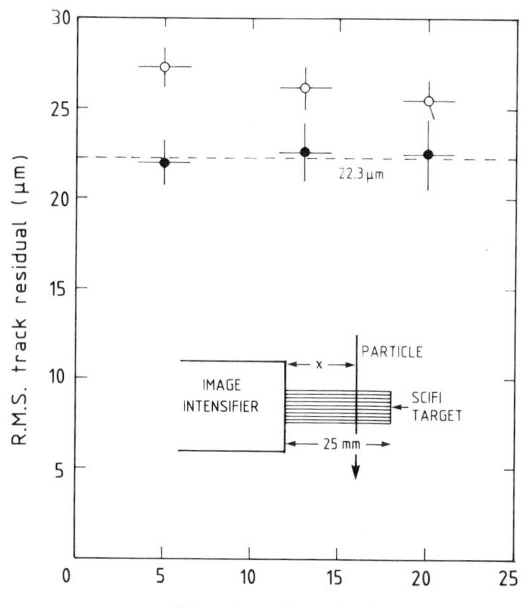

Fig. 36. R.m.s. track residuals vs. depth (x) into the glass SCIFI target (Ref. 10). The black dots show the original data (open dots) after correcting for background tracks.

Table 6

The supercollider experimental environment

	CLIC	LHC		SSC
Beam particles	e^+e^-	pp		pp
Luminosity	10^{33}	$1.4 \times 10^{33} \rightarrow$	10^{34}	10^{33} cm^{-2} s^{-1}
$2 \times E_{beam}$	2	16	16	40 TeV
Bunch spacing	170 μs	25	5	16 ns
Interaction rate	10^{-3} *	$1.4 \times 10^{8} \rightarrow$	10^{9}	10^{8} Hz
$\langle \text{Interactions} \rangle$ crossing^{-1}	$1.7\ 10^{-7}$ *	3.5	5	1.6
$\langle n_{ch} \rangle$ event^{-1} ($p_T > 0.5$ TeV/c)	~ 60	~ 100	~ 100	~ 100
$\langle n_{ch} \rangle$ cm^{-2} in jet core (r = 10 cm)	–	10	10	10
$\langle n_{ch} \rangle$ crossing^{-1}	–	~ 200	~ 260	~ 120
$\langle n_{ch} \rangle$ cm^{-2} s^{-1} (r = 10 cm)	–	$1.4 \times 10^{6} \rightarrow$	10^{7}	10^{6}
Radiation dose yr^{-1} (r = 10 cm)	–	10^{6}	10^{7}	10^{6} rads

* Single photon annihilation.

4. Tagging of heavy flavour production by the presence of a secondary decay vertex.

5. Measurement of z vertex positions and correct association of calorimeter energy with the corresponding signal or background vertex.

6. Independent measurement of jet directions and charged energy (redundancy, calibration and cross-check of calorimeter measurements, especially for rare events where the maximum event information is needed).

8.2 The need for a new approach

Present high-energy collider detectors, such as DELPHI or ALEPH, utilize giant superconducting solenoids of typical dimensions 7 m (length) x 5 m (diameter) and field strengths 1.2-1.5 T. The main factor which leads to these large dimensions is the requirement to measure momenta up to ~ 100 GeV/c. In order to achieve this goal, the central trackers, which comprise drift chambers or TPCs with measurement

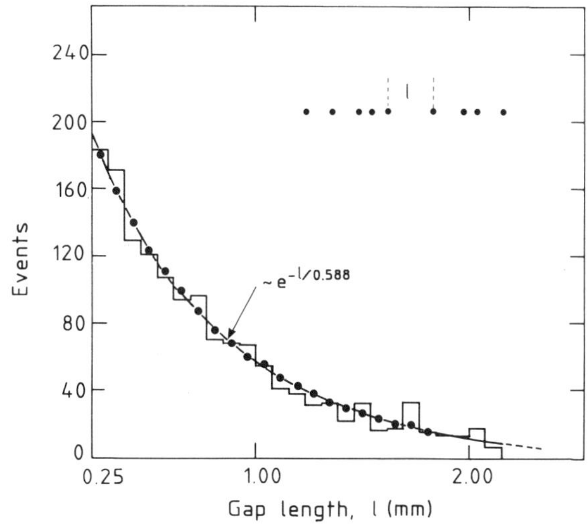

Fig. 37. The gap length distribution which is used to extract the spot frequency in glass SCIFI (Ref. 10).

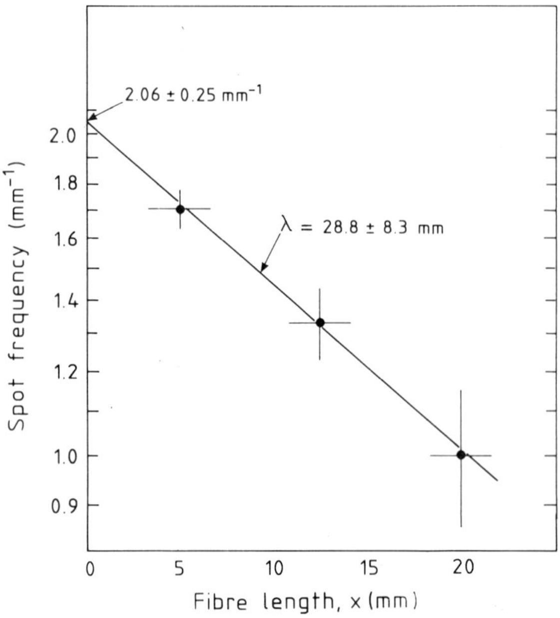

Fig. 38. The spot frequency vs. fibre length in glass SCIFI (Ref. 10).

precisions of 100–200 μm, must extend over a radial distance of 1–2 m. It is the large size of these inner systems which sets the enormous scale and cost of the overall detector. Future detectors which adopt this approach for the supercollider, where the momenta will be ≤ 1 TeV, will require even larger solenoids.

We argue that the combination of two new technologies can provide a dramatic reduction in the scale of future collider detectors. These technologies are as follows:

1. Precise micro-trackers such as SCIFI or Si micro-strips.
2. Superconducting solenoids with ultra-high (6 T) magnetic fields.

We can estimate the potential reduction in detector size. The momentum accuracy (σ_p/p) above a few GeV/c is proportional to $\sigma_s p/BL^2$, where σ_s is the sagitta measurement error, B is the field strength and L is the tracking lever arm. We will take the following representative values:

1. Today's detector (TPC or DC tracking): σ_s = 100 μm
B = 1.5 T
L = 100 cm
2. Tomorrow's detector (SCIFI tracking): σ_s = 5 μm
B = 6 T
L = L_{new}

In order to achieve the same performance for σ_p/p, tomorrow's detector needs a tracking lever arm of:

$$L_{new} = 100 \text{ cm} \sqrt{[\frac{5 \text{ μm}}{100 \text{ μm}} \times \frac{1.5 \text{ T}}{6 \text{ T}}]} = 11 \text{ cm} ,$$

i.e. a reduction by almost one order of magnitude.

A similar calculation can be performed for the low-energy region. Here multiple coulomb scattering (mcs) dominates the momentum precision and σ_p/p (mcs) $\propto \sqrt{X}/BL$, where X is the tracker thickness between the first and last measurement planes, expressed in radiation lengths (r.l.). We will assume for today's detector X = 2.5% r.l. and for tomorrow's detector X = 2.0% r.l. [equivalent to an 8 mm plastic scintillating fibre 'superlayer' or 6 layers of silicon]. This indicates a reduction in size by a factor of 4 is possible at constant σ_p/p (mcs).

We can summarize as follows:

- The combination of a precise SCIFI micro-tracker with an ultra high field superconducting solenoid can lead to a reduction in linear dimension of the central tracker by about one order of magnitude, while maintaining the same momentum measurement precision.
- This will result in a substantial reduction in the cost and size of subsequent detector elements (calorimeter solenoid, muon spectrometer, etc.).
- Furthermore the overall performance of the detector will be significantly improved with respect to:

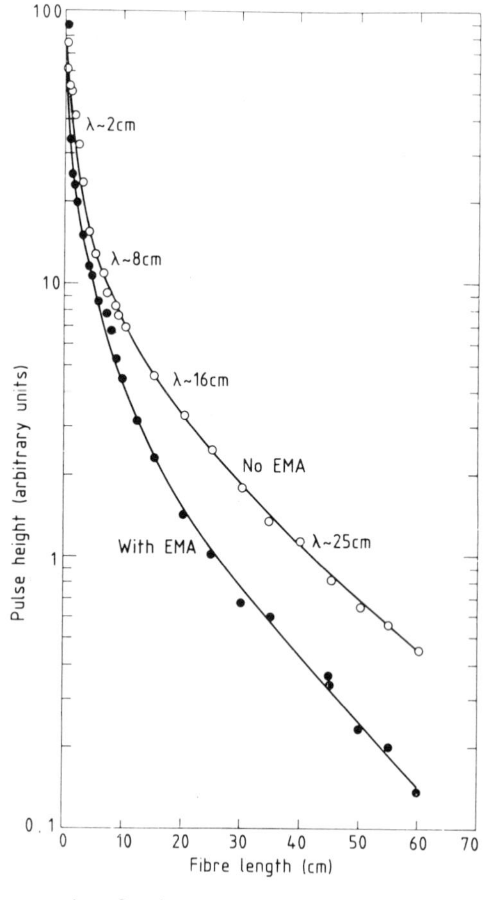

Fig. 39. Measurement of the attenuation of scintillation light in single multifibre bundles manufactured during production of the glass SCIFI targets 1 ('No EMA') and 2 ('With EMA'), shown in Figs. 17 and 18 (Ref. 10). The fibres are excited by a UV source.

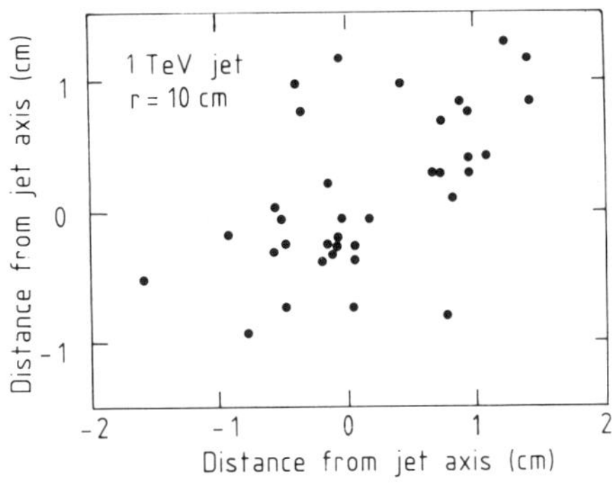

Fig. 40. Transverse distribution of charged particles in the core of a 1 TeV jet, at a distance of 10 cm from the interaction point (Ref. 51).

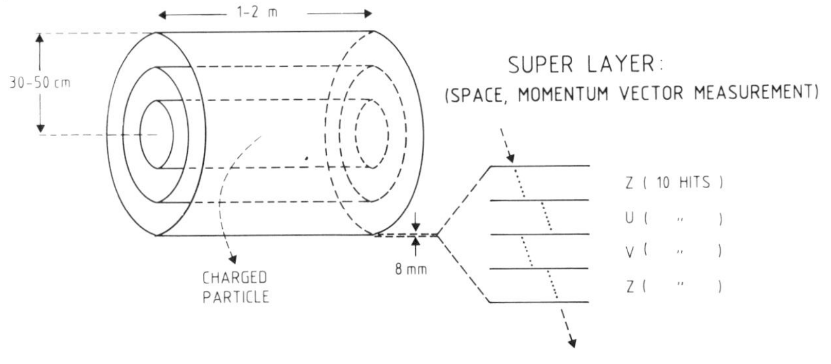

Fig. 41. Geometry of a SCIFI micro-tracker for a supercollider detector.

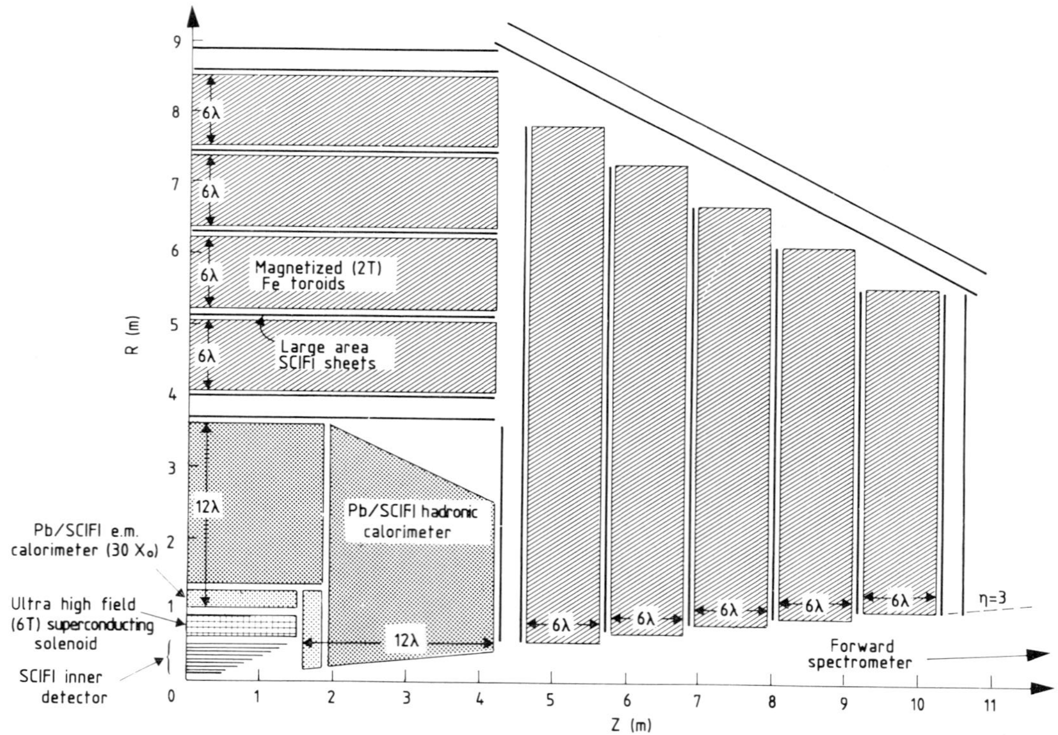

Fig. 42. ALFIE: All Fibre Experiment for a supercollider. The dimensions are chosen for operation at $\sqrt{s}$ = 20–40 TeV.

Table 7

SCIFI micro-tracker performance

Number of hits per track	120
σ_{xy} hit^{-1}	15 μm
Two-hit resolution	50 μm
σ_{xy} (impact parameter at origin)*	< 20 μm
σ_z (vertex) track^{-1}	100 μm
$(\sigma_p/p)^2$	$[5 \times 10^{-4}\ p\ (\text{GeV}/c)]^2 + (0.01)^2$

* p > 20 GeV/c

a) Calorimetry, since a deep uniform calorimeter and expensive materials become economically practical.

b) Electron identification, by the detection of synchrotron photons[51]. The key advantages of a strong B field are that the energy loss to synchrotron photons $\propto B^2LE_e^2$ and the mean photon energy $\propto BE_e^2$, where E_e is the electron energy. This is discussed further in Section 9.5.

8.3 Design criteria

The supercollider environment of high energy and high luminosity places the following requirements on a charged-particle tracking device:

1. Precise measurement accuracy (to measure momenta and secondary decay vertices).

2. Fine two-hit resolution/high granularity (to resolve particles in jets, to handle multiple events and multiple z vertices and to identify e^+e^- pairs).

3. Radial compactness/high density of measurements (hits) per unit path length (to reduce the overall cost and to maximize the performance of subsequent detector components).

4. Radiation hardness (both of the device and the local electronics).

5. Fast response/high rate capability.

6. Good pattern recognition capability (to minimize the computer time for event reconstruction).

SCIFI tracking is potentially capable of meeting all these design criteria, as we discuss below.

8.4 Design and performance

A SCIFI micro-tracker (Fig. 41) is built of three or more concentric cylindrical 'superlayers' constructed from coherent arrays of narrow (25 μm) diameter fibres. The outer radius is only 30-50 cm. Each superlayer consists of 4 layers, each 2 mm thick, made from fibres arranged in the directions z-u-v-z. The u and v fibres form a narrow angle stereo of ±50 mrad with respect to the z fibres, which lie parallel to the beam axis. The perpendicular material in each superlayer is 2% X_0 (plastic) - 8% X_0 (glass). With a yield of 5 hits per mm there are a total of 40 hits per track in each superlayer.

This geometry results in excellent track reconstruction since each superlayer measures local track vectors in space (three-dimensional position and angle) and in momentum (by impact parameter relative to the beam axis). For example, inside a superlayer the azimuthal angle precision, σ_ϕ = 0.6 mr, and the momentum precision, σ_p/p < 1% p. Unambiguous linkage of the track vectors between superlayers can therefore be made. In addition the vector stubs of (uninteresting) low energy tracks can be rejected at an early stage of event reconstruction.

In association with an ultra high field (6 T) solenoid, a compact SCIFI tracking system can achieve remarkable performance (Table 7). With a lever arm of 30 cm and assuming a measurement accuracy per hit σ = 15 μm and a systematic sagitta error of 10 μm, the momentum precision is

$$(\sigma_p/p)^2 = [5 \times 10^{-4}\ p(\text{GeV}/c)]^2 + (0.01)^2 .$$

This is sufficient for sign measurement at 1 TeV/c.

9. ALFIE: THE ALL FIBRE EXPERIMENT

9.1 Overview

SCIFI has important potential applications in a supercollider detector beyond the micro tracking system which we have just described. In order to illustrate these possibilities we have assembled a supercollider detector named ALFIE (All Fibre Experiment), based on the concepts presented in the previous section, which is completely read out with plastic SCIFI. The components of this detector (Figs. 42 and 43) are as follows:

1. A compact central detector, located inside an ultra high field superconducting solenoid of inner radius 60 cm, which comprises of:

   a) Micro-tracker.

   b) Charge-flow detector.

   c) z-vertex detector.

   d) Micro-converter.

2. A Pb/SCIFI electromagnetic and hadronic calorimeter.

3. A muon spectrometer comprising of magnetized iron absorber and large area SCIFI sheets.

We will now discuss each of these components in more detail.

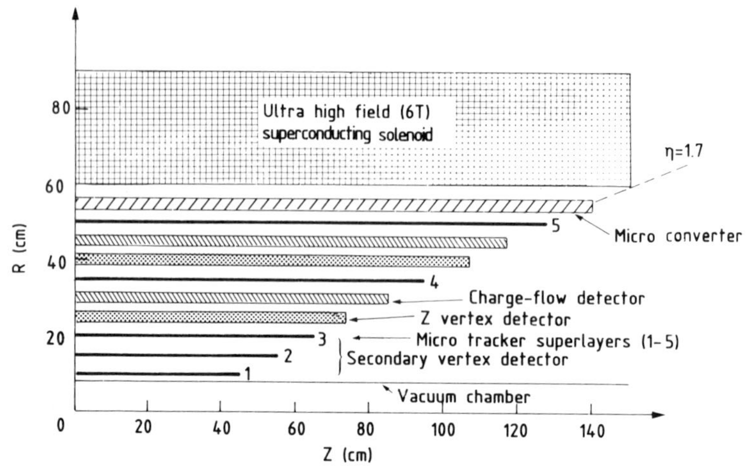

Fig. 43. The inner detector of ALFIE. Further representations of each element are shown in Figs. 41, 44-46.

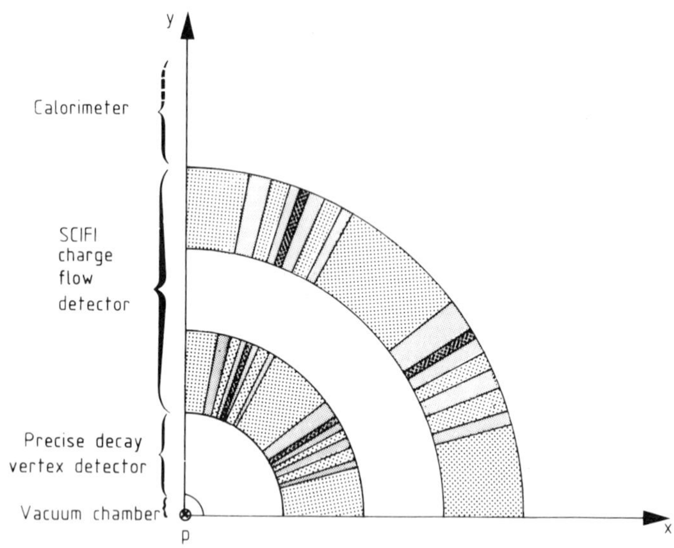

Fig. 44. The SCIFI charge flow detector.

9.2 Micro-tracker

This is constructed from plastic SCIFI in a similar fashion to the micro-tracker described previously. The device has five cylindrical superlayers with the first two comprising only of z fibres and the last three including u and v fibres. The function of the innermost layers is to identify the presence of heavy flavoured quarks or new long-lived particles by means of secondary decay vertices. The innermost fibre layer, which is situated at a radius (r) of 10 cm, receives a radiation dose of 10^6 rads per year and so must be replaced at yearly intervals. The situation quickly improves with increasing radius, e.g. the 3rd layer at r = 20 cm has a lifetime of four years.

9.3 Charge-flow detector

This detector (Fig. 44) measures the 'flow' of charged particles, without any attempt to resolve individual particles, in analogy with the measurements of energy-flow in calorimeters. The plastic SCIFI of diameter 1 mm are arranged parallel to the z axis in a cylindrical shell geometry.

The jets are measured by pulse height analysis of the SCIFI light output. In this way a simple yet relatively high resolution (in comparison with calorimetry) analysis of jet directions, widths and charged multiplicities can be made. The particular merit of this approach is the absence of any saturation of SCIFI response for essentially all jet flux densities, in contrast with the space charge limitations of wire chambers. (We remark here that, in an actual experiment, the charge-flow detector _or_ the micro-tracker would be used, but not both.)

9.4 z-vertex detector

The z measurement of the event vertex will be important since the _average_ beam crossing at a supercollider will contain 3-5 interactions. In these simultaneous interactions the only way to associate calorimeter cells with a common event and to reject background cells is by tracking back to the z vertex. This capability may be of crucial importance in the case of rare events which, for example, could be faked by the chance occurrence of two simultaneous high p_T interactions.

The z-vertex detector (Fig. 45) is constructed in a similar way to the charge-flow device but with the fibres arranged in circles, perpendicular to the beam axis. The readout face, which runs along the length of the chamber, is connected to the optoelectronic readout chain via non-scintillating coherent fibre bundles.

9.5 Micro-converter

A powerful potential application of SCIFI is an imaging pre-converter for photons and electrons (Fig. 46). The 'micro-converter' is made from plastic fibres of diameter 1 mm embedded in a medium of thin high-Z converters, e.g. Pb foils or Woods metal, of a total thickness ~ 3 X_0. The SCIFI/Pb volume ratio is ≥ 50% to ensure a very low threshold for photon detection. We can estimate this threshold to be ~ 5 MeV as follows. If we have a SCIFI/Pb volume ratio of 50% then the scintillator samples ~ $(dE/dx)_{SCIFI}/(dE/dx)_{Pb}$ = 1.7/11.7 = 14% of the total dE/dx. Therefore a photon (electron) of 5 MeV will deposit in the scintillator 5 × 0.14 = 0.6 MeV. The yield of plastic SCIFI is 1 pe per 35 keV deposited energy. Therefore a 5 MeV photon (electron) will produce a light pulse of 600/35 = 17 pe. In reality, the micro-converter has a finite granularity of 1 mm and so we must consider

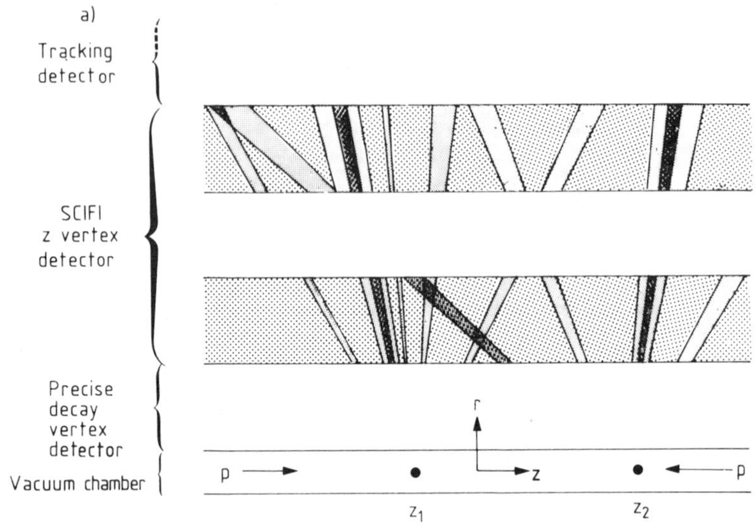

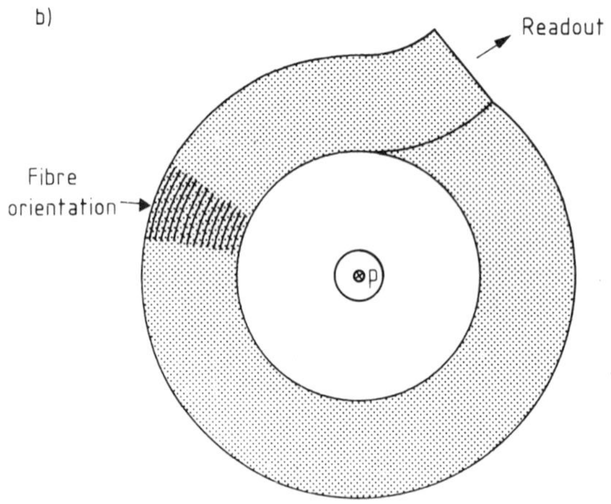

Fig. 45. The SCIFI z vertex detector. The fibres are oriented perpendicular to the beam axis.

the electron range. For a 2.5 MeV electron the range in Pb is ~ 2 mm so we expect at least one fibre crossing for both the electron and positron, i.e. a yield of 10 pe. At slightly higher incident energies the SCIFI micro-converter will image the photon conversions; the e^+e^- tracks will scatter in the converter and curl in the magnetic field.

The first application illustrated (Fig. 46a) for this device is to identify the presence of a nearby π and γ which could lead to false electron candidates in the relatively coarse-grained cells of the electromagnetic calorimeter. In addition to rejection of this background, the SCIFI micro-converter can make a positive electron identification in two ways (Fig. 46b):

1. Initiation of the electron shower in the thin pre-converter.

2. Identification of hard synchrotron photons radiated in the ultra-high magnetic field.

The characteristics of the synchrotron light radiated by an electron in a 6 T magnetic field over a path length of 30 cm are as follows: number of photons (independent of E_e) = 11, total synchrotron energy E (keV) = 14 E_e^2, critical energy E_c (keV) = 8 E_e^2 and mean photon energy E_γ (keV) = 1.3 E_e^2, where E_e is expressed in GeV. This shows the potential presence of a very powerful technique (in addition to calorimetry) for the identification of high energy electrons, namely the detection of hard (MeV) synchrotron photons which are located in a well-defined position along the tangent to the electron track. For a γ threshold of 5 MeV, we can see (Fig. 47) the threshold for electron identification is 30-40 GeV; lowering of the γ threshold would lead to a corresponding reduction in the electron identification threshold. The SCIFI tracking layers will be essentially transparent to photons of these energies ($\geq$ 5 MeV) and so most will reach the micro-converter. Photons which convert in the tracking layers will of course be detected with good efficiency and will aid the identification of electrons. This technique for electron identification is of particular importance since most of the alternative techniques, such as transition radiation or matching between the calorimetry energy and incident momentum, are of little benefit above ~ 100 GeV. At these energies, wherein lies the key physics, the only positive experimental signatures of an electron are longitudinal shower profile and production of hard synchrotron photons. Finally we comment that the radiation of photons in the material of the SCIFI tracker will increase the apparent synchrotron flux and aid the identification of the electron.

The final application of the SCIFI micro-converter (Fig. 46c) concerns the identification of single high p_T photons. Here the major background is from high energy π^o or η production which leads to two or more nearly-coincident photons. A two-shower resolution of 150 µm will allow π^o identification up to ~ 1 TeV.

9.6 Calorimeter

The calorimeter involves a matrix of parallel, 1 mm diameter plastic SCIFI embedded in Pb or Woods metal[12-14]. The volume ratio Pb/SCIFI is ~ 3/1 depending on the exact composition of the absorber and readout, as discussed below. The entire calorimeter is made from material of the same density. The calorimeter is divided longitudinally by cutting fibre blocks of the appropriate thickness, followed by machining and polishing of the entrance and exit faces. In the electromagnetic section (30 X_o) these blocks may have thicknesses of

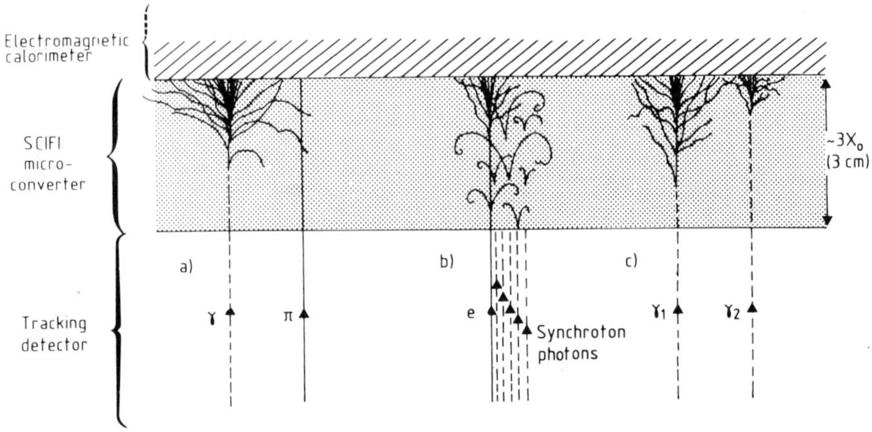

Fig. 46. The SCIFI micro-converter. The functions are a) to identify false electron signatures from $\pi\gamma$ overlap, b) to provide positive electron identification by means of showering and synchrotron photon identification and c) to identify false signatures of a hard single photon due to high energy π^0/η production.

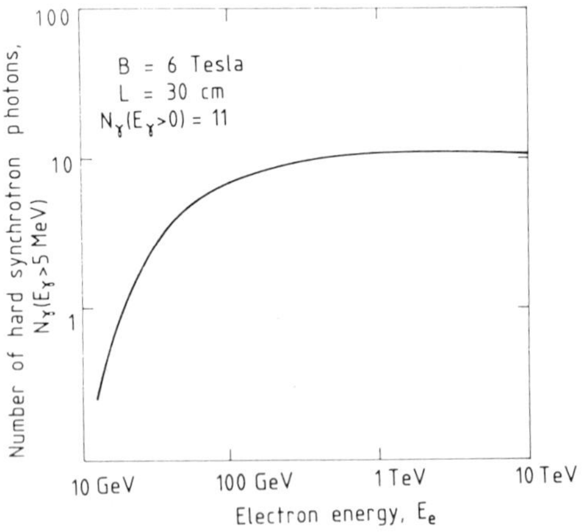

Fig. 47. The hard synchrotron photon flux, N_γ (E_γ > 5 MeV), radiated by electrons of energy E_e in a compact inner tracker similar to Fig. 43.

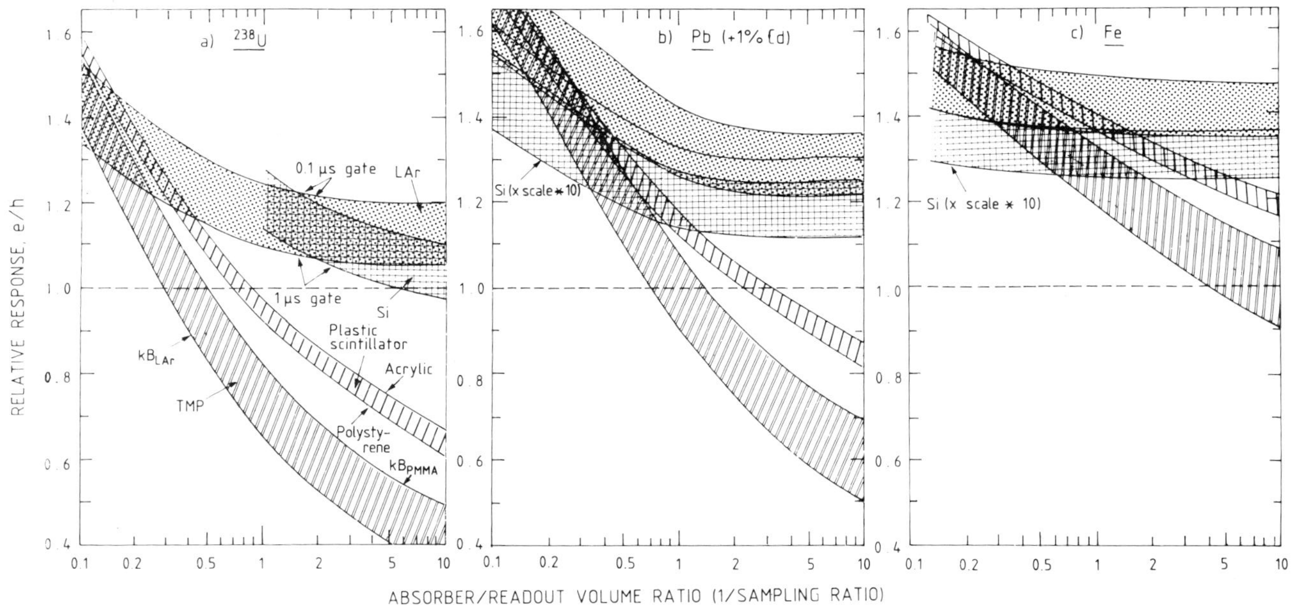

Fig. 48. The relative electron/hadron response vs. absorber/readout volume ratio for hadron calorimeters (Ref. 53). The absorber materials are a) uranium 238, b) lead (+1% cadmium, for improved capture of thermal neutrons) and c) iron. The range of possible responses of each readout material is indicated by a band. For TMP (tetramethyl pentane) the value of the Birks (recombination) constant, kB, is not yet known and so the values for LAr (liquid argon) and acrylic (PMMA) scintillator are used. Plastic scintillator responses for polystyrene and acrylic are indicated. The responses of Si and LAr are shown with the standard gate of 0.1 μs and a very long gate of 1 μs which captures essentially all of the thermal neutrons.

5 X_o, 20 X_o and 5 X_o in order to follow the shower profile and to check for (hadron) penetration and the exit. The hadronic section (a total of 12 λ) is also segmented longitudinally.

The fibre directions lie approximately radially in the barrel region and along z in the end cap region. In this geometry the fibres are approximately, but not exactly, pointing to the interaction region and so particles cannot travel the length of a single fibre and avoid the absorber material. The readout is accomplished with small waveshifter plates, each with two photodiodes attached to opposite edge faces (Section 10.2), which are arranged in a mosaic at the outer surface of each layer. With this approach the transverse granularity and cell shapes can be simply chosen or later modified, since the calorimeter medium itself has no fixed internal boundaries.

The advantages of this calorimeter are as follows:

1. High density. For a Pb/SCIFI volume ratio of 3/1 the radiation length X_o = 7.4 mm and the interaction length, λ = 21.3 cm.

2. Fast. The signal integration time is < 20 ns and the individual cell times can be determined to < 1 ns.

3. Hermetic and uniform. Internal dead regions or 'cracks' are absent.

4. Inexpensive materials and relatively simple construction compared with other calorimeters such as liquid argon or TMP.

5. Good electromagnetic and hadronic energy resolution.

6. Readout system which is compact, simple, flexible and stable in operation.

7. 'Compensated', i.e. relative electron/hadron response = 1.

8. Radiation hard to a level sufficient for operation in a supercollider detector.

We will elaborate further on the final three points. Wigmans[53] has recently clarified the mechanisms which influence the responses of calorimeters to electrons and hadrons. The responses depend on several parameters which include not only the absorber material, as was naively previously considered, but also the readout material and the relative amounts of each. The most important conclusion from this work is that compensating calorimeters can be made from absorbers other than uranium, a material which is expensive and difficult to handle. Compensation (e/π = 1) can be achieved by a variety of high Z absorbing materials, which suppress the electron response (by preferential absorption of photons in the Compton and photoelectric region) and boost the π response (by providing a good source of low energy neutrons which can be preferentially detected in the readout material). The situation is shown in Fig. 48. Here we see the general improvement in the e/π ratio towards unity as the Z of the absorber increases. However to be able to achieve $e/\pi = 1$ requires a readout material with a high cross-section for thermal neutrons, i.e. to contain free hydrogen atoms. This is the case for liquid TMP, and for plastic scintillator. In these materials, by altering the absorber/readout volume ratio, the neutron/minimum-ionizing particle ratio can be tuned through a wide range and compensation can be achieved. Specifically this will occur in a Pb/SCIFI calorimeter with a volume ratio of 3/1.

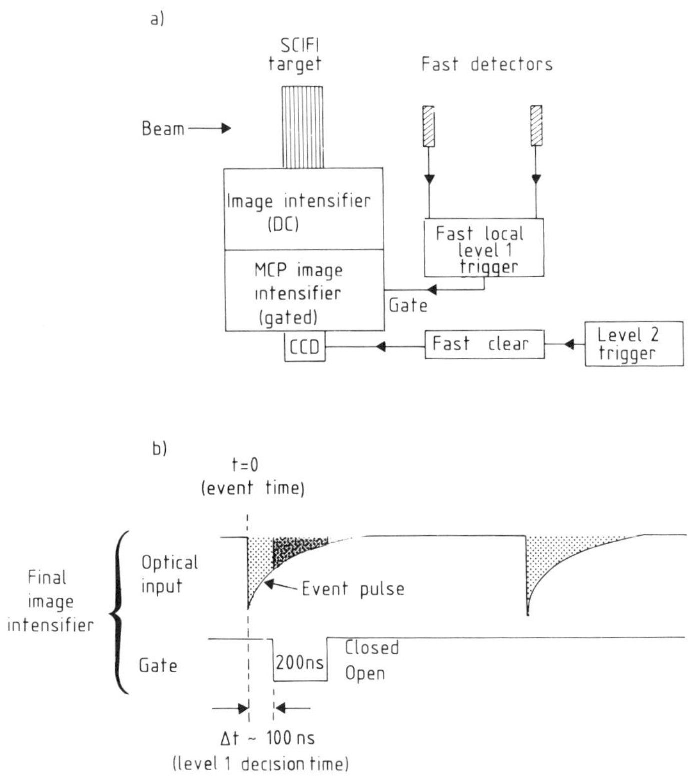

Fig. 49. a) Schematic and b) timing diagrams for the operation of SCIFI readout at beam rates up to ~ 10^6 Hz.

In order to have a good energy resolution in the electromagnetic section it is necessary to sample finely; 2 mm Pb will give an electromagnetic energy resolution, $\sigma_E/E = 12\%/\sqrt{E(\mathrm{GeV})}$. In a conventional approach, involving alternate plates of Pb and scintillator, this would require scintillator plates of thickness < 1 mm in order to maintain the required Pb/scintillator volume ratio of 3/1. Such plates are difficult to handle and have a poor transmission which would result in large transverse non-uniformities. (The transmission in very thin scintillator plates could, in principle, be improved by coating with a 'cladding' layer.) SCIFI, however, is well-matched to these simultaneous requirements of fine sampling thickness and uniform response (long attenuation length).

The radiation dose at the entrance to this calorimeter (r = 100 cm) is 10^4 rads year^{-1} (3×10^{11} minimum ionizing particles cm^{-2} year^{-1}). If we assume the mean photon energy is ~ 1 GeV then at shower maximum there will be ~ 10 particles per incident photon and the dose will be ~ 10^5 rads year^{-1}. On this basis we see that plastic SCIFI must perform reliably up to doses of 10^6 rads. This is a reasonable expectation based on present performances and new materials (Section 3.2) and considering the short fibre lengths involved. The radiation dose in the hadron calorimeter will be approximately an order of magnitude less (i.e. < 10^4 rads year^{-1}) because of the larger radial distance and the reduction in mean number of particles at shower maximum.

9.7 Muon spectrometer

The entire ALFIE inner tracking detector, solenoid and (12λ) calorimeter is contained within a radius of 3.6 m. Beyond this is located a muon spectrometer. The dimensions of the muon system are enormous and incompressible since it is necessary to have 4-5 m magnetized Fe to absorb sufficiently the multi-TeV jets and to provide a 15% measurement of the muon momenta. Tracking chambers made of SCIFI sheets are used in the muon spectrometer. The purpose of this exercise is not to argue that the SCIFI is better than conventional large area drift chambers but simply that it may be competitive.

The SCIFI sheets are made from plastic fibres of diameter 1 mm and length 6 m. This will provide a position measurement accuracy $\sigma \sim 300$ μm. We assume the bulk cost of a single image intensifier/CCD readout channel to be 25 KSF and (as for UA2) 2000 fibres to be handled by a single channel. These fibres cover an area of $2000 \times 6 \times 10^{-3} = 12$ m^2. So the total cost of a SCIFI muon chamber, including mechanical costs is ~ (25 + 5) KSF/12 ~ 2.5 KSF m^{-2}. This is identical to the typical cost of large area drift chambers. The particular advantages of the SCIFI chambers are speed, compactness and the ability to be wrapped in circles to match optimally the field direction in the forward toroids. The triggering capability of this system will be discussed in the next section.

10. READOUT AT THE SUPERCOLLIDER

10.1 Image intensifier/CCD systems

The high speed operation of an image intensifier/CCD system is paced by the CCD. The maximum rate at which a CCD may be exposed is several x 10^5 Hz. This is set by the fast clear time of 1 μs which would, for example, lead to a 50% dead time at an exposure rate of 10^6 Hz. The readout times are typically 10 ms, reducing to 1 ms for

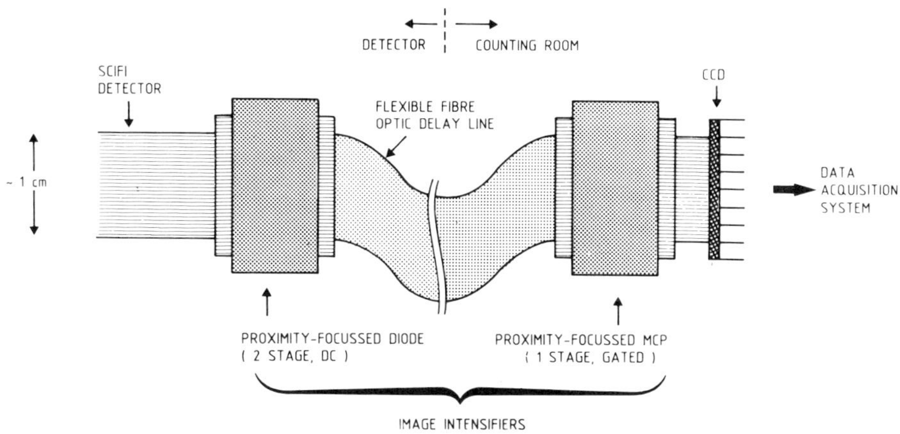

Fig. 50. SCIFI readout assembly for supercollider operation. The coherent fibre optic cable provides sufficient time delay for formation of a level 1 trigger which is then used to gate open the final MCP image intensifier.

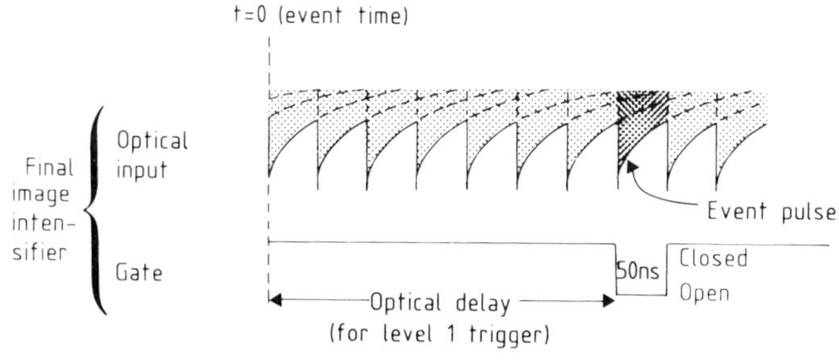

Fig. 51. Timing diagram for the final MCP image intensifier of Fig. 50, at an input event rate of 0.2 10^8 Hz.

the highest speed contemporary CCDs. With these slow readout times (which are due to the high degree of multiplexing) it is clear that detailed CCD information cannot participate in the trigger until level 3 ($\leq 10^3$ Hz) where the events are analysed in parallel microprocessors. It is possible, however, to obtain fast OR'd SCIFI pulses which can participate in level 2 triggers ($\leq 10^5$ Hz). These pulses are generated at the anodes of the final image intensifier stages and correspond to the sums of the total SCIFI pulse height in regions of area ~ 1 cm^2.

The operation of an image intensifier/CCD system at low rates $\leq 10^6$ Hz is illustrated in Fig. 49. Here the decision time for a level 1 trigger is less than mean time between events. In these cases the phosphor decay time of the DC first stages can be suitably chosen to store the event at the input of the gated MCP stage while the trigger decision is taking place (Fig. 49b). The gate is opened by a successful level 1 trigger and the CCD is exposed. Data is then read out of the CCD. The readout is aborted and the CCD is fast cleared if the level 2 trigger is satisfied.

At higher rates it is necessary to provide an optical delay line upstream of the gated image intensifier stage (Fig. 50). The first two stages are proximity focussed photodiodes mounted directly at the edge of the SCIFI detector. Since these devices do not have a MCP they have a high quantum efficiency and operate in longitudinal magnetic fields of any strength. The photodiodes serve to amplify the optical signal by a factor of 100 and to shift the light to a longer wavelength for improved optical transmission. The fastest available phosphors are used (P47, τ = 35 ns). The output is connected to a flexible coherent fibre bundle which acts as the optical delay line. The length of this delay line must be sufficient to allow the level one trigger to open the gate of the final stage before the arrival of the optical data, e.g. a level 1 decision time of 350 ns would require a 60 m delay line. This situation is shown in Fig. 51 for a period of 50 ns between crossings, which corresponds to the minimum system integration time. In order to preserve the lifetime of the final stage image intensifier, the device and CCD may be cooled and the gain reduced. (Suitably cooled, CCD's can record and shift signals of < 10 electrons.)

The ability of a SCIFI micro-tracker to handle the high rates of the supercollider (Table 6) is due to its extremely high granularity: the full system comprises ~ 5 x 10^7 fibres or ~ 10^7 'cells' (defined by the 50 µm image intensifier spot size). During an integration time of 50 ns approximately five events will occur, with a total multiplicity of 100 (high p_T event) + 4 x 40 (minimum bias events) = 260 charged particles. However, the resulting occupancies are low. For example, in the innermost layer at r = 10 cm there are $2\pi 10/(50 \times 10^{-4}) = 1.3 \times 10^4$ cells, and so the occupancy per trigger is $260/1.3 \times 10^4 = 2.1\%$.

It would clearly be an advantage to build a device which replaced the coherent fibre optic delay line. Figure 52 shows some conceptual schemes which involve a) vacuum and b) gas-filled devices. The vacuum device is basically a proximity focussed image intensifier with a long drift space in a small electric field (10 cm with ΔV_{drift} = 11 V provides a 100 ns delay). A strong longitudinal magnetic field provides focussing. The gas filled device requires a photocathode which is compatible with exposure to gases. A candidate is TMAE, the photosensitive component used in RICH detectors. Here we require photons to be absorbed immediately after traversing the input fibre optic window. This may be achieved by heating or pressurizing the TMAE

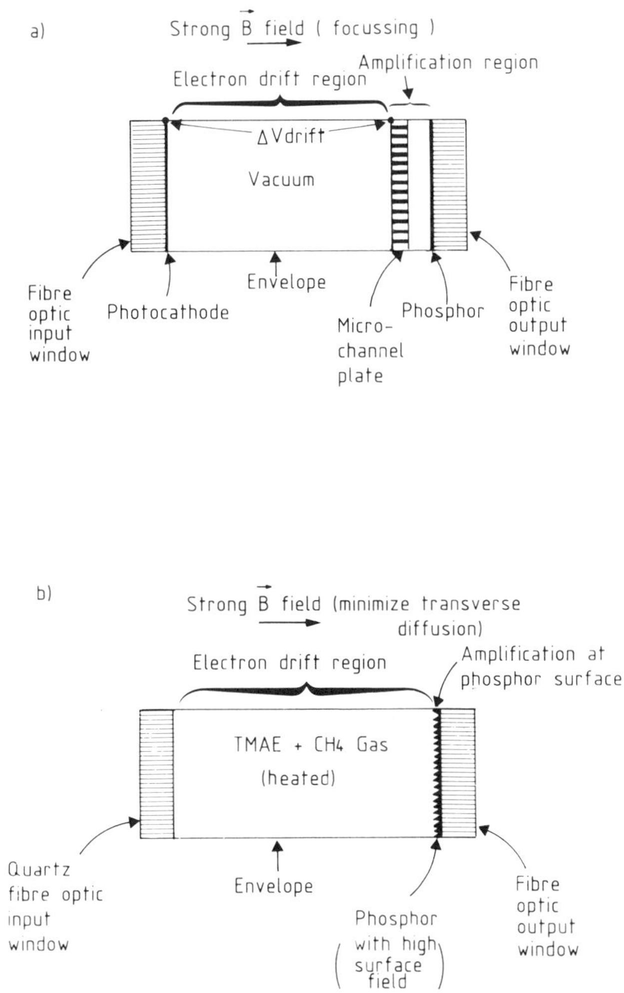

Fig. 52. Conceptual schemes for a coherent optical delay line involving a) vacuum and b) gas-filled devices.

or, perhaps preferably, by arranging for a liquid TMAE photocathode at the inner face of the input window. Since TMAE is only sensitive to photons with wavelength λ < 220 nm, it will be necessary to precede this device with a converter stage involving a suitable (BaF_2?) phosphor and quartz fibre optics. Finally, after drifting the appropriate distance it should be possible to achieve some electronic gain by avalanche multiplication at a suitably-designed phosphor surface.

10.2 Phototubes and Si photodiodes

Single or multi-anode phototubes and Si photodiodes can read out both SCIFI tracking detectors with moderate diameter (~ 1 mm) fibres and SCIFI calorimeters. Phototubes (which have been described in Section 5.1) have the merits of robustness, low noise, high gain and high rate capabilities, but cannot be used in magnetic fields.

Probably the best devices to read out a supercollider SCIFI calorimeter are Si photodiodes. The simplest construction involves 'fluorescent flux concentrators'[54]: small waveshifter plates of dimensions ~ 30 x 30 x 1 mm^3 which have a Si photodiode glued to each of two opposite narrow faces. The dimensions of the plates are chosen according to the desired calorimeter cell size, e.g. here we chose $\Delta\phi \times \Delta\eta \simeq 0.03 \times 0.03$ for the electromagnetic calorimeter. A mosaic of these plates covers the outer surface of each longitudinal section of the calorimeter. This readout is very compact and provides a simple and effective technique for longitudinal readout, with a flexible choice of cell sizes and shapes. Two photodiodes ensure the identification and rejection of spurious pulse height due to tracks which cross a Si wafer.

The light output can be estimated from the present measurements of ~ 10,000 pe GeV^{-1} in a Pb/SCIFI block with a 1/1 volume ratio and bialkali phototube readout. We will assume a Pb/SCIFI ratio of 3/1, a photon collection efficiency in the waveshifter plate of 0.2 and a relative quantum efficiency (Si/phototube) = 2. The expected yield is then 10,000 x 1/3 x 0.2 x 2 = 1300 pe GeV^{-1}, or a noise equivalent per photodiode of ~ 1 GeV. With a little improvement this is probably acceptable for supercollider operation. The advantages of Si photodiode readout are well-known; the absence of drift (no gain), compactness and compatibility with magnetic fields are of particular note. Finally we remark that it should be possible to calibrate each calorimeter cell to < 1% simply from the singles pulse height distribution. Each 3 x 3 cm^2 cell at 1 m radius receives 10^5 Hz charged particles. This constitutes a constant energy source which can be used to inter-calibrate all cells and to monitor drifts with high statistical accuracy. It should be straightforward to follow the slow degradation of light output of a SCIFI calorimeter, which does not have short time scale problems associated with gas and liquid impurities.

11. CONCLUSIONS

SCIFI detectors have advanced rapidly since their recent 'rediscovery' to the stage where they are now being incorporated in a few pioneer experiments. Current work has demonstrated their striking capacity for electronic 'imaging' which leads to a broad variety of unique applications in tracking and calorimetry. With some further development it appears that SCIFI detectors are well-matched to the harsh experimental environment of the supercollider. Moreover, the use of SCIFI in tomorrow's supercollider detectors can simultaneously reduce their size and cost while substantially improving the performance.

Acknowledgements

I would like to thank many colleagues for discussions, in particular, M. Bourdinaud, J.P. Fabre, J.M. Gaillard, C.G.A. Hill, H. Leutz, L. Linssen, S. Majewski, R. Mead, R. Ruchti, H. Schönbacher, J.C. Thévenin and my colleagues in the CERN-MPI(Munich)-RAL Collaboration. I also wish to to thank A. Zichichi and the Directors of this Workshop, G. Charpak and F. Villa, for having organized such an enjoyable and stimulating meeting. Finally, my thanks to S. McCreadie for typing this paper so efficiently.

REFERENCES

1. See, for example, the review by G.T. Reynolds, IRE Trans. Nucl. Sci. NS-7, 115 (1960).
   D.A. Hill et al., Proc. 2nd Symp. on Photo-Electronic Devices, London (1961) 475.
   H.C. Burrows et al., Nuclear Electronics I, (1962) 153.
   D.G. Anderson et al., British Scientific Instrument Research Association (SIRA) AERE/EMB/PR1301 (1961); Proc. 6th Int. Congress on Glass (1962) 429. (The latter work resulted in the development of GS1 scintillating glass.)

2. S.R. Borenstein et al., Proc. 1981 Isabelle Summer Study (1981) 1438; Phys. Scripta 23, 550 (1981); IEEE Trans. Nucl. Sci. NS-29, 402 (1982); IEEE Trans. Nucl. Sci. NS-31, 396 (1984).

3. L.R. Allemand et al., Nucl. Instrum. Methods 225, 522 (1984).
   H. Blumenfeld et al., IEEE/NS-33, No. 1, 54 (1986).

4. D.R. Potter, IEEE/NS-29, No. 1 (1982).

5. R. Ruchti et al., IEEE Trans. Nucl. Sci. NS-30, No. 1, 40 (1983); IEEE/NS-31, No. 1, 69 (1984); IEEE/NS-32, No. 1, 590 (1985); IEEE/NS-33, No. 1, 151 (1986).

6. J.M. Gaillard et al. (UA2 Collaboration), Proc. DPF Conf., Eugene, Oregon (1985) 912.

7. W.R. Binns et al., Nucl. Instrum. Methods A251, 402 (1986).

8. A. Konaka et al., Proc. 23rd Int. Conf. on High-Energy Physics, Berkeley (1986).

9. A. Bross, Nucl. Instrum. Methods 247, 319 (1986).

10. M. Atkinson et al., Nucl. Instrum. Methods A254, 500 (1987). The results using CCD readout are presented in a paper which will presently be submitted to Nuclear Instruments and Methods.

11. D.R. Potter, Proc. Workshop on New Solid State Devices for High-Energy Physics, Lawrence Berkeley Laboratory (1985).

12. H. Blumenfeld et al., Nucl. Instrum. Methods 225, 518 (1984); 235, 326 (1985).

13. H. Burmeister et al., Nucl. Instrum. Methods 225, 530 (1984).

14. DELPHI Technical Proposal, CERN/LEPC/83-3, 159 (1983).

15. J. Fent et al., Nucl. Instrum. Methods 211, 315 (1983).
H. Fessler et al., Nucl. Instrum. Methods 228, 303 (1985);
Nucl. Instrum. Methods A240, 284 (1985).

16. M. Albrow et al. (UA1 Collaboration), CERN-EP/86-131 (1986), submitted to Nuclear Instruments and Methods.

17. L. Bachman et al., Nucl. Instrum. Methods 206, 85 (1983).

18. R. Ruchti et al., Proc. 1982 DPF Summer Study on Elementary Particle Physics and Future Facilities, Snowmass, Colorado (1982) 373.
M. Atkinson et al., Nucl. Instrum. Methods 225, 1 (1984).

19. I.M. Chiang et al., AGS Proposal E787 (1986).

20. K. Pretzl, private communication.

21. D. Binnie et al., Proc. 1984 Summer Study on the Design and Utilization of the Superconducting Super Collider, Snowmass, Colorado (1984) 593.
M. Atkinson et al., Nucl. Instrum. Methods 237 (1985) 505.
J. Kirkby, Proc. 5th Topical Workshop on Proton-Antiproton Collider Physics, St. Vincent, Aosta Valley (1985) 672.

22. P. Sonderegger, CERN-EP/86-214 (1986), submitted to Nuclear Instruments and Methods.

23. J.B. Birks, Theory and practice of scintillation counting, Pergamon Press, Oxford (1964).

24. I.B. Berlman, Handbook of fluorescence spectra of aromatic molecules, Academic Press, New York and London (1971).

25. Nuclear Enterprises Ltd., Bath Road, Beenham, Reading, Berkshire, England.

26. Bicron Corporation, 12345 Kinsman Road, Newbury, Ohio 44065, USA.

27. C. Aurouet et al., Nucl. Instrum. Methods 169, 57 (1980).

28. M. Bourdinaud and J.C. Thévenin, Physica Scripta, Vol. 23, 534 (1981).

29. J.C. Thévenin et al., Nucl. Instrum. Methods 169, 53 (1980).

30. P.L. Mattern et al., IEEE Trans. Nucl. Sci. NS-21, 81 (1974); NS-22, 2468 (1975).

31. H. Schönbacher and W. Witzeling, Nucl. Instrum. and Methods 165, 517 (1979).

32. Y. Sirois and R. Wigmans, Nucl. Instrum. Methods A240, 262 (1985).

33. G. Marini et al., CERN/85-08 (Yellow Report, 1985).

34. J.P. de Brion et al., Saclay preprint DPhPE 86-07 (1986).

35. J.C. Thévenin, private communication.

36. C.L. Renschler and L.A. Harrah, Nucl. Instrum. Methods A235, 41 (1985).

37. S. Majewski, University of Florida, Gainesville, private communication.

38. W.R. Binns and A. Bross, Workshop on New Solid State Devices for High-Energy Physics, Lawrence Berkeley Laboratory (1985).

39. A.R. Spowart, Nucl. Instrum. Methods 135, 441 (1976); 140, 19 (1977); 150, 159 (1978).

40. Levy Hill Laboratories Ltd. (C.G.A. Hill, Director), 5 Sheffield House, Fieldings Road, Cheshunt, Waltham, Hertfordshire EN8 9TJ, England.

41. Collimated Holes Inc. (R.W. Mead, President), 460 Division Street, Campbell, California 95008, USA.

42. Optical characteristics of cathode ray tube screens, Electronic Industries Association, 2001 Eye Street N.W., Washington, DC 2006, USA. TEPAC Publication No. 116 (revision of JEDEC publication No. 16-C) (1980).

43. Hamamatsu Photonics K.K., 1126 Ichino-cho, Hamamatsu City, Japan.

44. Philips International BV, Elcoma Division, PO Box 218, 5600 MD Eindhoven, The Netherlands.

45. R. Meunier, CERN, private communication.

46. Delft Electronische Producten, Postbus 60, 9300 AB Roden, The Netherlands.

47. Thomson-CSF, Division Tubes Electroniques, 38 rue Vauthier, BP 305, 92102 Boulogne-Billancourt Cedex, France.

48. S. Reynaud, CERN/EF/INSTR 86-1 (1986).

49. Proxitronic, Robert Bosch Strasse 34, D-6140 Bensheim, Fed. Rep. Germany.

50. Fibre Optics Development Systems Inc., 427 Olive Street, Santa Barbara, California 93101, USA.

51. Report of the Task Force on Detector R & D for the Superconducting Super Collider, SSC-SR-1021 (1986).

52. A. Seiden, these proceedings.

53. R. Wigmans, CERN/EF/86-18 (1986), submitted to Nuclear Instruments and Methods.

54. E. Lorenz et al., Nucl. Instrum. Methods A249, 235 (1986).

Discussion J. KIRKBY

G. LUTZ

How do you propose to get the signals out of the proposed super-collider detector. Do you plan to lead all signal fibres to the outside?

J. KIRKBY

There will be an effective multiplexing into groups of ~ 10^5 chambers near the detector. This is equal to the number of pixels or, equivalenty, the number of fibres, per CCD. Immediately at the exit of the scintillating fibres there will be a couple of DC image intensifier stages, followed by the presently-undeveloped optical delay, followed by a gated MCP image intensifier. In order to avoid the high radiation environment the CCD's will be attached to the exit of the MCP image intensifier via flexible coherent optical bundles. The total cross section of the fibres is a few 100 cm^2, so the electrical data can be handled by a ~100 cables.

F. SAULI

Most results obtained so far with scintillating fibres have been obtained using image intensifiers with phosphors having long (msec) decay times. Faster phosphors exist, but have lower efficiency. Is there any evidence that one can make a fast intensifier (for high rate applications) still preserving a good single photoelectron imaging efficiency (without too much noise)?

J. KIRKBY

The fastest phosphors used in image intensifiers have decay time constants τ ~ 60 ns, such as P^{46}. We have assembled a multistage image intensifier system wich utilizes P^{46} phosphors throughout. In the laboratory we measure the efficiency of this phosphor to be about one third of a P^{20} phosphor which is among the most efficient available. We believe this system will allow us to detect single photons with good efficiency. This will be experimentally checked in some beam tests we have planned for october 1986. Tube noise is no problem since the gating times we will use are extremely short (few 100 ns).

THE INFN ELOISATRON PROJECT: EUROPEAN LONG INTERSECTING STORAGE ACCELERATOR

C. Aglietta, K. Alberini, G. Badino, G. Bari, M. Basile, M. Bassetti, J. Berbiers, A. Bertin, E. Boschi, V. Braginski, R. Bruzzese, N. Cabibbo, G. Cara Romeo, R. Casaccia, C. Castagnoli, A. Castellina, A. Castelvetri, L. Cifarelli, F. Cindolo, M. Civita, A. Contin, G. D'Ali, M. Dardo, C. Del Papa, V. de Sabbata, L. Ferrario, W. Fulgione, S. Galassini, P. Galeotti, M. Gasperini, P. Giusti, R. Goldoni, G. Iacobucci, I. Laakso, A. R. Leo, M. Leo, G. Luches, G. Maccarrone, A. Marino, T. Massam, V. N. Melnikov, R. Meunier, F. Motta, R. Nania, V. Nassissi, G. Navarra, F. Palmonari, G. Papini, G. Passotti, P. Pelfer, G. Pocci, G. Prisco, M. Puglisi, M. Ricci, G. Rinaldi, C. Rizzuto, F. Rohrbach, P. Rotelli, O. Saavedra, N. Sacchetti, G. Sartorelli, G. Soliani, M. Spadoni, M. Steuer, G. Susinno, S. Tazzari, K. Thorne, G. Torelli, G. C. Trinchero, P. Vallania, G. Venturi, S. Vernetto, F. Villa, A. Vitale, L. Votano, M. Willutzky and A. Zichichi

Presented by A. ZICHICHI

ABSTRACT

Past experience, present knowledge, and a straightforward forecast in the extreme energy limits for subnuclear physics are reviewed, in order to put forward the case for a multi – TeV intersecting hadronic machine, whose basic features are described.

1. INTRODUCTION

1.1 A new era

A new era in modern scientific thought has been opened up.

New ideas, new concepts, new phenomena make physics of just twenty years ago seem as old as millenia.

The quadridimensional space – time of Einstein seemed to be a conquest beyond which no one would be able to go. This, however, seems to be an incredibly narrow outlook for two reasons, both of them fundamental: the number of dimensions, and the properties of those dimensions. No one had thought, before the sixties, that there could exist space – time dimensions with 'FERMIONIC' properties. Those of Einstein are 'BOSONIC'. This is how the new concept of superspace was born, and with it, SUPERPARTICLES and SUPERMATTER. In Table 1 is given a synthesis of the present status of our knowledge of matter (quarks, leptons, photons, strong and weak gluons, Higgs particles) and supermatter.

The world in which we live and the matter of which we are made could have their roots in a *'bosonic'* superspace with 10 dimensions, plus the 32 *'fermionic'* ones. And this is not all. The concept of *'point'* that has held its position for centuries and centuries falls by the wayside. In its place is the *'superstring'* – a unidimensional entity with a pointless structure in a 42 – dimensional superspace.

In this extraordinary progress of our knowledge, the winning parameter is ENERGY and the objective is the unification of all the forces of nature: Figure 1 shows the present status of our knowledge.

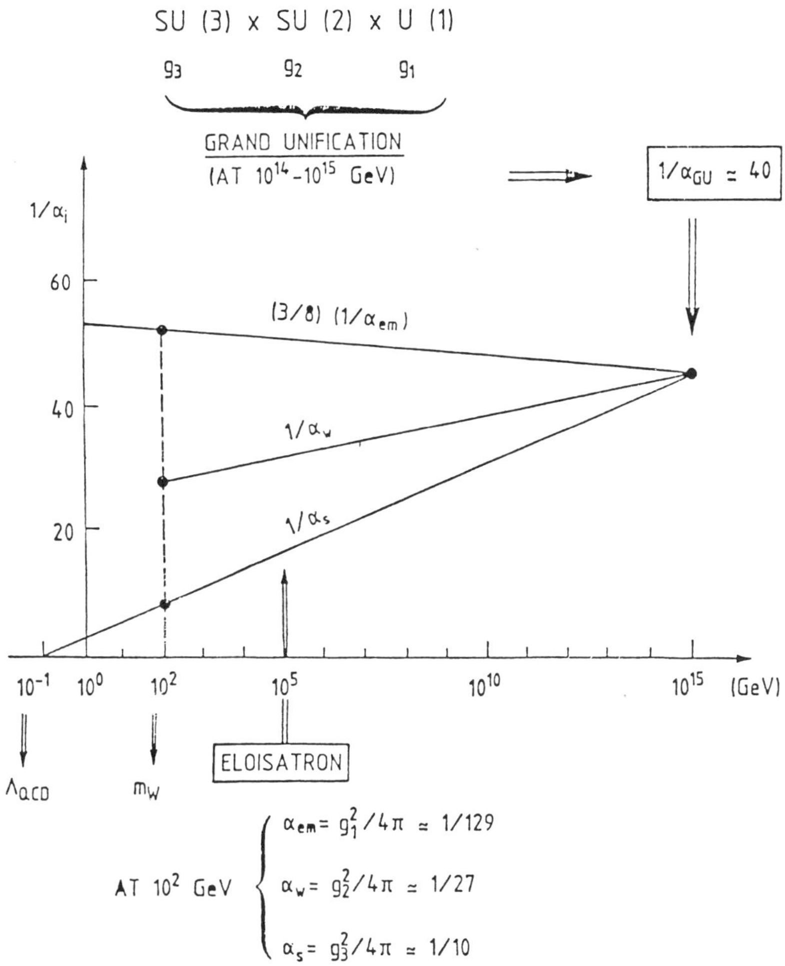

Fig. 1 All the fundamental forces of nature should be generated by a unique force. The convergence of the three fundamental constants g_1, g_2, g_3, is the basis for GUT.

Table 1

Present status of World and Superworld

World R = 0	**Superworld** R $\neq$ 0		
Spin 1	Spin ½		
Gluons Photons $(W^{\pm}, Z^0)$	Gluinos Photinos $(W^{\pm}, Z^0)$-inos	Gauge particles	FORCES
Spin ½	Spin zero		
Quarks Leptons	Squarks Sleptons	Matter particles	MATTER
Spin zero	Spin ½		
Higgs	Shiggs	Gauge breaking	SSB

SSB ≡ Spontaneous supersymmetric breaking

1.2 The position of Italy

Italy, with its intellectual forces engaged in this advanced field of frontier physics, has been able to reach an outstanding position in Europe and the world, which is why this unprecedented scientific development did not find the country unprepared. If Italy keeps faith with its programs, it will find itself in a position of absolute primacy in the most advanced sector of fundamental science.

1.3 The energy level and the related technological development

The 'ELOISATRON' project will be the proof of this development.

First of all, what is the Eloisatron? It is a machine in which protons collide with protons.

The level of energy is 100 TeV (10^{14} electronvolts). That is to say, one hundred times the level at which we work today at CERN and in the USA (at Fermilab). The circumference of the machine is 250 km: its cost would be of the order of 5×10^9 US dollars.

An essential characteristic of the ELOISATRON is that it has no weak points. Even though it will be necessary to do a lot of research and development (R&D), there is no doubt of its realizability. Its construction will provide for technological development in six leading sectors of High – Technology, namely:

1) cryogenics,

2) superconductivity,

3) high vacuum,

4) ultra – fast electronics,

5) magnetic fields,

6) super computers,

as well as the possibility of other research, e.g. superconducting RF cavities, and new techniques for detecting subnuclear particles.

1.4 The basic steps of the project

Let us look at the basic steps of the ELOISATRON project.

In 1979 the INFN (Istituto Nazionale di Fisica Nucleare) again took up the project to build a superconducting cyclotron in Italy (the CS project). Besides the scientific importance for studies in nuclear physics, the aim of the CS project was to develop around it a scientific undertaking of high technology affecting three Italian industries: one specializing in magnets (Ansaldo), another in superconducting cables (LMI), and another in cryogenics (Zanon).

This first step was crowned with success. In 1981 the construction of the CS commenced.

The following year the second phase started with the participation of Italy in the construction of the most delicate part of the German project HERA (Hadron – Electron Ring Accelerator): that is, the *superconducting magnets for the proton ring* at DESY, Hamburg. The collaboration between Italian industry and Italian and German scientists has given encouraging results.

In fact, Italy will construct 50% of the superconducting magnets for HERA, and production has already started.

Table 2

The great projects of INFN — Steps towards the ELOISATRON

- 1979 - **CS project** (Superconducting Cyclotron) resumed.
- 1981 - **The construction of the CS** (Ansaldo-LMI-Zanon) begins.
  The HERA project is presented to the INFN Council.
- 1982 - The excavations at the Gran Sasso begin.
  First meeting of the HERA designers with Ansaldo-LMI-Zanon.
- 1985 - **The construction in Italy** of the great magnets for **HERA** begins.
- 1986 - **Beginning of R&D works** for the **LSCM** (Long SuperConducting Magnet) prototypes for ELOISATRON general test (10%).
- 1987 - Beginning of the experiments at the Gran Sasso.
  End of the construction of the HERA magnets.
  Beginning of the construction of the **ELOISATRON magnets (10%).**
- 1988 - **Beginning of the excavations for the ELOISATRON tunnel** (250 km).
- 1990 - **End of the LSCM construction: (10%) ELOISATRON.**
  Installation in the LEP tunnel and beginning of tests. **Experiments at (10-16) TeV.**
- 1991 - **Beginning of the LSCM construction for the ELOISATRON.**
- 1992 - **Completion of the ELOISATRON tunnel.**
- 1995 - **End of tests** and dismantling of the **(10%) ELOISATRON** prototype installed in the LEP tunnel.
- 1997 - **Completion of the installation of magnets in the ELOISATRON tunnel.**
- 1997-2000 - **Beginning of the 10^2 TeV physics experiments: 100% ELOISATRON.**

At this point it is clear that collaboration between scientists and large Italian industries is possible.

Table 2 shows the essential phases of the ELOISATRON project from 1979 to 2000.

1.5 Why put the ELOISATRON in Italy?

Italy is not a superpower. Instead of trying to do many things, and not doing them very well, it is considered decidedly better to do a few things and do them excellently in a competitive way and at the level of the superpowers.

On occasion it has been said that Italy, geophysically speaking, could not be suitable for such a large machine. This is why the Italian National Institute for Geophysics (ING) has made a series of studies. *Conclusion:* at least seven sites are suitable for the ELOISATRON ring. Figure 2 shows the results of the ING studies.

This is what the ELOISATRON project should be: Italy should promote the undertaking of the construction, in its own country, of the largest subnuclear machine in the world in collaboration with those countries that share its interests in a field so vital for scientific, technological, and cultural progress.

It should also not be forgotten that we are working on a scientific project whose goals are peace and civil progress. Only if pacific high technology is made competitive, will the possibility of victory over the war technology become a reality. Science to be utilized for peaceful ends can only become a reality through concrete projects.

At the Versailles Summit Meeting called by President Mitterand, the Italian government put forward the proposal of the ELOISATRON to stress the importance that Eu-

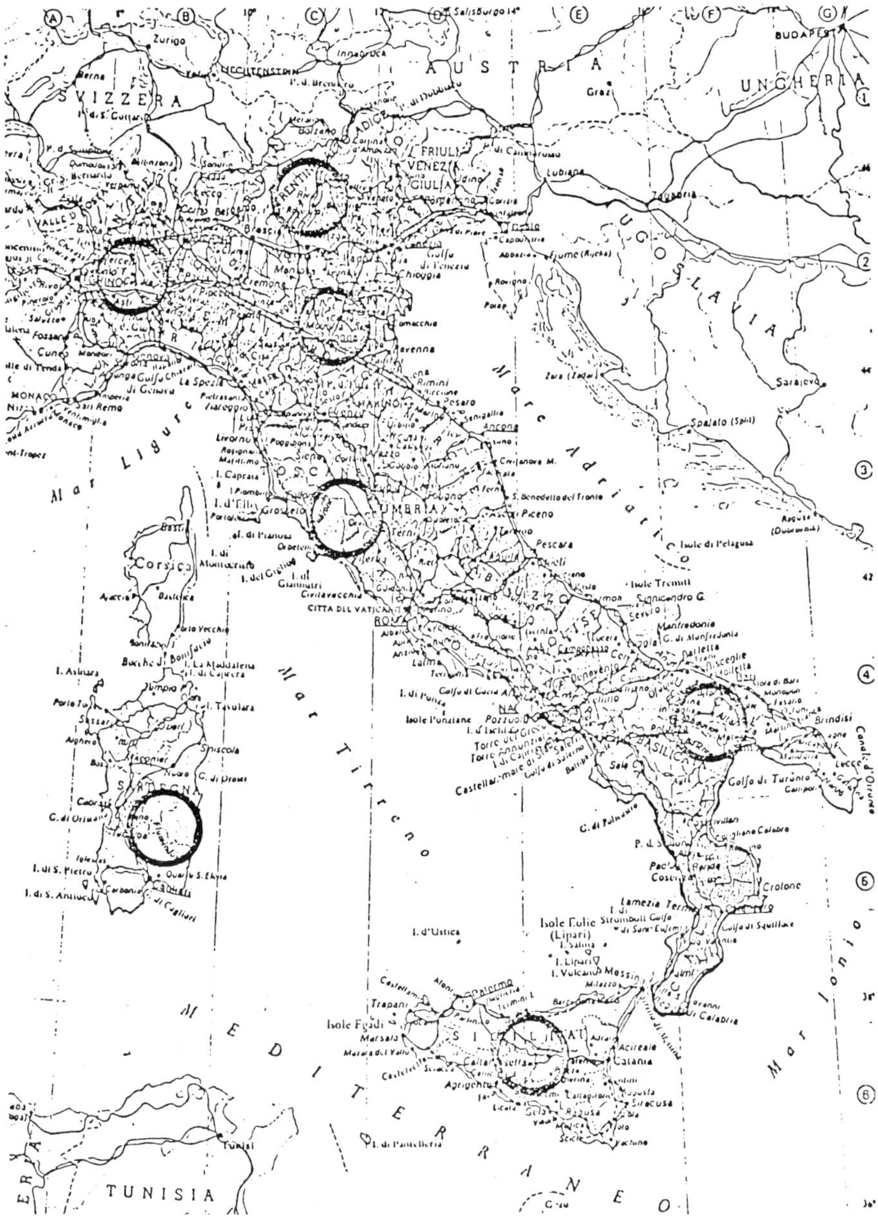

Fig. 2 Possible sites for the large ELOISATRON ring [Study made by the Italian National Institute for Geophysics (ING) 1985].

rope should not abandon the field of subnuclear physics. The 'I' in ELOISATRON stands for *'intersecting'*. At that time Americans were in favour of a machine with only one ring, for protons and antiprotons. Today, they have returned to the idea of a machine with 'intersecting' rings.

1.6 Objectives of the project

The primary objective of the ELOISATRON project is to stop Europe from falling back into a secondary position, with respect to the superpowers, in this crucial field of modern physics.

In fact the true authentic innovations stem from the discovery of new fundamental laws of nature, of new phenomena. The source of these novelties has always been there, where new ideas and new concepts are born. If Galilei were alive today, he would be at CERN fighting for the ELOISATRON.

1.7 The third phase of the project and the ELOISATRON schedule

This great European undertaking, after completion of the CS and HERA phases, needs a subsequent phase: a general test at the 10% level. This test must be made at the minimum cost. We have therefore to use the best structures existing today in Europe, which is why the 10% test of the ELOISATRON should be done at CERN.

The schedule for the ELOISATRON (as shown in Table 2) foresees that the excavations for the 250 km tunnel will start in 1988, with completion of the machine in 1997. If the ELOISATRON follows this schedule, then physicits in the year 2000 will have the most powerful instrument in Europe to help them understand how this world is made.

He whose ideal is an independent Europe must never forget that the prime factor for independence in the modern era is Science.

2. WHY START NOW TO DISCUSS A MULTI – TeV MACHINE IN EUROPE?

Since by 1990 both LEP and HERA will be operating, and if, as many of us believe, subnuclear physics is to be pursued with vigour and without a break of many years, then it is appropriate to start work now on a new hadronic machine in the 10^2 TeV range.

3. FROM NEW IDEAS TO REALITY

Some people might say that it would be better to wait for new ideas. But past experience has shown that to wait is not a wise choice. In fact, a long time is needed in which to transform new ideas to reality.

Two examples should suffice:

i) Superconducting high−field magnets, first proposed in 1961, became 'reality' in 1983 (the Tevatron): 22 years were required.

ii) Collective field accelerators were proposed by Veksler, Budker and Fainberg in 1956: 30 years later, there is still no practical design for a high−energy machine based on these ideas.

The ELOISATRON design, as will be seen in Section 5, is in fact based on extrapolations from known facts and technologies.

4. PHYSICS

In recent years a fashionable approach to the extreme energies has been the 'desert', i.e. nothing should exist in the range from a few 10^{-1}TeV up to $\sim 10^{12}$TeV.

The 'desert' would be a serious obstacle if there were no problems whatsoever in this field of extreme energies.

The high−energy limit of our present knowledge has two frontiers. One is in the domain of experimental physics. At present the known limits on the inverse radii of all known leptons (e,μ,τ) are of the order of 10^{-1} TeV. This means that point−like stuctures of the presently existing particles are already in the multi−TeV range. The other is in the theoretical domain. Here we have the family and the hierarchy problems, which, together with the proliferation of the Higgs sector and of its associated large number of parameters, make the multi−TeV range overcrowded with problems to be understood.

But even if everything would look 'perfect', we should not forget the lesson we can learn from past experience.

4.1 The lesson from the past

The most famous is the 'desert' of Lord Kelvin, who stated in 1897: 'There is nothing to be discovered in physics now. All that remains is more precise measurements'. A few months later J.J. Thomson announced the discovery of the electron.

Some people could say: "This example is too old; these are 19th century stories". For this reason we propose to review more recent facts.

Let us start with the 30 GeV proton synchrotrons at CERN and BNL, the original motivations being the (πp) and (pp) scattering and phase–shift analysis as well as tests of Isospin and T invariances.

What did we get from these machines?

- New particle states, which produced the celebrated global flavour symmetry SU(3) of Gell–Mann and Ne'eman.
- The first measurement of the ($\omega - \phi$) mixing angle, a crucial test for resolving the puzzle of the vector meson masses.
- The measurement of (e^+e^-) and ($\mu^+\mu^-$) production in hadronic interactions, which was started in 1964 at CERN and resulted in the discovery of the J particle at BNL in 1974.
- The first proof that the nucleon had an important electromagnetic structure in the time–like region.
- The discovery of the existence of antinuclei: the antideuteron.
- The discovery of two kinds of neutrinos: $\nu_e \neq \nu_\mu$.
- The discovery that $\nu_\mu \neq \bar{\nu}_\mu$.
- The discovery of C and P violation.
- The discovery of CP and T violation.
- The discovery of the weak neutral current.

All the above findings had nothing to do with the original motivations. Let us look at some more examples:

SLAC – The original physics aims were to study the electromagnetic form factors of the nucleons, the electromagnetic transition form factors ($N-N^*$), and quantum electrodynamic (QED) checks. *Found:* the proof that point–like structures exist inside the proton. Moreover, in the field of electroweak physics, the first high–energy measurement of parity violation in purely electromagnetic interactions was performed.

ADONE – The Italian $(1-3)$ GeV (e^+e^-) machine. Here the motivations were of two types. *The conventional ones* were the checks of QED and of radiative corrections; the (μe) electromagnetic equivalence; the electromagnetic form factors of pions, kaons, and protons; and the study of the tails of the vector mesons. It is probably interesting to recall that these vector mesons (ρ,ω,ϕ) were theoretically proposed as 'gauge' particles in order to understand the conserved hadronic currents associated with isospin, hypercharge, and baryon number. *The unconventional* proposals were the search for heavy leptons using a new technique based on the acoplanar (μe) pairs, and the search for leptonic quarks. The totally unexpected discovery was that the ratio of hadronic to muonic cross–section was found to be much higher than the theoretically predicted value, based on the tails of the three vector mesons ρ,ω,and ϕ.

SPEAR and DORIS – The "heavy lepton", looked for at ADONE, was found at SPEAR using the (μe) technique invented at Frascati. SPEAR started at 3 GeV (the energy limit of ADONE) with great enthusiasm because of the ADONE discovery of the high hadronic cross–section ratio mentioned above. However, a series of great new things were there to be discovered: the J/ψ family of new particles and the open charm states.

PETRA and LEP – So far nothing unexpected has come from these machines; the evidence for gluons is in the aera of 'expected' results.

ISR – The ISR is a special case. It was a machine of the highest technical performance. Had the experimental apparatus been allowed to be of the same level as the machine, the result would have been tremendous: J/ψ, Υ, new heavy–flavour states, high–p_T jets, and direct photon production would have been found in the first years of ISR operation. Unfortunately, the trend was to encourage 'first generation' experiments, i.e. very simple and powerless set–ups.

THE 400 GeV CERN and FERMILAB SUPER PROTON SYNCHROTONS – The discovery of the 9.5 GeV Υ states by Lederman and collaborators is an outstanding example of an unexpected result.

THE 540 GeV CERN $p\bar{p}$ COLLIDER – Too recent to give unexpected results; it has, however, provided the expected $W^{\pm}$ and Z^0 weak bosons.

What lesson can we derive?

Firstly, when the construction of a new machine is started, the experimental set–ups must also be started, and with the same emphasis as that placed on the machine construction. The strategy of first– and second–generation experiments, where 'first' means 'powerless', corresponds to wasting all the money and effort: the ISR is an example.

Secondly, watch the energy gaps. The maximum energy of ADONE was 3 GeV. SPEAR started at 3 GeV, but then jumped to higher energies and was, for some time, bound to miss the J/ψ. SPEAR's maximum lay at 9 GeV, whereas PETRA started at above 10 GeV. Lederman's Υ was at 9.5 GeV. Apart from the machine point of view, the energy gap is the reason why the 10% ELOISATRON step is needed.

Thirdly, when unexpected discoveries are made, the expected ones appear as a 'desert' of imagination.

We now start discussing the 'desert' of present – day physics and its problems.

4.2 The 'desert'

This physics is based on the assumption that there are no new gauge forces from the presently accessible 0.1 TeV to an upper energy E_{max}.

The renormalization group extrapolation shows that the effective couplings of the three gauge forces $SU(3)_c$, $SU(2)_{ew}$, and $U(1)_{em}$, converge to the same value at the same energy E_{max}. This energy is found to be

$$E_{max} \sim 10^{12}\text{TeV} .$$

In order to unify all gauge forces, including gravity, the natural scale becomes

$$E_{max} = E_{Planck} = 10^{16}\text{TeV} .$$

Figure 3 shows the dangerous range of the 'desert'.

However, the key question is the following: Are there problems still to be solved in this theoretical picture of very high energy physics? The answer is positive and threefold. The problems whose solutions are open to competition are those already mentioned: the hierarchy, the family, and the proliferation. For a review of these problems, see Ref. [1]. The solutions of these problems call for experimental investigation in the multi – TeV range.

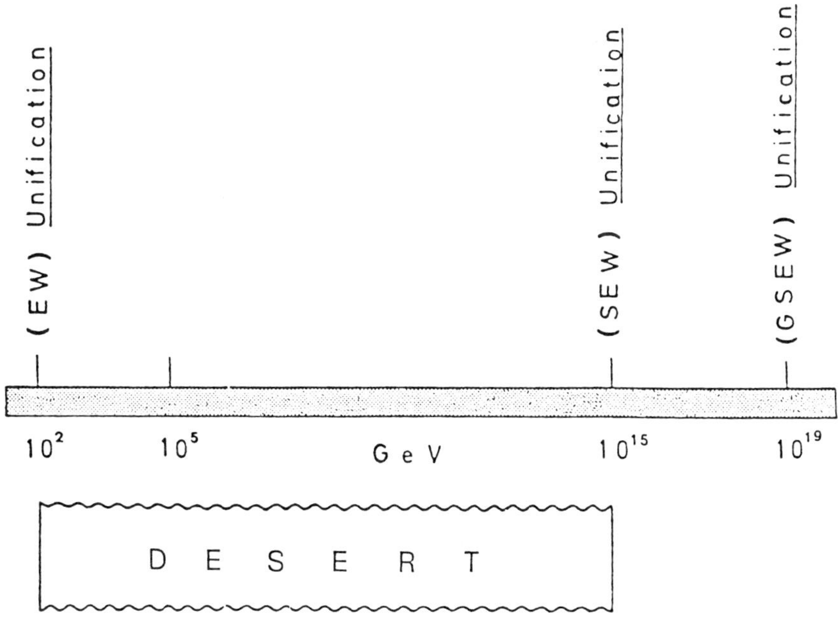

Fig. 3 The 'desert' range.

Table 3

Ist family	2nd family	3rd family	4th family
$\begin{pmatrix}\nu_e\\ e\end{pmatrix}$	$\begin{pmatrix}\nu_\mu\\ \mu\end{pmatrix}$	$\begin{pmatrix}\nu_\tau\\ \tau\end{pmatrix}$	$\begin{pmatrix}\nu_L\\ L\end{pmatrix}$
$\begin{pmatrix}u\\ d\end{pmatrix}$	$\begin{pmatrix}c\\ s\end{pmatrix}$	$\begin{pmatrix}t\\ b\end{pmatrix}$	$\begin{pmatrix}t_h\\ b_h\end{pmatrix}$

Table 4

Luminosity ($cm^{-2}\,s^{-1}$)	E_T for $\sqrt{s} = 10$ TeV (TeV)	No. of experiments which reach 1 TeV scale for $\sqrt{s} = 40$ TeV
10^{30}	0.1	2 out of 7
10^{32}	1	5 out of 7
10^{34}	2	All

Energy (TeV)	Luminosity ($cm^{-2}\,s^{-1}$)
5	3×10^{32}
15	3×10^{31}
The same q^2 is reached in the two cases	

4.3 The unification problem

If the gravitational force is to be unified with other forces of nature, the four-dimensional Einstein space-time has to be enlarged, in the number dimensions and in the properties of the dimensions. In addition to the Einstein bosonic dimensions, fermionic dimensions are needed. This is how superspace comes in, and supersymmetry (SUSY) as a byproduct. Supersymmetry generates an immense number of particles, and nobody knows how heavy the masses of these particles are: this is just a matter of theoretical speculation. A possible version of the supersymmetric particle states is due to Farrar and Fayet and it is given in Table 1; notice that the quantum number R is zero for all known standard states.

The energy level where the supersymmetric partners of known and yet-to-be-discovered particles can be produced, is open to theoretical speculations.

The present status on the long road towards unification is shown in Fig. 1, where the coupling constants α_s,α_W, and α_{em} are plotted against energy. Notice that at 0.1 TeV the expected increase in α_{em} and the expected decrease in α_s and α_W have already been observed.

4.4 Extrapolation from present knowledge

There are three known families of quarks and leptons (see Table 3).

Five quarks (u, d, c, s, b) have been discovered so far. A sixth quark, the top quark (t), is required by the ABJ anomaly cancellation, and its natural location would be the up-like member of the third family. There are, however, two arguments which suggest the existence of a fourth family:

i) supersymmetry needs a very massive quark ($m_{quark} \sim 10^2$ GeV) in order to have the gluino mass compatible with the experimentally known limits;

ii) the model of Glashow et al. says that if beauty decays $\geq$ 95% into charm, and if $\tau_b \geq 10^{-12}$ sec, then $m_t > 50$ GeV.

However, it could very well that $m_t < 50$ GeV. In this case the simplest way out is to introduce a fourth family. If we believe in SUSY, this should be the last family.

Without any reference to sophisticated theoretical arguments, let us look at few experimental data on quark-mass estimates. They say that

$$m_c/m_s \simeq 4, \quad m_b/m_s \simeq 10. \tag{1}$$

Suppose [1] that

$$m_c/m_s = m(\text{up-like})/m(\text{down-like}),$$

$$m_b/m_s = m(\text{family } N+1)/m(\text{family } N).$$

In other words, suppose that the already known quark-mass ratios (1) stand for the ratio (up-like) over (down-like) and for the ratio of the next-family quark mass over the preceding one, then:

$$m(\text{up-like})/m(\text{down-like}) \simeq 4,$$

$$m(\text{family } N+1)/m(\text{family } N) \simeq 10.$$

In this case the top mass would be in the 25 GeV range.

Concerning the fourth family, we would have

i) for the down-like member (heavy beauty), $m_{b_h} \simeq 55$ GeV;

ii) for the up-like member (heavy truth), $m_{t_h} \simeq 220$ GeV.

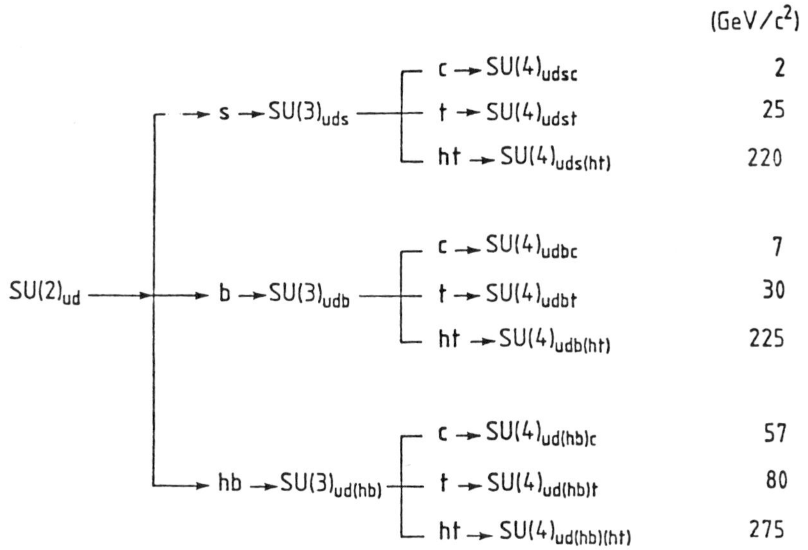

Fig. 4 $SU(3)_{uds}$ repeated with 'beauty' and 'heavy beauty' replacing 'strange', and $SU(4)_{udsc}$ repeated with 'top' and 'heavy truth' replacing 'charm'.

So even if the 'desert' is there, there could be a lot of standard particles yet to be discovered.

In fact, we could imagine, regardless of the drastic mass differences in the heavy flavours, that global symmetries such as $SU(3)_f$ go on like the 'up – down – strange' symmetry repeated with beauty and heavy beauty, and the $SU(4)_f$ global symmetries of 'up – down – strange – charm' repeated with top and heavy truth symmetry, as shown in Fig. 4.

If these global symmetries are there, the diagrams of Figs. 5, 6, and 7 show the structure of the new states, obtained from quark x = up – like and quark y = down – like, combined with the $SU(2)_f$ of (u, d) quarks.

If the global symmetries indicated in Fig. 4 do not exist, the number of states obtained by combining the eight quarks of the four families would be even greater.

In any case, the study of how the various quarks combine among themselves to produce new particle states is crucial to the understanding of what is going on in the multi – TeV range of quark interactions.

In this respect it is interesting to extrapolate our present knowledge of the cross – section to high energy and high masses. We will use two simple criteria: dimensionality and scaling.

From dimensionality we have $\sigma \sim (1/m^2)$; scaling dictates that the cross – section cannot depend on s, but on the ratio (s/m^2).

We therefore assume that

$$\sigma(m,s) \sim 1/m^2 \; f(s/m^2)$$

and extrapolate, from the 'strange' or 'charm' cross – sections, the cross – sections for heavier flavours in the multi – TeV range. The basic formula is

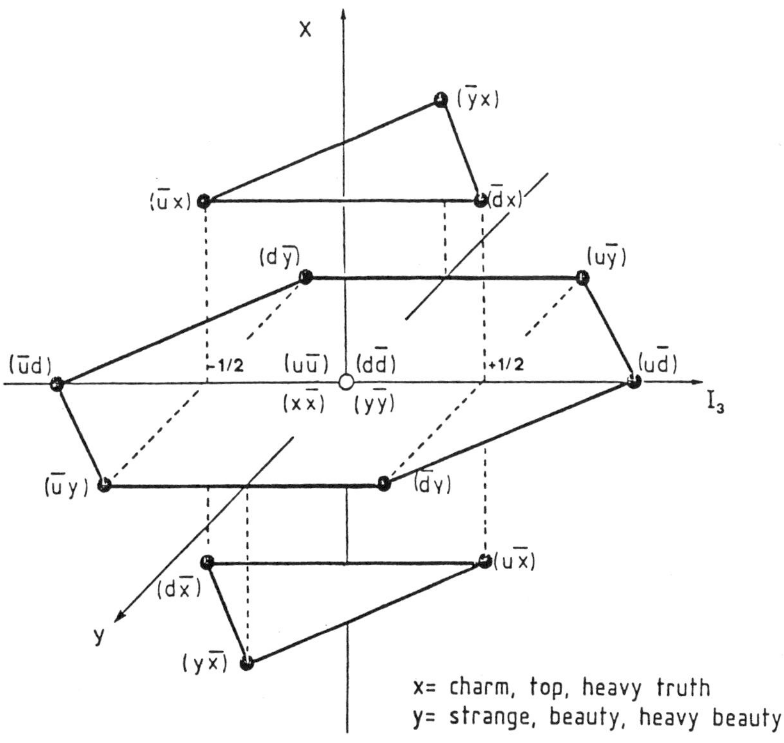

Fig. 5 The $SU(4)_{udxy}$ structure for $J^P = 0^-$ and $J^P = 1^-$ meson states.

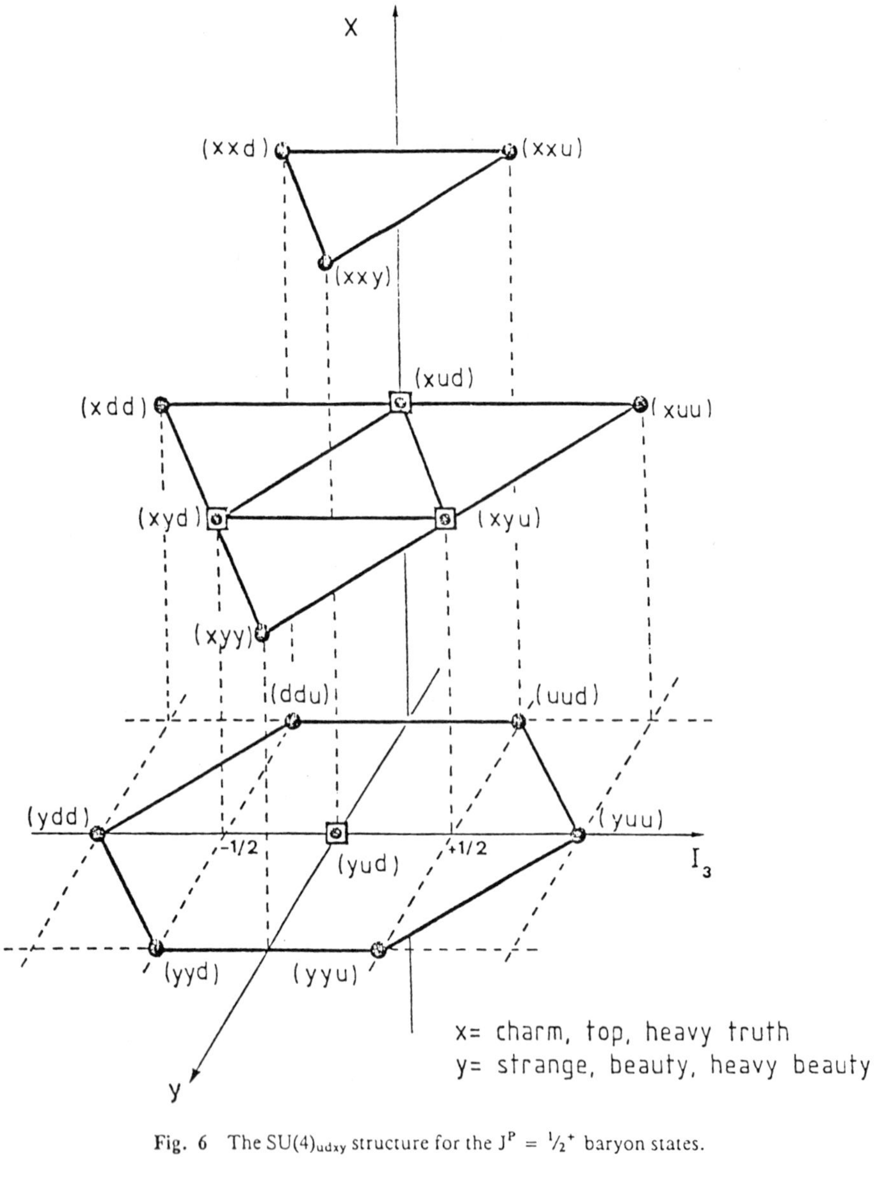

Fig. 6 The $SU(4)_{udxy}$ structure for the $J^P = \frac{1}{2}^+$ baryon states.

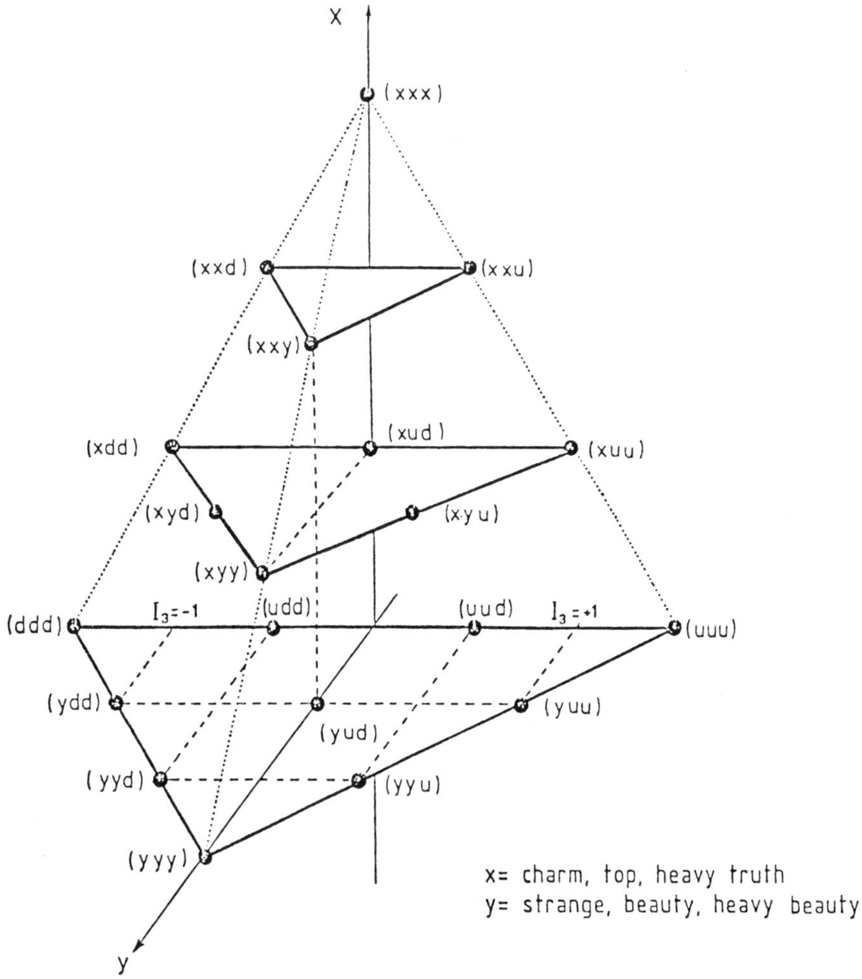

Fig. 7 The $SU(4)_{udxy}$ structure for the $J^P = {}^3\!/_2{}^+$ baryon states.

$$\sigma_{fi}[(\sqrt{s})_{pp} = E_i] = (m_j/m_i)^2 \cdot \sigma_{fj}\,[(\sqrt{s})_{pp} = E_j = (m_j/m_i)\,E_i],$$

where i and j are the quark – flavour indices.

The result is (Figs. 8 and 9) that the order of magnitude for the cross – section values is in the range $(10^{-1} - 10^{0})$ μb, at $(\sqrt{s})_{pp} = 50$ TeV.

All this does not take into account new, harder interactions. If they would be included, then our cross – section estimates would all look pessimistic.

To conclude (irrespective of any sophisticated theoretical argument), by extrapolating present physics knowledge, using dimensionality and scaling, we find that a multi – TeV machine gives access to very large mass objects, as shown in Fig. 8 and 9.

4.5 (pp) versus (e^+e^-)

Proton – proton machines are less expensive than circular (e^+e^-) machines. In fact the cost of (pp) machines varies with energy E, whilst for (e^+e^-), owing to the steep increase in synchrotron radiation losses with E (which causes the economically optimum radius to be proportional to E^2), the cost varies with E^2.

On the other hand, as shown by the BCF (Bologna – CERN – Frascati) group [1], the multihadronic final states produced in (pp) collisions can be understood in simple terms once the leading effects are correctly accounted for. This means that (pp) machines can be powerful tools for the study of QCD effects.

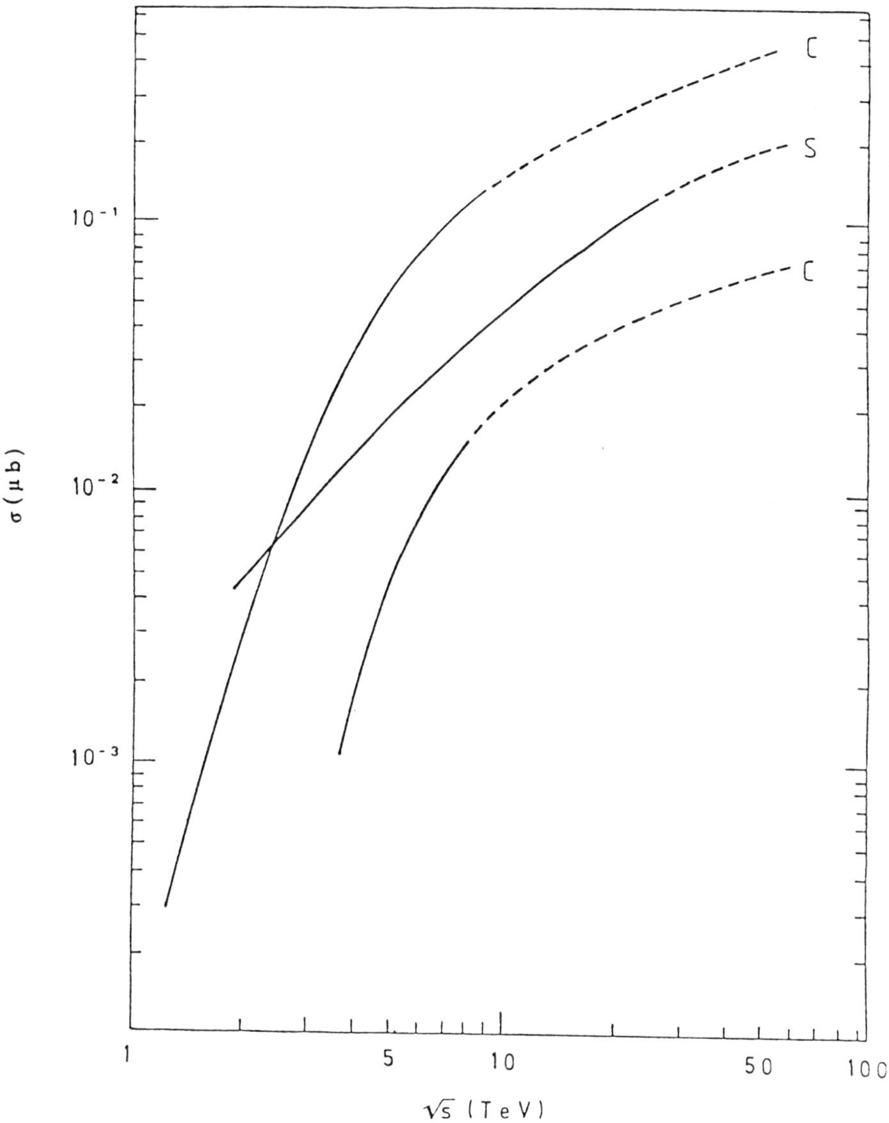

Fig. 8 The cross-section for production of 'heavy truth' derived from 'strange' (curve s) and 'charm' (curves c) cross-sections following formula (3). Notice the width, due to the experimental uncertainties relative to the charm data.

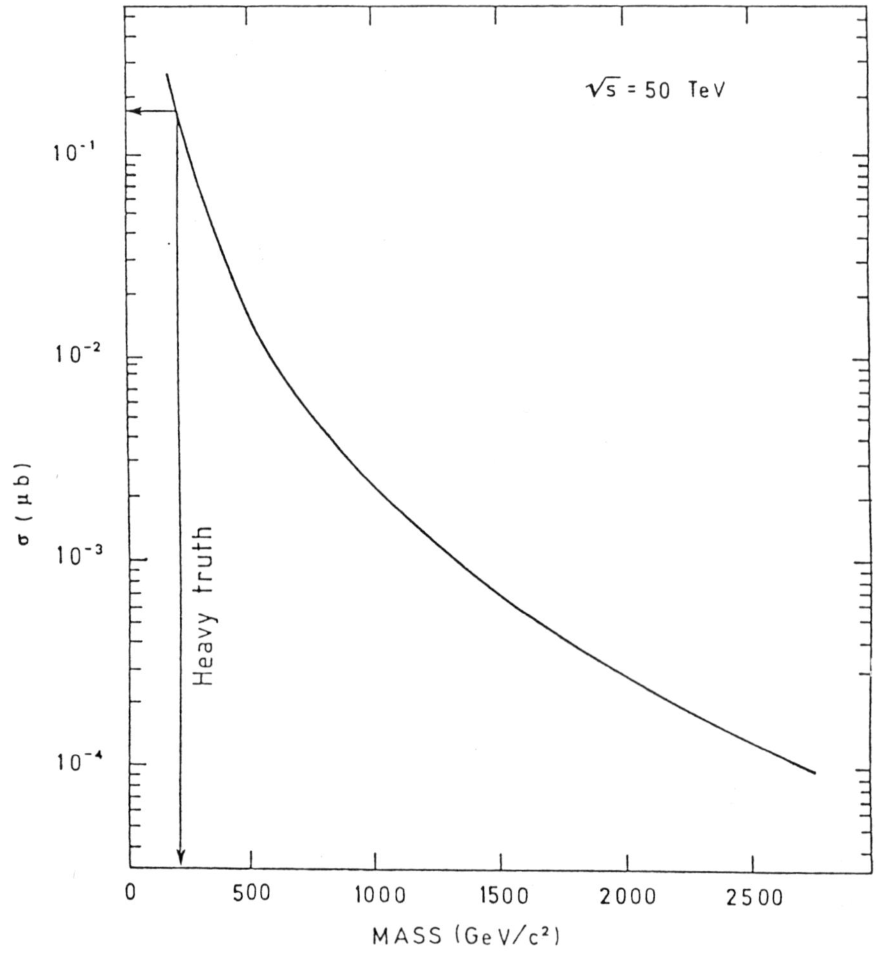

Fig. 9 The production cross-section at $\sqrt{s}$ = 50 TeV as a function of the quark mass.

4.6 A basic difference between (pp) and (p$\bar{p}$) machines

The main problem with (p$\bar{p}$) machines, as compared with (pp) intersecting rings, is the luminosity.

To show the importance of luminosity, let us look at the results obtained [2] in the analysis of the performances achieved by seven typical experiments as a function of machine energy and luminosity (see Table 4).

For the case of high$-p_T$ π^0 production, a factor of 3 in energy is compensated by a factor of 10 in luminosity.

However, if interactions much harder than QCD should exist, then the basic quantity becomes energy.

There is no question but that, in the design of a new machine, we should try for the highest energy and the highest luminosity. Any 'rare' process to be investigated needs the highest luminosity. This means that (pp) collisions win.

5. THE MACHINE

5.1 General remarks

Two extensive design studies of multi$-$TeV proton colliders have been carried out: the LHC in Europe, at CERN, and the SSC in the United States.

For the LHC it is proposed to utilize the LEP tunnel: a 9 TeV per beam machine can be built assuming that 10 T magnets can be obtained from industry after a three to four years R&D program.

The SSC project is not constrained by the site and is more ambitious: a 20 + 20 TeV collider is proposed. For this machine a detailed cost and performance comparison was made between three different magnets. Two were medium field (5.5 – 6.5 T) magnets, the technology for which is practically already available in industry; the third was a 3.5 T 'superferric' magnet. The 6.5 T two – in – one magnet proposed by BNL, a straightforward extrapolation of the HERA magnets that are being produced now by Italian and German industry, was finally chosen. In particular it was concluded that the advantage of the simple design for the 3.5 T 'superferric' magnet was not enough to compensate for the additional complication and cost introduced by the very long circumference required.

Innovative machine component and system engineering is what is mostly needed to make these machines reliable and cost effective.

5.2 Parameters for a 50 TeV ring

5.2.1. The maximum useful luminosity is limited by beam – beam effects and by maximum allowable number of (unresolvable) interactions per bunch crossing that the experiments can tolerate.

The maximum luminosity allowed by the beam – beam limit is

$$L_M = N_b f \, k\gamma \, \xi_M/(\beta^* r_p), \tag{2}$$

where N_b is the number of particles in a bunch, f the revolution frequency, k the number of bunches, ξ_M the maximum allowed linear tune-shift per crossing, β^* (= 1 m) the Twiss beta function at the crossing point, and r_p the classical proton radius. Experience with the SPS shows that ξ_M is given by

$$\xi_M = 0.02/J, \tag{3}$$

where J (= 8) is the number of crossing points around the ring.

Let $<n>$ be maximum number of events per bunch crossing that the experiment can tolerate. In order for the apparatus to be able to resolve one bunch crossing from the next, bunches should be at least t_b away from each other in time (or $s_b = ct_b$ in space). The maximum tolerable luminosity L_n is given by

$$L_n = <n> k\, f/\Sigma, \tag{4}$$

where Σ is the pp total cross-section. In the energy interval we are considering it can be approximated by

$$\Sigma_{(mb)} = 45 \ln E_{(TeV)} \tag{5}$$

The maximum allowable luminosity will of course be the smallest, between (2) and (4). Also note that as the radiation damping time (τ_d) becomes important, the time behaviour of the beam emittance and of the luminosity have to be studied in detail. This has not been done when computing the first-order parameters given below.

5.2.2 Let B be the bending field, the overall radiated power per unit circulating current is then given by

$$P_s/i = (4\pi/3) r_p B c \gamma^3; \tag{6}$$

then the radiated power linear density is given by

$$(p/i) = P_s/(i2\pi\rho) = (2/3)er_pB^2\gamma^2/m_p. \tag{7}$$

The latter quantity is very important when dealing with 'cold bore' magnets. Should it become too large, non−conventional cryostats with special power absorbers at higher temperatures or expensive 'warm bore' magnets have to be envisaged.

In additon there are the cryostat system thermal losses. As a first approximation we assume they are proportional to the overall magnet lenght through the constant p_0, the thermal power loss per unit lenght. The overall power dissipation is then given by

$$P_{tot} = (4\pi/3)r_pc\gamma^3(2i)B + p_0(2\pi\gamma m_pc)(eB)^{-1} \tag{8}$$

The radiated power is calculated for two beams, and thermal losses other than those in the bending magnets are neglected.

It is seen from expression (8) that there is a value of B that minimizes the overall power loss. It occurs when the radiation linear power density is equal to p_0:

$$\begin{aligned} B^2{}_{opt} &= 3m_p(2er_p\gamma^2)^{-1}(p_0/2i) \\ &= 3m_p(2er_p\gamma^2)^{-1}(p/2i) \end{aligned} \tag{9}$$

Note that p_0 is a constant that one would wish to be small. Also note that, since the value of B_{opt} depends on the circulating current and therefore on the value of $\langle n \rangle$, it has to be compted in a self−contained way.

By substituting B_{opt} in Eq. (8) one obtains the corresponding overall power dissipation:

$$P^{(o)}_{tot} = (8\pi e/3\varepsilon_0)1/2(2i)1/2\gamma^2, \qquad (10)$$

where ε_o is the vacuum dielectric constant. Let F be the fraction of the arcs filled with bending magnets, the average arc radius is then

$$R = \rho/F = \gamma m_p c/(eBF), \qquad (11)$$

and the circumference corresponding to B_{opt},

$$C \approx 2\pi\gamma m_p c/(eB_{opt}F). \qquad (12)$$

5.2.3 The results obtained from the above considerations strongly depend on the assumed set of parameters: in particular $<n>$, p_0, F, and the injected beam emittance (which enters into the formulae for L and i) have to be studied in detail. However, a tentative set of parameters is useful for understanding the order of magnitude of the enterprise.

A parameter list for 50 TeV, $\gamma\varepsilon = 5 \times 10^{-6}$mrad, and the optimized field (9) is given in the first two columns of Table 5. We have assumed F = 0.8 for the filling factor and a cryogenic efficiency of 1/300 (the latter may be too optimistic if the magnets are operated at around 2 K). For p_0 we have assumed 0.5 W/m, the same as for the SSC but lower than assumed for the LHC (2 W/m). If one allows the radiation linear power density to increase, higher fields than B_{opt} can be used. In the same table a list of parameters is given for B = 10 T and the same numerical constants as above.

It is worth noting that the overall power loss increases rather slowly with $<n>$ and with the field. A detailed performance and cost analysis has therefore to be performed in order to decide between the proposed solutions.

Table 5

$\langle n \rangle$	1	4	1	4
B (T)	5.9	4.2	10	10
R (km)	35.5	50.2	20.8	20.8
C (km)	223	315	131	131
k	29735	42052	17453	17453
N_b	1.2×10^{14}	3.4×10^{14}	7.1×10^{13}	1.4×10^{14}
L ($cm^{-2} s^{-1}$)	2.2×10^{32}	9×10^{32}	2.2×10^{32}	9×10^{32}
P_s (kW)	89.2	126.2	152.1	304.2
P_t (300) (MW)	54	76	61	107

Table 6

Estimated cost in US dollars ($\times 10^6$)

(Extrapolated from results of the Cornell Workshop;
to be achieved through a vigorous R&D program and innovative engineering)

	Low-field	High-field
General facilities Injector systems Experimental areas	1000	1000
Magnets	1200	2000
RF, controls Refrigeration Other facilities and installations	800	600
Tunnel, roads Power distribution Service buildings	800	300
Contingency	800	800
Total	4600	4700

5.3 The cost

The cost estimate depends on many factors and is, of course, the major problem. If one chooses an optimistic approach to see where one stands (assuming there will be a vigorous R&D program and innovative engineering), the figure arrived at, for the total cost, is about 5 x 10^9 US dollars.

A breakdown of cost into the various items is given in Table 6. This estimate could change according to the results of the large R&D needed before the executive project is worked out.

5.4 Specialized working groups

The ELOISATRON project is divided structurally into the following working groups:

1) Machine theoretical
2) SC magnets
3) Injector
4) Civil engineering
5) Experimental areas
6) Refrigeration power plants
7) Standard plants
8) Vacuum
9) Radiofrequency
10) Controls
11) Anti – 'desert' physics
12) Data handling and supercomputers

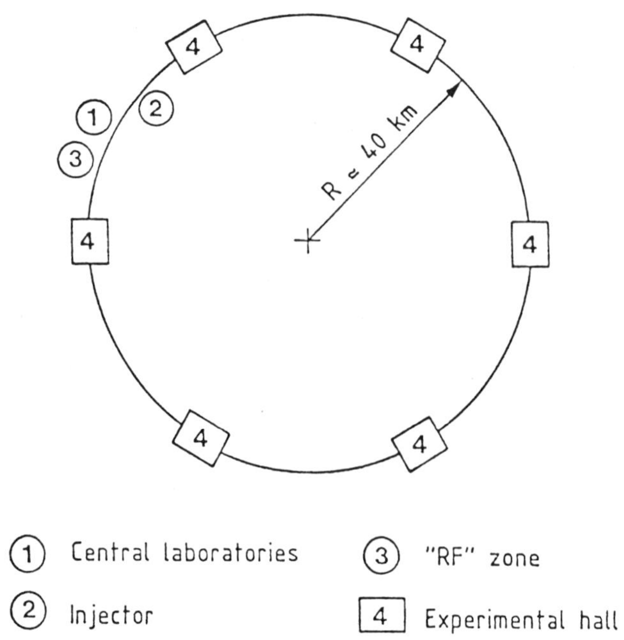

Fig. 10 General layout of the ELOISATRON ring.

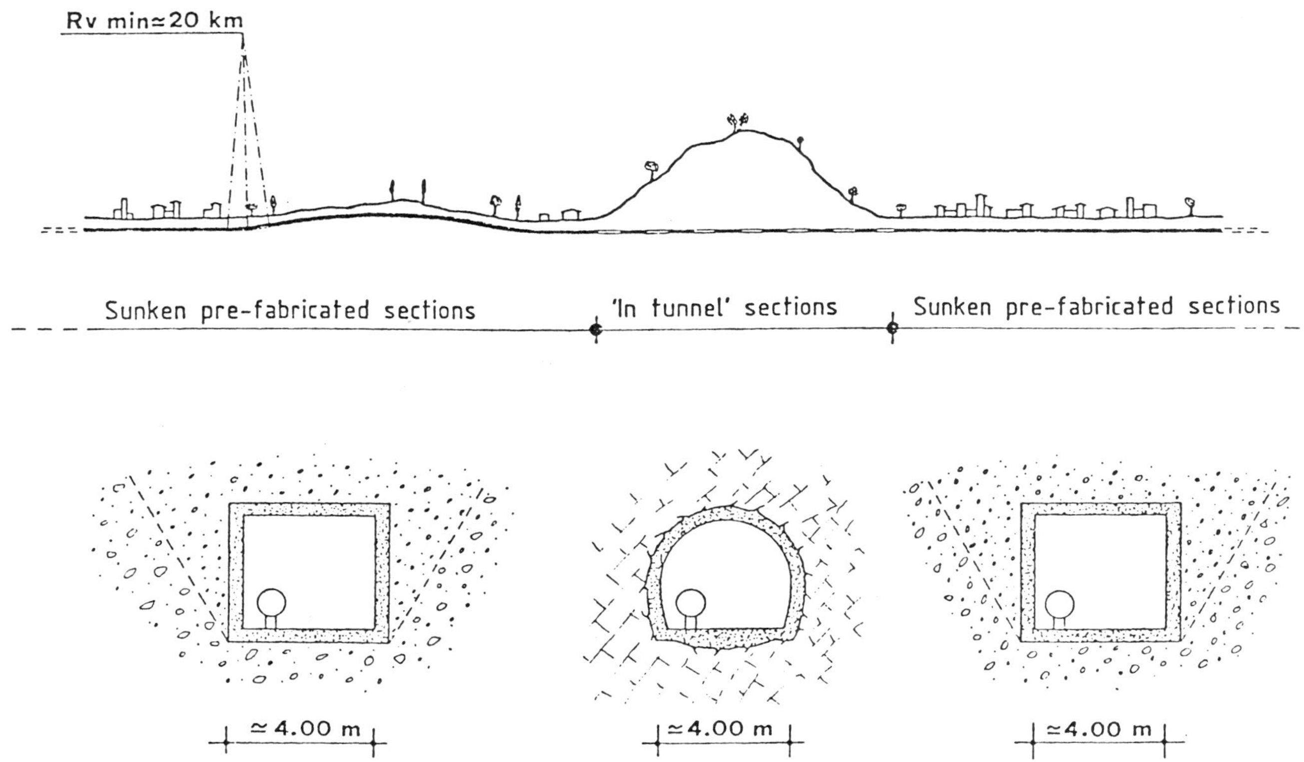

Fig. 11 A longitudinal section of the ELOISATRON. Notice the sunken prefabricated sections and the 'in tunnel' sections.

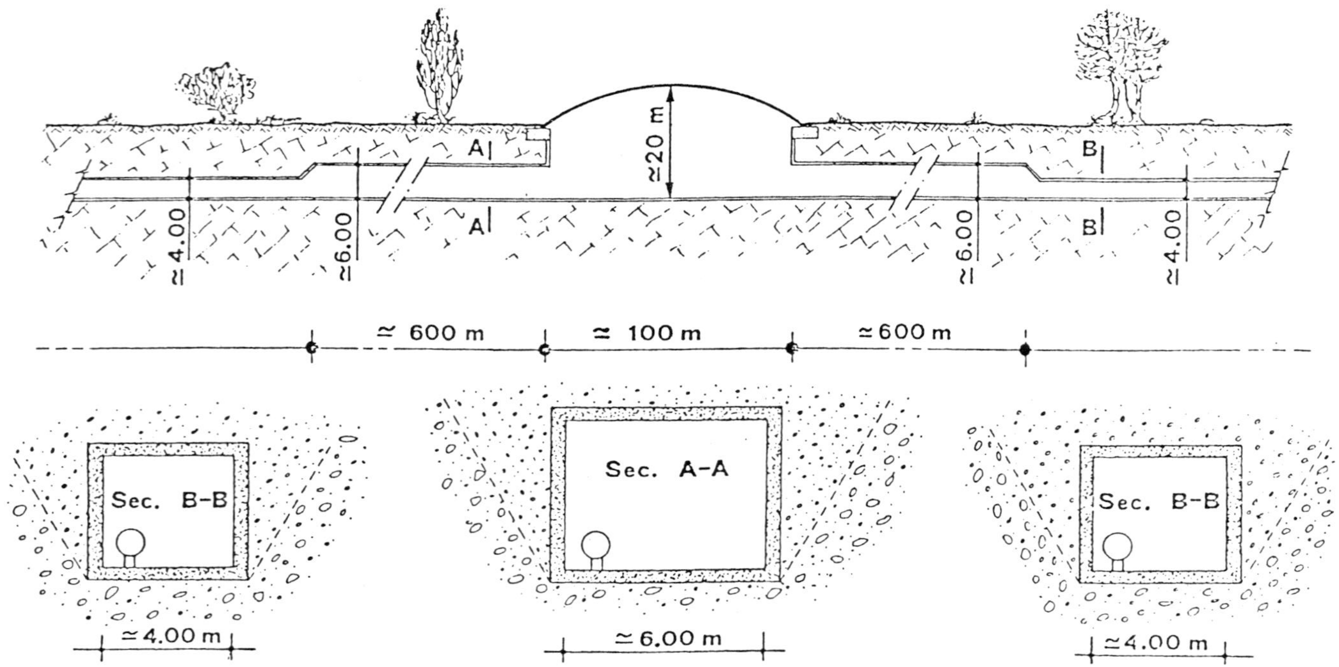

Fig. 12 The schematic longitudinal section of an experimental hall.

5.5 Where and how to build the ELOISATRON

The ELOISATRON should be installed in Italy where – as mentioned earlier – there are many good regions that are seismologically very stable. Figure 10 shows an example of the general layout of the machine.

Depending on the flatness of the ground, the machine could be built, in part, as sunken prefabricated sections (6m deep, 4 x 4 m^2 in cross – section) and, in part, 'in tunnel', as shown in Fig. 11. The cost has been estimated to be not higher than for the pipe – line solution.

Experimental areas should be large, for example at least 100 m wide x 20m high. Their lenght could be either 40m or 100 m, depending on the physics program to be pursued. (see Figs. 12 and 13).

6. CONCLUSIONS

This project is open to the widest collaboration, both national and international. It should not interfere with the other European projects at present under way.

The first priority is therefore to find new sources of money.

The second priority is to solve the technical and engineering problems; for example: superconducting magnets and antiproton sources, if antiprotons are wanted.

Finally, a close and effective collaboration should be established among the national and international centres where there is knowledge of and competence in the various fields of interest to the ELOISATRON.

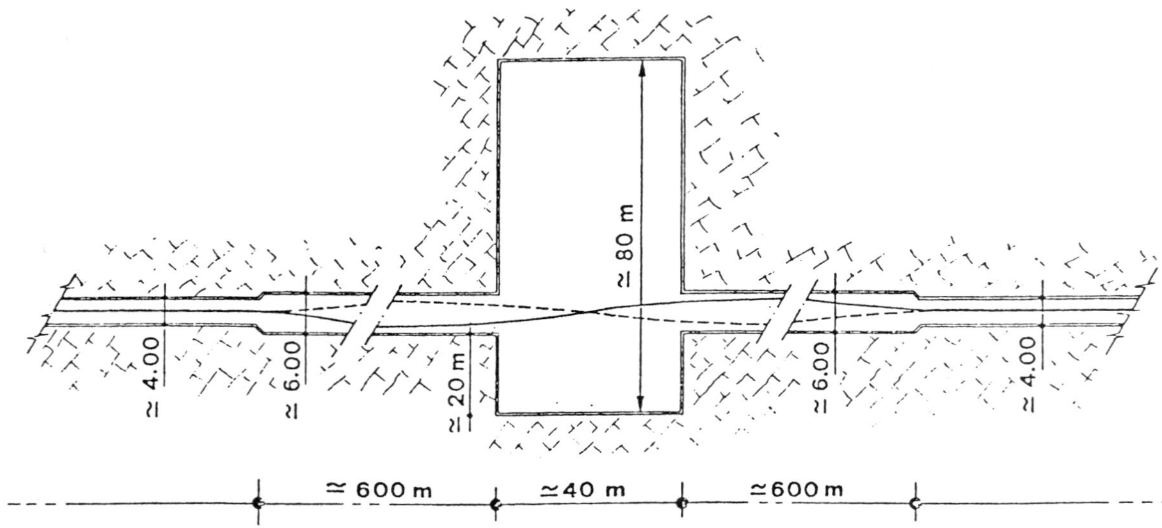

Fig. 13 The schematic plan view of an experimental hall.

References

[1] M.Basile et al., Report INFN/AE– 83/7 (1983).

[2] G. Kane, private communication.

Discussion A. ZICHICHI

P.WEILHAMMER

You mentioned only one place specifically to implement the (10%) Eloisatron, namely the LEP tunnel, around 1990. In case of opposition of LEP users at that time in which LEP experiments will do major installation work, what are other possibilities for a (10%) – Eloisatron installation?

A. ZICHICHI

There are three proposal for the 10% project: one at DESY, one at CERN and one, coming from C. Rubbia, to start from scratch in Italy. In my opinion the third solution is probably too expensive, but the discussion is still open and we don't have yet a solution.

W. SCHMIDT – PARZEFALL

At present, the HEP – community does not seem to completely agree, whether it is more promising to do the next step towards experimentation at higher energies with pp or with e^+e^- collisions. You skipped this point in your presentation.

A. ZICHICHI

The 100 Tev physics , in term of constituents is like 10 Tev e^+e^- because 50% are gluons, the rest being quarks. Since at the moment there is even no idea how to build a 1 Tev e^+e^-, the comparison, from the technical point of view, is in favour of a hadron machine. There are however also physical reasons for the choice. At high energies the pp interaction is probably more rich than an e^+e^- annihilation which is forced to go through the intermediate photon. Even ep interaction could be more important at high energies as in Hera, the first machine probably capable to show point – like interactions between leptons and quarks.

J. KIRKBY

I think there is one R&D group missing from your list: a marketing group. The task of this group is to develop the case that the giant Eloisatron project creates more money than it costs, by creation/stimulation of industry ect... I can think of two current examples where marketing has transformed attitudes: the Olympics and the Channel tunnel. Both these are now welcomed as money – machine ventures where before they were seen as giant money sinks.

A. ZICHICHI

There is no specific group since in my intention this would be an engagement for all people involved in the project. I agree completely with you and in effect, as a consequence of the work performed in Italy on this specific points, now Italy fully supports CERN and is willing to support our new projects, like Eloisatron. I hope that more other countries can follow our example.

G. LUTZ

You mentioned a period of 20 to 30 years from putting a new idea on accelleration into reality. Don't you think that this is partly the result of not putting enough effort into it, due to the lack of a specific project for which it might be used? What is your opinion on how much such a development period could be shortened by increasing the effort.

A. ZICHICHI

We had a workshop on this precise aspect and if new ideas will come we will put all our efforts on them. However the 10% project can be built with present technologies and at reasonable costs: why wait? For the 100% Eloisatron there is enough time for new developments.

ELOISATRON PROJECT

This project has been presented at:

- **VERSAILLES Summit Meeting (France)**	**June 1982**
- **SANREMO (Italy)** Nobel and Galilei Celebrations	**2 May 1983**
- **ROME (Italy)** Nobel and Galilei Celebrations	**9 May 1983**
- **CASTIGLIONE DELLA PESCAIA (Italy)** 2nd Pisa Meeting on Advanced Detectors	**2 June 1983**
- **ERICE (Italy)** Ettore Majorana Centre for Scientific Culture International School of Subnuclear Physics 21st Course	**8 August 1983**
- **COMO (Italy)** 3rd International Conference on Physics in Collisions	**2 September 1983**
- **WASHINGTON (DC, USA)** National Science Foundation Working Group Meeting on High Energy Physics	**3 October 1983**
- **SALERNO (Italy)** University of Salerno	**4 December 1984**
- **POTENZA (Italy)** University of Potenza	**3 March 1985**
- **BOLOGNA (Italy)** University of Bologna	**14 March 1985**
- **LIVERMORE (California, USA)** Lawrence Livermore National Laboratory	**9 April 1985**
- **BOULDER (Colorado, USA)** Universities of Colorado — Conference on World Affairs	**11 April 1985**
- **CASTELGANDOLFO (Italy)** National Conference on the INFN (Italian Institute for Nuclear Physics)	**10 July 1985**

- **ERICE (Italy)** **13 August 1985**
  Ettore Majorana Centre for Scientific Culture
  International School of Subnuclear Physics
  23rd Course

- **ERICE (Italy)** **3 September 1985**
  Ettore Majorana Centre for Scientific Culture
  International Conference on Advanced Solid
  Earth Geophysics

- **TRIESTE (Italy)** **8 October 1985**
  21st Congress of the SIF
  (Italian Physical Society)

- **GENEVA (Switzerland)** **11 November 1985**
  CERN — Senior Staff Consultative Committee

- **STRASBOURG (France)** **20 November 1985**
  Council of Europe — Parliamentary Assembly
  Committee on Science and Technology

- **BOLOGNA (Italy)** **15 December 1985**
  Academy of Sciences — Invited Plenary Address for
  the opening of the academic year

- **MOSCOW (USSR)** **29 January 1986**
  Presidium of the USSR Academy of Sciences

GENERAL DISCUSSION

G. CHARPAK

I was quite impressed by the results shown by BOURQUIN: 35 μm resolution in a gas detector, considering that you can get considerable number of measurements, imply that you reach an accuracy on the vertex which is probably comparable with the one achieved with solid state detectors. For these reasons it is a pity that you have given up so soon because of the ageing problems found with isobutane. There is a lot of experience on this gas and probably you can find a way to solve your problems keeping the isobutane.

M. BOURQUIN

In fact we have not given up. The decision of replacing isobutane was dictated by the necessities of the running experiment, but studies on other possible gas mixtures are still under way.

J. KIRKBY

It doesn't seem to me there should be a competition between wire chambers and solid state chambers. If you start searching for the best precision you of course will choose solid state detectors because of the proved 5 μm resolution and very good double hit resolution. But, you cannot build a tracking device entirely with silicon, because of the multiplexing problem which gives you too long read – out time. So you are forced to back – up your silicon with a wire chamber which is fast, usable in a trigger etc..... On the other hand if you want to push wire chamber resolutions to 5 – 10 μm you will require a high density of wires with all the related mechanical and electronical problems: you will loose completely the advantage of a fast device. So in my opinion the two devices must cohexist, each one providing complementary information.

F. PIUZ

We have a certain experience with both type of detectors and we believe that the 200 μm double hit resolution achievable with silicon detector is the most important parameter for events with high density of particle. The 1 mm obtained with wire chambers will not be sufficient.

R. HORISBERGER

In my opinion the relevant quantity for a vertex detector is the resolution /% radiation lenght. Extrapolating from the Walenta data, he is able to have two 5 μm measurements in 1.6% radiation lenght while, from the Delphi vertex chamber proposal, we can get two 5 μm measuremt in 1.0% ration lenght. Increasing the preassure will not be helpful since it will require to renforce the mechanic and, as a consequence, increase the radiation lenght. Another point, as already mentioned, is the double track resolution: from our experience when you go at 8 cm, 50 μm pitch is the maximum limit you can go without loosing physics because of double track separation. I don't see how one can do as well with a wire chamber.

F. SAULI

Altough wire chambers can provide high space resolution, in general this goes at the expense of the rate capability. With induction chamber it seems to me unlikely that one can build large systems with time resolutions much better than 1 μs (due to the amplifiers bandwidth). On the other hand, silicon detectors do not suffer from this limitation. So we must be fair and compare the devices not only on accuracy but together with rate capability.

D. RITSON

We must underline that as you go closer to the interaction point, segmentation and an easy paking are the essential requirements. In my opinion, silicon detectors are a valid solution while wire chambers, with their increasing complexity, will probably have a hard life.

M. ATAC

I agree for silicon being used for the innermost part on huge Collider detectors, and we must encourage research on few relevant items: i) Gallium Arsenide development because of its radiation hardness; ii) multiplexing capability; iii) radiation resistive analog devices. As a final remark I would like to point out that the future of experimentation will rely on total energy flow, i.e. calorimetry. But to distinguish jets, electrons etc... you can't go too near the interaction point and you will need π^0 recontruction: wire chambers have their future role as complementary devices for calorimetry.

A.H. WALENTA

Let us go to more practical items. I try to extrapolate the DELPHI proposal to HERA: for a 1 meter long, 3 layer silicon detector I ended with a cost of 18 MDM. This comes essentially because with wire chambers you can go from 5 cm to 1 meter long wires with the same accuracy, while for silicon you can't go above 5 cm because of noise and capacity problems. So for 1 or 2 meter long chamber you will have reasonable proposals only with wire chambers.

G. CHARPAK

The costs problem are foundamental. Can we check with the DELPHI estimate?

R. HORISBERGER

The DELPHI estimate for the silicon vertex detector is 450 KSFR. The detector is 25 cm long (8 cm beam pipe radius), 2 layers. The price includes only the silicon.

G. CHARPAK

So probably if you include everything, Walenta was optimistic.

D. RITSON

Probably is possible to go down to 3 – 4 cm with the inner radius and reduce the costs to reasonable values. Our experience had shown that background problems are often over weighted while reality is more favorable.

R. HORISBERGER

I want to point that the overall costs of the DELPHI vertex is around 1 MSFR, including a $\mu-$VAX ect... The read – out cost is around half of the silicon cost.

G. LUTZ

Also for the ALEPH vertex detector the total cost is around 1 MSFR. However, you cannot discuss silicon detectors against gas detectors without having any specific purpose in mind. It will be quite different if it is in the context of present experiments or of future experiments.

F. PIUZ

I would like to make two points:

i) in a multi tracks event the resolution of the first and the second track measured with silicon are comparable, while normally, in a wire device, the accuracy for the second track is very poor;
ii) from the first point, even if with wire chambers you can have more layers, not all points will be usable when you have very near tracks and, a consequence, even few silicon layer will give comparable number of informations.

K. MORGAN

To have a good accuracy is not so important if, later on, you are not able to associate two 5 μm apart spots to true tracks.

G. LUTZ

With one layer, I agree, is difficult to associate points to tracks. But if you use 2 layers, each with 10 μm precision, you can define a direction and require matching with your candidate. Also pattern recognition could be easier with Silicon detector due to their 100% efficiency.

F. SAULI

I have a question on silicon detectors. To measure a coordinate I imagine you integrate the signals and make a center of gravity. How long is your integration time?

R. HORISBERGER

In our test we used 140 ns integration time.

F. VILLA

Could you tell me the time for two successive tracks to be distinguished.

R. HORISBERGER

140 ns, adding, more storage capacities to the Microplex chip.

D. RITSON

Walenta chamber achieves very good results but, being far from the center of the beam pipe, due to multiple scattering he will get information only in a statistical way and not event by event. To go really on an event measurement you have to go to small radia and only Silicon detector, at the moment, seem to be able to do this.

J. JAROS

I would add that we made some studies on possible mixed devices for MARK II and it was clear that, due to multiple scattering, as you go down in radius, you effectivelly gain using Silicon detector. You just need few very precise points near the vertex and a less precise device (a wire chamber) with a reasonable lever arm to obtain the optimum.

G. CHARPAK

If you combine the two dimension readout CCD with the tubes presented by Sauli, even if the read−out is slow, you will have such a density of information that you can handle easily ambiguities and not pointing spurius tracks.

J. JAROS

That is right, but again CCD makes sense with small radia, otherwise multiple scattering will complicate life.

F. SAULI

CCD can be used at SLC because of the very low crossing rate. We need to wait until they can be gated before using in any other Collider, othewise they will be all black.

G. CHARPAK

But the pixels are so small that they will not be all black.

G. LUTZ

In effect they have already being used in fixed target experiment.

J. KIRKBY

Multiple scattering is one point, but double hit resolution is the other main point since we will face very dense events. Only silicon detectors give good results in this respect.

J. JAROS

We have also to compare results from the physics point of view. I have shown that 100 μm double tracks resolution obtained with our drift chambers garantee efficiency for 90% of tracks in hadronic events. This is enough to make physics.

F. VILLA

This at low energy; at high energies you will require more.

M. ATAC

Is much easer to think to Silicon in e^+e^- machine. At Tevatron, with 1 m long intersection region, you have different and most serious problems related to heat, cables, space...... The solution could be in an hydrid system (1 silicon layer + wires) but it is a quite complicated solution.

R. HORISBERGER

I would like to comment on double tracks resolution. If you want to detect only the existence of some displaced tracks from a secondary vertex, I agree, this parameter is not so crucial. But if you want to reconstruct completely an exclusive B decay on an event by event basis, then even 90% for 5 tracks or more may not be acceptable and only silicon devices will give you a sufficent reconstruction efficiency.

J. JAROS

With multiple scattering and actual resolution, we are not yet in the possibility to reconstruct all vertices and associate all particles to each one of them. Today we can hope to tag a decay, to tell the flavour and reconstruct few tracks with larger impact parameter. A complete picture on event by event basis is, in my opinion, not for the present generation of detectors.

F. VILLA

There is also a space problem in the game. To cover a certain area with silicon you need a certain amount of space for the Microplex and electronics associated. Which are its dimensions?

R. HORISBERGER

The Microplex is about 6.5 mm wide and 5 mm deep for 128 channels. In total, at the moment 11 cm of strips detectors is accompained by 5 mm of read – out system, so about 1/22. But if you try to read also the back side of silicon detectors, then this fraction will increase.

M. ATAC

Going back to the intrinsic speed of silicon detectors, from my experience with 300 μm the rise time should not exceed 10 – 20 ns. So why were we talking of 140 ns gate width to read – out these strips.

G. LUTZ

The problem is in electronic: you can have low power, low noise electronics but not so fast. Going to 10 mW probably you would reach 30 ns rise time.

V.M. KELLY

From the numbers quoted at this meeting it seems that the cost of a silicon detector is ~ 500 k$ and the overall associated electronic costs is ~ 200 k$. You cannot expect with the present demand a great reduction in costs. Probably integrating the detectors with the amplifiers in the same wafer you could reduce the price, but it is not something for the near future.

F. VILLA

I hope the prices will go down, but in effect there are examples of things whose price has not being reduced, like solar cells for example. On the other hand if there is really a need for a costly detectors, then is up to us to find the right way to money.

G. CHARPAK

I have a question for Walenta. You told us the price for a silicon detector in HERA, but which is the cost for solution you are now proposing?

A.H. WALENTA

It is of about 600 KDM.

G. LUTZ

Going back to Kelly proposal for an integration of the detector with the electronics, we are presently investigating this possibility. However there are certain technical problems and expecially the yield on a wafer, which today is already only 50%, will probably go down for detectors with electronics in monolithic way.

INDEX

Absorption
  SCIFI, 232, 239
  waveshifters and, 243
ACCs, MARK-J detector, 91
Acetaldehyde, 163
Acetic acid, 163
Acrylic scintillators, 232, 233, 234
ADC, 63
Advanced computer programs, (ACP), Fermilab, 170
Aging, 90 (see also Wire chamber aging
Alcohol-gas mixtures
  and chamber aging, 154-157, 160, 163, 164
  MARK-J detector, 85
ALFIE, 276, 279-288
  calorimeter, 283, 285-288
    advantages, 286-288
    electron/hadron response, 285
    structure, 283, 286
  charge-flow detector, 281
  components, 279-280
  micro-converter, 281-283
  microtracker, 280-281
  muon spectrometer, 288
  z-vertex detector, 281, 282
Alignment
  MAC vertex chamber, 130, 131
  one TeV system, 6, 15
Alternate doped multilayer, 216, 224
Alumina plates, TEC, 98-99
Aluminized mylar, see MAC vertex chamber
Aluminum wires
  aging, 156, 157, 159, 161, 163, 164
  PEP vertex chamber, 39-40
Amplification, nondestructive, 216-219, 220
AMY, 115
Analog readout, 206
Angular multihit capability, 25
Angular resolving power, one TeV jet system, 5, 16
Annealing, SCIFI, 235
Anode signals, MARK-J detector, 72
Anode wire (see also Wire chamber aging
  MAC vertex chamber, 118
  MARK II drift chamber, 59
  MARK-J detector, 73, 75
  multidrift tubes, 102, 103, 104
  PEP vertex chamber, 39-40
Anomalous decay vertices, 44
Anthracene, 232, 233, 234
Antiblooming diodes, 256
Argon-carbon dioxide, 87
Argon-carbon dioxide-methane, 117
Argon-carbon dioxide-xenon, 117
Argon-ethane
  MARK II drift chamber, 55-56
  PEP vertex detector, 69
  streamer plateau for single electrons, 117
Argon-ethane-ethanol, 154, 157

Argon-ethane-isopropanol, 154-155
Argon-ethane-methanol, 154, 156, 157
Argon-hydrocarbon gases
- MAC vertex chamber, 117-118
- MARK II, 86, 87

ARGUS, 16, 35
Asymmetry, charge, 30
Attenuation, scintillator, 239
- glass, 242, 243, 244, 245, 246, 274
- plastic, 232, 235, 236

Attenuation length, 231, 234, 245, 264
Axial drift cells, 39
Axial vector couplings, 19, 29-30
Azimuthal angle
- MARK II, detector, 42
- MARK-J detector, 75, 76

B hadron lifetimes, 115, 145
B mesons
- lifetime of, 27-28
  - MAC vertex chamber, 141-145
  - one TeV system, 6
  - PEP vertex chamber, 45-51
- one TeV system, 12, 16
  - magnetic field, 6
  - reconstruction of, 4, 5
  - resolution, 16
  - tracking, 10-11
- tagging, 4, 5, 24-27, 44 (see also Tagging)

B mesons at PEP and SLC, 19-33, 45-47, 49-51
- physics issues, 19-20
- signal isolation, 21-24
  - numbers for PEP, 23, 24, 21-24
  - tagging numbers for MARK II at SLC, 24-27
- sources, 20
- tagged B samples, 27-33
  - lifetimes, 27-28
  - mixing and b-->u transition, 28-29

B mesons at PEP and SLC (continued)
- tagged B samples (continued)
  - production characteristics at SLC, 29-33

b quarks, 1
- fragmentation function, 19
- hadronization of, 145
- lepton spectrum, 31-32
- lifetimes, 38
- from new particles, 20

Background, 2
- charm signals, 178
- MAC vertex chamber, 120
- supercollider, 268

Back-to-back events, Bhabha tracks, 42, 44
b$\bar{b}$ events
- Higgs decay and, 33
- MAC vertex chamber, 142, 143, 145
- MARK II PEP vertex chamber, 47
- Si detectors, 16
- vertex displacement, 23

B$\bar{B}$ events
- like-signal leptons, 16
- multitrack resolution and, 16
- third generation detectors and, 39

BBQ, 233
Beam centroid, MAC vertex chamber, 142, 143
Beam gas events, PEP vertex chamber, 40
Beam pipe
- MAC vertex chamber, 121-122
- MARK II drift chamber, 53, 68
- MARK-J detector, 72, 77, 90
- time expansion chamber, 100

Beam test, MAC vertex chamber, 119-120, 121
Beam trigger event, MARK-J detector, 85
Bhabha events
- MAC vertex chamber, 116
  - calibrations, 129-130, 132, 133

Bhabha events (continued)
MAC vertex chamber (continued)
chamber efficiency, 135
cross-talk, 129
drift corrections, 130-131
miss distance, 139
multihadron events, 140
residuals, 137-138
MARK II PEP vertex chamber, 41-44
MARK-J detector, 82, 83, 84, 85, 87
Bias, MAC vertex chamber, 130
Bias voltage, strip detectors, 209
Biased MOS structure, 207
BINOS, 80
Birks constant, 242
bis-MSB, 233
Bottom, see b quarks
Branching ratios, 16, 192
B decays, 19, 27
semileptonic, 28
Bunch spacings, supercolliders, 270
Buried channel CCD, 200
Butane, 117-118

Cabibbo-suppressed hadronic decays, 192
Calibration
MAC vertex chamber
channel-to-channel, 131-134
off-line, 129-130
thresholds, 120
time-to-distance, 130-131, 132
MARK-J detector, 89
Calorimetry
MARK-J detector, 83
one TeV jets, 1, 16
SCIFI, 229, 230
supercollider, 278
CAMAC, 75
Carbon dioxide gas mixtures
MAC vertex chamber, 117, 118
MARK II drift chamber, 55
Carbon dioxide gas mixtures (continued)
MARK-J detector, 86, 89
and wire aging, 163
Cathode wires
carbon dioxide and, 163
MAC vertex chamber, 116, 118
MARK II drift chamber, 59
MARK-J detector, 73
PEP vertex chamber, 39-40
$c\bar{c}$ events
background
B tagging at SLC, 26
reduction of, 23
MAC vertex chamber, 142, 145
Si detectors, 16
CCD, see Charge-coupled devices
CD algorithm, 134
Cell design
SLC drift chamber, 53-55
time expansion chamber, 96
Center of gravity method, 72
Central detector, MAC
efficiency, 135-137
pattern recognition, 134-135
resolution, 137-140
Centroid measurement, 100
Cerenkov counters, 169
Cerium, 241, 242
CERN-MPI-RAL SCIFI readout, 257
CESR, 20, 27, 28, 29, 115
Chamber, MAC
design, 122-124
performance, 136-136, 137
Chamber resolution
MAC vertex chamber, 137
PEP detector, 68
Channel guide, 210, 211
Channel number, 36
Channel stops, 213
Charge asymmetry, 30
Charge coupled devices (CCD)
as buffer before trigger, 35
fully depleted, 209-216
multi image, 216-217
properties, 204-205
SCIFI readout, 253, 254, 255, 256, 257, 258

Charge coupled devices (continued)
semiconductor principles, 200-201, 202
supercollider, 289-292
Charged particles
B decays into, 16
gluon and quark jets, 2, 3
MAC tracking paramaterization, 146-147
one TeV jets, 1, 2
supercollider tracking, 278
tau decay into, 44
Charge-flow detector
ALFIE, 279, 281
SCIFI, 229
Charmed particles, 19
lifetime measurement, 181-185
MARK II PEP vertex chamber, 38, 44
signal extraction, 178-181
tagging at Z^o, 26-27
Charm photoproduction
charm signal extraction of, 178-181
charmed meson lifetime measurement, 181-185
event reconstruction, 173-178
track finding, 173-174
track fitting, 174-176
vertex fitting, 176-178
Fermilab experiment E691, 168-170
computing, 170
data sample, 170
tagged photon spectrometer, 168-170
future plans, 190
silicon microchip detector design, 171-173
silicon microchip detector performance, 186-189
detector, 186
mass resolution, 189
noise, 187
per-plane efficiency, 187
per-track efficiency, 188
Charm photoproduction (continued)
silicon microchip detector performance (continued)
two-track resolution, 188-189
Cladding interface, SCIFI, 239
CLEO, 16, 115
Collimator, MAC chamber, 120, 121, 122
Collision rate, SSC tracking detectors, 35
Colour centres, 241
Computation, MARK-J detector, 75
Conduction by electrons, 196
Conduction by holes, 197
Coordinates, measurement precision, 12-14, 35
Copper, aging, 158, 160, 162
Cosmic ray triggers, MARK-J detector, 80-81, 87, 88, 89
Cost
multidrift tubes, 113
SCIFI, glass, 245
Couplings, vector, 29-30
Covariance matrices, 147
CP violation
B mesons, 19
K^o, 28
Cross sections, at one TeV, 1
Crossings, supercolliders, 270
Cross-talk, 240
Bhabha tracks, 42, 44
MAC vertex chamber, 129
SCIFI, 239
Current
discharge, and wire aging, 164
PEP vertex detector, 40-41, 68-69

D mesons, 16, 19, 50
event reconstruction, 28, 30
lifetimes, 45, 46
mean flight distance, 22
one TeV system, 12

D mesons (continued)
  tagging, 26-27
Dark current, 205
Data acquisition
  Fermilab experiment E691, 170
  MARK-J detector, 91
Data compression, pp collider, 15
$d\bar{d}$ events, 23, 26
Decay length
  B tag efficiency versus, 25
  MARK II PEP jet vertex decays, 41, 47, 48
  tau, 44-45
Decay time constant, SCIFI, 234, 245
Decay vertices (see also specific particles and detectors)
  B decays, 21, 22
  MARK II vertex chamber, 38
Delrin, 123, 124
Density
  SCIFI, glass, 245
  supercollider calorimeter, 286
Density of measurements, supercollider, 278
Depleted mode, 198
Depleted P channel metal oxide semiconductor, 218
Depletion, diode, 203
DEPMOS, 218-219
DESY, MARK-J, see Mark J detector
Detection efficiency, MARK II, 22
Detection gap, MARK-J detector, 74, 75, 80
Detector-amplifier structure, 216-219, 220
Deuterons, see D mesons
Diameter
  SCIFI, 231
  MARK-J detector chamber, 73
Di-jet system, 2
Dileptons, like-sign, 29
Dimethyl ether
  MAC vertex chamber, 117
Dimethyl ether (continued)
  multidrift tubes with, 105-112
  time expansion chamber, at HERA, 95, 98
Diode strip detectors, 199-200
  double sided readout of, 206-209
  properties, 204
Diodes, semiconductor drift chambers, 196, 197, 202-204
Discharge current, and wire aging, 164
Discrimination threshold
  MAC vertex chamber, 121
  multidrift tubes, 106
Discriminator circuit, MAC vertex chamber, 124, 126
Distortions, SCIFI readout, 253
Doping, 199, 224
DORIS, 27
Double diode structure, 203
Double sided readout strip detectors, 206-209
Double track resolution, see Two-track resolution
Drift, MAC vertex chamber calibrations, 130-131
Drift cells, PEP vertex chamber, 39
Drift chambers, 205
  MAC, see MAC vertex chamber
  MARK II, 51-65, 68
  MARK-J, see Mark J detector
  semiconductor, 196, 199, 205
    fully depleted CCD detectors, 209-216
    structures, 202-204
  tagged photon spectrometer, 168
  tilted cells, 16
  time resolution, 35
Drift distance
  MAC vertex chamber, 117, 119, 120
    multihadron events, 141
    residuals, 137-138

Drift distance (continued)
  MARK II drift chamber, 55-56
  MARK-J detector, 75, 76, 85
Drift field, MAC vertex chamber, 116
Drift length, MARK-J detector, 84
Drift lines, time expansion chamber, at HERA, 96
Drift time
  MAC vertex chamber, 119
    calibrations, off-line, 129-130
    measurement system, 127
  MARK-J detector, 72
  multidrift tubes, 105, 107, 108, 108
  time expansion chamber, at HERA, 97
Drift tubes, see Multidrift tubes
Drift velocity
  isopropyl alcohol and, 161, 163
  MAC vertex chamber, 127-128
  MARK II drift chamber, 51-53, 57
  MARK-J detector, 72, 73, 78-80, 89
    gas system modification and, 87
    operation experience, 78, 79, 80
DRPs, MARK-J detector, 91
Dynamic RAM, 224
Dynode, SCIFI readout, 255

Efficiency
  MAC vertex chamber, 135-136
  MARK-J detector
    deterioration of, 85
    gas system modification and, 86, 87
    wire number versus, 88
  MARK II, 22
  multidrift tubes, 106-107
  polar angle versus, 83
  silicon microchip devices, 187, 188
Electric field
  MARK-J detector, and drift velocity, 78, 79, 80
  time expansion chamber, at HERA, 97
Electrodes, MARK II drift chamber, 59
Electromagnetic calorimeter
  ALFIE, 279, 285-288
  MARK-J, 83
  SCIFI, 230
Electromagnetic spray, PEP, 40
Electron drift, 1 TeV silicon strip detectors, 14
Electron identification, supercollider, 268, 278
Electron traps, 241
Electron-hadron response, ALFIE, 286
Electron-hole pair, 198
Electronic noise, microchip detectors, 187
Electronics
  MAC vertex chamber
    design, 124, 125, 126
    drift corrections, 130-131
    performance, 127-128
  MARK II drift chamber, 63
  MARK-J detector, 75, 91
  multidrift tubes, 105
  one TeV system, 15
  TEC at HERA, 99, 100
Electron-positron collisions, 19-20 (see also specific detectors)
  B physics, 20, 22
  MARK-J detector, 72
Electron-positron pairs, see Bhabha scattering
Electrons
  conduction by, 196
  heavy neutrino coupling, 48, 49
  radiation damage, see Radiation damage
  supercollider, 268, 278
Electroweak interactions, 1
ELOISATRON, 115, 145

ELOISATRON (continued)
basic steps ofof project, 302-304
energy level and related technology, 301-302
location of project, 304-305
objectives, 306
present status of fundamental knowledge, 298-300
third phase and schedules, 306-307
EMA, 239, 240, 264, 274
Emission
SCIFI, 232, 239, 244
waveshifters and, 243
End effects, MARK II, 58
Endplates
MAC vertex chamber, 123
multidrift tubes, 102, 103-104
Energy loss, semiconductors, 198-199
Energy resolution
ALFIE, 286
SCIFI, 229
ENERTEC, 193
Epitaxial growth, alternate doped multilayer, 224
Epoxy, 117
Error
MAC vertex chamber, 145
time expansion chamber, at HERA, 100
Ethane gas mixtures
MAC vertex chamber, 117-118
MARK II drift chamber, 55-56
PEP vertex detector, 69
Ethanol
MARK-J detector, 86, 87
PEP vertex chamber, 41, 69
and wire aging, 163
Event rates 20
Event reconstruction
B decays, 28
Bhabha tracks, 42, 43, 44
D, from B decay, 30
MARK-J detector, 83, 85
multidrift tubes, 108
one TeV jets, 4, 5, 6, 16
Event reconstruction (continued)
secondary vertices, 35
track pair effects, 64

f quarks, 30
Factor enhancement, 24
FADC readout, 69, 91
Fast clear, readout, 219, 220
Fast gas, 51-53
Fast memory chip, 63
Fastbus module, 63
Fermilab, see Charm photoproduction
Fermions, Z^o couplings, 29-30
Fibre-optic coupling, SCIFI, 253
Field cage design, SLC, 58-60
Field effect transistors, 196, 216-219, 220, 224
Field lines
MARK-J detector, 74, 75
MARK II drift chamber, 53, 54
Fine spatial resolution, scintillators, 228
Fitting functions, 130, 133
FNAL, 16
Formaldehydes, 163
Fragmentation, 19
Fully depleted CCDs, 214-216

Gain, 69
gas, 116, 117, 118
PEP vertex chamber, 40
silicon microchip devices, 190
Gap length distribution, SCIFI, 271
Gas gain, 116, 117, 118
Gas systems
MAC vertex chamber
design, 124-125, 126
drift corrections, 130-131
MARK II drift chamber, 51-53
prototype results, 55-56
wire positions and, 63
MARK-J detector, 72, 73, 74, 85, 89
and drift velocity, 78, 79, 80, 87

Gas systems (continued)
MARK-J detector (continued)
modifications, 86-87
multidrift tubes, 105-112
PEP vertex detector, 41, 69
TEC at HERA, 95, 98
and wire aging
argon-ethane-isopropanol, 154-155
argon-ethane-methanol, 156, 157
Gate structure
CCDs, fully depleted, 214
nondestructive readout device, 219, 220
Glass capillaries with liquid scintillator, 248-249
Glass scintillators, 240-248
experimental status, 264-267, 269, 271, 272, 274
light yield and attenuation length, 242, 244, 245, 246
manufacture, 249-253, 246, 248
mechanisms, 240-242
properties, 242, 243, 244, 245
radiation hardness, 246, 248
trapping efficiencies, 227
Glow discharge, 68-69
PEP vertex chamber, 40-41, 68-69
and wire aging, 164
Gluons, 1
fragmentation, 19
quark-gluon di-jet system, 2
track separation angle, 2, 3
Gold wire, aging, 164
Grid planes, MARK II, 53
Grid wires
MARK II drift chamber, 53, 55
MARK-J detector, 73
Guard wires, MARK II, 58, 59

Hadron calorimetry
ALFIE, 279, 285-288
SCIFI, 229, 230
Hadron collider, readout for, 36
Hadron decays
Cabbibo-suppressed, 192
SCIFI microtracker and, 268
Hadron supercollider environment, 268
Hadronic events
MARK II drift chamber, 64
MARK-J detector, 83, 85
multiple, 40, 140, 141, 142
SCIFI, 234, 245, 268
Hadronization, b-quark, 145
Hadrons
multi-hadronic events, 40, 140, 141, 142
supersymetric partners, 1
Heat cycling, SCIFI, 235
Heavy flavour
SCIFI, 229
Si detectors, 16
tagging, 268
Heavy neutrinos, PEP, 47-49
Heavy quarks (see also specific detectors)
MARK II vertex chamber, 38
new, third generation detectors and, 38-39
HERA, time expansion chamber, 95-99
Hermeticity
ALFIE, 286
SCIFI, 229
Higgs
decay products, 20
standard, decay products, 33
third generation detectors and, 39
Hitachi chip, 63
Holes
CCD devices, 212, 213, 215
conduction by, 197
one TeV silicon strip detectors, 14
3-Hydroxyflavone, 233, 239, 243

Image intensifier
  SCIFI readout, 253, 254, 255
  supercollider, 289-292
Impact parameter resolution
  B decays, 21-22
  existing and proposed detectors, 17, 18
  MAC vertex chamber, 139
  MARK II PEP vertex chamber, 40, 41, 44
Impact parameters
  leptons from B semi-leptonic decay, 45-47
  MAC vertex detector, 142, 145, 149
  MARK II PEP vertex chamber, 45-47, 51
  silicon versus drift chamber detectors, 68
  time expansion chamber, at HERA, 100
Impurity, 241
Information storage, 15, 35
Insulation, strip detectors, 206
Integrated luminosity, 47
Interaction rates, supercolliders, 270
Invariant mass, 26
Ion tails
  MARK-J detector, 72
  multidrift tubes, 107
Ionization
  and fake rate, 57-58
  multidrift tubes, 105-106
  PEP vertex detector, 68-69
  time expansion chamber, 98
ISIT, 255
Isobutane
  MAC vertex chamber, 117
  MARK II drift chamber, 55
  MARK-J detector, 89
    and drift velocity, 78, 79, 80
    replacement of, 85
Isochrony, MARK II, 53, 54, 55
Isopropyl alcohol, and wire aging, 154-155, 160, 161
Jet cells
  MARK II drift chamber, 53
    field cage design, 59
    mechanical construction, 60-61
  versus straws, 116
Jets
  MARK II PEP vertex chamber, 47, 48
  one TeV, 1-15
  reconstruction, 37
  SCIFI, 229
  supercollider, 268
JETSET, 115
Junction field effect transistor, 216-219, 220

Kapton circuitry, 66-62
Killers, 241
K-M matrix element, 28
KM mixing angles, 38
K27, 233

LABEN 5242 preamplifier, 124
Landau fluctuations, 206
Lattice defect, 241
Layered detectors
  number of layers, 68
  one TeV jets, 7, 16
Leakage, 193
LEAR, 115
LEP, 16
  B production, 20
  Higgs decay, 33
  third generation detectors, 38
Lepton impact parameter, 46-47, 51
Lepton spectrum, 31-32
Leptons
  couplings, measurement of, 30
  decays, see specific detectors
  high transverse momentum, 31
  like-signal, 16
  sign of, at 2 TeV, 6
  tau, see Tau
Lepton-tagged events
  at MAC, 143, 144

Lepton-tagged events (continued)
  at PEP, 23-24, 46-47
Lifetimes
  B mesons, 19, 27-28
    MAC vertex chamber, 141-145
    MARK II PEP vertex chamber, 49-51
  bottom hadron, 115
  charmed meson, 181-185
  MAC vertex chamber, 137, 141-145
  MARK II PEP vertex chamber, 44, 45-46, 49-51
  new particles decaying to B, 31
  at one TeV, 1
  third generation detectors, 38
Light quarks
  decay into, 31-32
  MAC vertex chamber, 142, 145
  vertex displacement, 23
Light yield, scintillator
  glass, 242, 244, 245, 246, 264
  plastic, 232, 235
Like-sign dileptons, 29
Liquid scintillators, 248-249
  glass capillaries, 248-249
  Teflon tubes, 249
  trapping efficiencies, 227
Localization accuracy, multi-drift tubes, 105-106 109, 110, 111
Lorentz angle, 14
Low-field drift velocity, MARK-J detector, 72
L-3 TEC, 91
Luminosity
  for Higgs decay, 33
  MAC vertex chamber, 141
  MARK II PEP vertex chamber, 47
  supercolliders, 270

MAC vertex chamber
  beam test, 119-120, 121
  B lifetime, 141-145
MAC vertex chamber (continued)
  B lifetime (continued)
    event selection, 141-142
    measurement of, 142-145
  design, 120-126
    beam pipe and pressure vessel, 121-126
    chamber, 122-124
    gas system, 124-125, 126
    radiation shielding, 120-121
    signal processing, 124, 125, 126
  gas gain, 117, 118
  motivation and strategy, 116
  pattern recognition, 134-140
    algorithm, 134-135
    efficiency, 135-136
  performance, 127-134
    channel-by-channel calibration, 131-134
    electronics, 127-128
    global alignment, 130, 131
    off-line calibration, 129-130
    operating point, 128-129
    time-to-distance calibration, 130-131, 132
  radiation damage, 117-119
  tracking parameterization, 146-147
Magic Gas, 117
Magnetic field
  MAC vertex detector, 146-147
  one TeV jet system, 4, 6, 11
  SCIFI, 229
  time expansion chamber, at HERA, 98
Malter discharges, 41
MARK II detectors
  B decays
    signal isolation, 21, 22
    tagging numbers, 24-27
  impact parameter resolution, 37, 38
  one TeV jet system versus, 4
    pattern recognition, 10-11
    resolution, 6
    tracker parameters, 7

MARK II detectors (continued)
PEP vertex chamber, 39-51
B lifetime measurement, 49-51
design, 39-40
operating experience, 40-41
performance, 41-44
physics results, 44-49
SLC drift chamber, 51-65
cell design, 53-55
electronics, 63
field cage design, 58-60
mechanical construction, 60-63
performance, 64, 65
prototype results, 55-58
MARK III, 115
MARK-J detector
design, 72-75
measurements in test beams, 75-78
operation experience, 78-88, 89
Bhabha events, 82, 83, 84, 85
cosmic ray triggers, 80-81, 87, 88, 89
data acquisition, 91
deterioration, 86-87, 90
drift velocity, 78-80
gas system changes, 85, 86-88
hadronic events, 83, 85
Mass
Higgs, standard, 33
invariant, 26
t quark, 32
Mass resolution, silicon microchip detectors, 189
Matrix element
B decays, 19
K-M, 28
Mean flight distance, 22
Mean refractive index, SCIFI, 234
Measurement accuracy, supercollider, 278
Measurement resolution, MAC vertex chamber, 138
Memory chip, MARK II drift chamber, 63
Mesons
B, *see* B mesons
charmed, *see* Charm photoproduction
Metal-oxide-semiconductor structure, 196, 197
Methane, 117, 118
Methanol, 156, 157, 163
Micro-channel plate (MCP), 253
Micro-converter, ALFIE, 279, 281-283
Microstrip detectors, 192
charm photoproduction, *see* Charm photoproduction
MARK II drift chamber, 51
third generation detectors, 39
Micro-tracker
ALFIE, 279, 280-281
SCIFI, 278-279
supercolliders, 273, 275, 276, 277
Mimimum-ionizing particles
multidrift tubes, 106-107
SCIFI, glass, 245
Mixing
B physics, 19, 20, 28-29
third generation detectors, 38
time resolution, 35
Momentum
B from t decay, 32, 33
B tag efficiency versus, 25
charmed particle, 179
leading particles in jet, one TeV, 2
one TeV jet particles, 2, 4, 5
supercollider, 278
transverse, 4, 5, 21, 22
Momentum errors, MARK II PEP vertex chamber, 44
Momentum resolution
B decays, 21, 22

Momentum resolution (continued)
one TeV jet system, 6, 7
PEP/PETRA detectors, 37
silicon microchip detectors, charm photoproduction, 189
tagged photon spectrometer, 168
Monte Carlo
B lifetime, 50-51
B mixing, 29
B tagging, 24
jet, 23
time expansion chamber, 100
MOS, CCDs, 200-201, 202, 214
MOSFETS, nondestructive, 216-219, 220
MSD, 193
Multidrift tubes, 101-112
construction, 102-105
dimethyl ether, results with, 105-111
experimental setup, 105
future developments, 111-112
Multihadron events
MAC vertex chamber, 140, 141, 142
PEP vertex chamber, 40
Multihit capability
B tag efficiency versus, 25
MAC vertex chamber, 116
Multi-image CCD, 216-217
Multi-layer detectors
number of layers, 68
Multiple coulomb scattering, vertex chambers
MAC, 116, 138, 146-147
PEP, 40
Multiple scattering, 192
charm signals, 176
MAC vertex chamber, 137, 138
MARK II PEP vertex chamber, 40, 44
one TeV system
and momentum resolution, 7
and vertex reconstruction, 16
silicon versus drift chamber detectors, 68
Multiple scattering (continued)
time expansion chamber, at HERA, 100
Multiplexing, readout, 253, 296
Multiplicity, gluon and quark jets, 2, 3
Multitrack resolution, and $B\bar{B}$ identification, 16
Multiwire proportional chamber, 106
Muons
heavy neutrino coupling, 48, 49
MAC vertex chamber, 142
MARK II PEP vertex chamber, 49, 50
supercolliders, 268
Muon spectrometer, 268, 279, 288
Mylar drift tubes, see MAC vertex chamber

Naphthalene, 232, 233, 234
N-doping, 210, 211
Negative space charge region, 198
Neutrinos, 30
MARK II PEP vertex chamber, 47-49
new, third generation detectors and, 39
New particles, 1, 20, 30-31
t versus b', 31-32
third generation detectors and, 38-39
Nickel wire, aging, 158-159, 161, 163, 164
Noise
CCDs, 201
multidrift tubes, 106
silicon microchip devices, 187, 190

One TeV jets, 1-15
tracking system design, 4, 6-15
Lorentz angle, 14
magnetic field, 6

One TeV jets (continued)
  tracking system design (continued)
    pattern recognition requirements, 7-12
    tracker parameters, 7
    z measurement, 12-14
  tracking system requirements, 2-4, 5
Operating point, MAC, 128-129
Optical delay line, 291
Organic scintillators, 230, 231, 232, 233, 234
Oxidation, 164
Oxygen, 72, 86, 90, 235, 237
OXYSORB, 72, 86, 90

Packing fraction, 229
Paired layers, 9
Paired tracks, 13, 64, 65
PANAMETRICS, 80
Particle fluxes, supercollider, 268, 270
Pattern recognition
  MAC vertex chamber, 130, 134-140
  one TeV system, 7-14
  supercollider, 278
Pauli interference effects, 181
PBD, 233
PEP vertex chamber, 115
  B physics, 20
    lifetime measurement, 49-51
    mixing, 29
    signal isolation, 21, 22
  design, 39-40
  operating experience, 40-41
  performance, 41-44
  physics results, 44-49
  reliability problems, 68
  resolution of, 37
PETRA, 37, 83
Phantom tracks, 174
Phase space, MARK II, 53
Philips tube, 255
Phosphor, SCIFI readout, 253
Photoelectron yield, SCIFI, 231
Photon beam, 169
Photon conversion, MAC, 116
Photon detection, supercollider, 278
Photon gain, SCIFI readout, 253
Photon yield, SCIFI, 231, 234, 245
Photoproduction, charm, see Charm photoproduction
Photosensitivity, MAC, 117
Phototubes, SCIFI, 292
Pions, three-charged, 44
Plastic scintillators, 230-240
  experimental status, 257-264
  light yield and attenuation length, 332, 235
  limitations and cures, 237, 239, 240
  mechanisms, 230-232
  properties, 232, 233, 234
  radiation hardness, 235-237, 238
  trapping efficiencies, 227
Plastics, see Polymers
Platinum, aging, 158, 160, 162, 164
Plessey SP9687 dual comparator, 124
PMOS, 218
Polar angle, wire efficiency versus, 83
Polarization, b system, 30
Polymers
  MAC vertex chamber, 117
  SCIFI, see Plastic scintillators
  and wire aging, 164
Polystyrene, 232, 233, 234
Polyvinyl toluene, 232, 233, 234
POPOP, 233, 239, 243
Power dissipation, one TeV system, 15
pp colliders, 15
$\bar{p}p$ events, 29
  B production, 20
  information storage, 15
Preamplifiers
  MARK-J detector, 75

Preamplifiers (continued)
silicon microchip detector, 172
Pressure
MAC vertex chamber, 117, 11, 126
MARK-J detector, 73
Pressure vessel, MAC, 121-122
Primary ionization statistics, multidrift tubes, 105-106
Primary vertex, leptons from, 31
P-type material, 198, 207
Pulse height
aging and, 152, 153, 154, 155, 162
multidrift tubes, 113
Pulse width, MARK-J detector, 85
Pulsed laser beam, multidrift tubes, 107, 108

Quark decays, see specific detectors and particles
Quark gluon di-jet system, 2
Quark jets
track separation angle, 2, 3
vertex displacement, 23
Quark mass, t, 32
Quarks, new, see New particles
Quenching, glass SCIFI, 241, 242

Radial drift
MAC vertex chamber, 116
semiconductor detectors, 204
Radiation
shielding, MAC chamber, 119-120, 121
supercolliders, 268, 270
synchrotron, 16, 120, 128, 278
Radiation damage, 193
MAC vertex chamber, 116, 117-119
one TeV system, 7, 15
Radiation damage (continued)
semiconductor detectors, 205
Radiation hardness
ALFIE, 286
SCIFI, 229, 231, 245
scintillators
glass, 245, 246, 248
plastic, 235-237
supercollider, 278, 286
Radiation length, SCIFI, 231, 234
Radius
minimum, vertex detector, 21
one TeV system, 7, 9
RAMs, 224
Random access nondestructive readout pixel device, 219, 220
Random access to simple cells, CCD with, 224
RCA 8850 quantacon phototube, 242
Readouts, 35, 36
MARK II drift chamber, 63
MARK-J detector, 75, 91
SCIFI, 253-257
semiconductor devices, 205, 215
CCDs, 201
double sided, 206-209
nondestructive, 216-219, 220
supercollider, 289-292
Readout electrons, 15
Reconstruction, see Event reconstruction
Reliability, drift chambers, 68
Residuals, MAC vertex chamber, 137
Resolution
charmed mesons, 181
MAC vertex chamber, 116, 119, 120, 137-140
electronics and, 127-128
voltage, pressure, and discriminator thresholds, 121
MARK II drift chamber, 56-58

Resolution (continued)
MARK II vertex chamber, 38, 41, 44
MARK-J detector
azimuthal angle and, 75, 76
double-track, 75, 76
gas system and, 86, 87
one TeV jet system, 2-3, 4
angular resolving power, 5
detector alignment, 6
tracker parameters, 7, 9
z measurement, 13
silicon versus drift chamber detectors, 68
SMD performance, 186
time expansion chamber, at HERA, 98, 100
Reverse bias, 197
microchip detectors, 171
of strip detector, 208, 209

Sampling interval, MARK-J detector, 75
Scattering, see Multiple scattering
Scintillating fibre detectors
ALFIE, 279-288
calorimeter, 283-288
charge flow detector, 281
micro-converter, 281-283
microtrcker, 280-281
muon spectrometer, 288
z-vertex detector, 281, 282
applications, 229-230
characteristics of, 226-229
experimental status, 257-267
glass, 264-267, 269, 271, 272
plastic, 257-264
manufacturers, 249-253
microtracker concept, 267-279
design, 278-279
need for, 270, 273-278
operation principles, 225-226
readout, 253-257
scintillators, 230
Scintillating fibre detectors (continued)
scintillators (continued)
glass, 240-248
liquid, 248-249
plastic, 230-240
supercollider, readout at, 289-293
supercollider environment, 268-288
Second coordinate, 35
Secondary vertices, 21-22
mass and sphericity cuts, 24
one TeV jet system, 4
reconstruction efficiencies, 35
supercolliders, 268, 270, 278
Self-absorption, SCIFI, 239
Self-quenching streamer mode, 116
Semiconductor detectors
charge coupled devices, 200-201, 202
drift chambers, 202-204
new concepts, 214-220
fully depleted CCDs, advanced, 214-216
multi image, 216-217
nondestructive readout and amplification, 217-219, 220
new developments, 206
double sided readout of strip detectors, 206-209
fully depleted CCDs, 209-214
properties
of devices, 204-205
of semiconductors, 198-199
strip, 199-200
structures, 196-198
Semi-leptonic decay, 28, 45-46
Sense wires
MAC vertex chamber, 116
MARK II drift chamber, 53, 55
MARK-J detector, 73
time expansion chamber, 97

Separation angle, one TeV system, 9, 10, 11
Shielding, MAC, 120-121, 122
Sign
  b, measurement of, 30
  like-sign dileptons, 29
  MAC vertex chamber, 142, 143
Signal processing
  ALFIE, 286
  B decays, 20, 21-23
  charm modes, 178
  MAC vertex chamber, 124, 125, 126
Signed impact parameter, MAC, 142, 145, 149
Signed residuum, MARK-J detector, 84, 89
Silicon detectors (see also Semiconductor detectors)
  charm photoproduction, see Charm photoproduction
  heavy flavour decays, 16
  microstrip, 171
    MARK II drift chamber, 51, 68
    supercolliders, 273
    third generation detectors, 39
  one TeV jets, 1-15, 16
  time resolution, 35
Silicon intensifier target, 255
Silicon photodiodes, 292
Single-photon signal levels, SCIFI readout, 253
SLC drift chambers
  B physics, 20
    signal isolation, 21, 22
    tagged sample production, 29-33
    tagging numbers for MARK II, 24, 27
  cell design, 53-55
  electronics, 63
  field cage design, 58-60
  mechanical construction, 60-63
  performance, 64, 65
  prototype results, 55-58
SLC drift chambers (continued)
  third generation detectors, 38
Sodium iodide, 241, 242
Softening point, SCIFI, 234
Solenoids, superconducting, 273, 279
Space charge region, 198
Spatial resolution
  MAC vertex chamber, 116, 117
    beam test, 119, 120
    electronics and, 127-128
    multihadron events, 140
    observed, 137-138
    operating point, 128
    voltage, pressure, and discriminator threshholds, 121
  MARK-J detector, 72
  time expansion chamber, 98
SPEAR, 115
Speed, CCDs, 201
Spot frequency, 271, 272
Spot profiles, 266
Spot size, 267
Spurious track, 11
$s\bar{s}$ events
  B tagging at SLC, 26
  vertex displacement, 23
SSC, 20, 115, 270
Stablohm-800 resistive wire, 152-154, 162, 163
  argon-ethane-isopropanol gas, 154-155
  argon-ethane-methanol gas, 156, 157
Stainless steel wires, 158, 160, 162, 164
Standard Model, 19
  couplings, measurement using charge asymmetry, 30
  CP violation for K^0, 28
Stokes shift, SCIFI, 232, 239
Straw tube vertex chambers, 21 (see also MAC vertex chamber; Multidrift tubes)
Streamers, MAC, 117

Strip detectors, 196, 206-209
diode, 199-200, 204, 206
fully depleted MOS-CCD combinations, 215
MARK II drift chamber, 51
one TeV jets, 1-15
third generation detectors, 39
Supercolliders
ALFIE, 279-288
calorimeter, 283, 285-288
charge-flow detector, 281
components, 279-280
micro-converter, 281-283
microtracker, 280-281
muon spectrometer, 288
z-vertex detector, 281, 282
readout at, 289-292
image intensifier-CCD systems, 289-292
phototubes and silicon photodiodes, 292
scintillating fiber detectors, 268-288
design and performance, 278-279
design criteria, 278
need for new approach, 270, 273-278
need for tracking, 268-270
Superconducting solenoids, 273, 276
Supersymmetric hadronic partner, 1
Synchrotron radiation
heavy flavour decays, 16
MAC vertex chamber, 120, 128
supercollider, 278

t quarks
decay products, 20, 31-32, 33
lepton spectrum, 31-32
SLC, production at, 30-31
Tagging, 4, 5, 27-33
B mesons
at PEP, 23-24, 25, 64, 65
at SLC, 24-27, 38, 44, 47
lepton tagging cuts, 23-24
Tagging efficiency
B mesons, 24-25
silicon versus drift chamber detectors, 68
Tantalum collimators, 121, 122
Tau, 16, 38, 44
heavy neutrino coupling, 48, 49
mean flight distance, 22
Teflon
MAC vertex chamber, 117
scintillator tubes, 249
TELEDYNE, 80
Temperature
and drift velocity, 78, 79, 80
and glass SCIFI, 245
p-Terphenyl, SCIFI, 232, 233, 234
TEV I, 20
TEV II, 20
TeV jets
one, 1-15
supercollider, 268
Tevatron, 169
Threshold
MAC vertex chamber, 120, 128
SLC chamber operations, 69
Thrust, and signed impact parameter, 149
Time expansion chamber
L-3, 91
MARK-J detector, see MARK-J detector
modified, for ZEUS at HERA, 95-99
description, 96, 97, 98
design, 98-99
time scale, 99
Time resolution
charmed mesons, 181
SSC tracking detectors, 35
Time-marker signals, MARK-J, 80
Time-to-distance relationship, MAC, 119
Time-to-voltage converters, MAC, 127
Timing resolution, MAC, 127
Titanium liner, MAC, 121, 122

Top quark, see t quarks
Topology, reconstruction of, 35
TPC, 21, 22, 115
Track finding, microchip detector, 173-174
Track fitting, microchip detector, 174-176
Track pairing
  one TeV jets, 13
  MARK II drift chamber, 64, 65
Track reconstruction, 2, 4, 5
Track residuals, in glass SCIFI, 269
Track separation angle
  $B\bar{B}$ identification, 16
  one TeV jets
    measurement precision, 10, 11
    requirements, 2-4, 5, 9
Tracking parameterization, MAC, 146-147
Track-pair resolution, see Two-track resolution
Trajectories
  B physics, vertex configuration, 22
  MAC vertex detector, tracking parameterization, 146-147
  one TeV system, pattern recognition, 10, 11
Transverse momentum
  B decays, 4, 5, 21, 22
  leptons, 31
Trapping efficiencies, 227
Triggers
  fully depleted MOS-CCD combinations, 215
  MARK-J detector, 80-81, 87, 88, 89
TRISTAN, 38, 115
$t\bar{t}$ events, 30
Tungsten wire aging, 163
Two-hit effects, 2-3, 10-11
Two-track resolution
  MARK-J detector, 72, 77
  MARK II drift chamber, 57-58, 64, 65
Two-track resolution (continued)
  one TeV jets, 9
  SCIFI, 229, 277, 278
  silicon microchip devices, charm photoproduction, 188
  silicon versus drift chamber detectors, 68
  time expansion chamber, at HERA, 100

u quarks, 19, 30
UA1, 20
UA2
  readout systems, 255-257
  SFD, 257-261
$u\bar{u}$ events
  B tagging at SLC, 26
  vertex displacement, 23

VC algorithm, 134
Vector couplings, Z^0 to fermions, 29-30
Vector segments, one TeV system, 7,8, 9, 10-12
Vertex chambers, see MAC vertex chamber; PEP vertex chamber
Vertex displacement, B physics, 23
Vertex finding, 192
Vertex fitting, charm signals, 176
Vertex measurement, MARK II vertex chamber, 38
Vertex reconstruction, one TeV system, 16
Vertex resolution
  MAC vertex chamber, 139, 140, 141
  silicon microchip devices, 190
Vertex separation, charmed particles, 179, 181
Vertex topology reconstruction, 35
VLSI electronics, SCIFI, 255
Voltage
  MAC vertex chamber, 117

- Voltage (continued)
  - MAC vertex chamber (continued)
    - operating point, 128
    - and resolution, 121
  - MARK II drift chamber, 55
  - MARK-J detector, 78, 79, 80
  - PEP vertex detector, 41, 68-69
  - TEC, at HERA, 98
- $V_{b-->u}V_{b-->c}$, 29
- Wavelength shifters, SCIFI, 230, 231, 232, 233, 234, 239
- Wire chamber aging
  - aluminum wire, 156, 157, 159, 161
  - carbon dioxide, 163
  - copper, 158, 160, 162
  - isopropanol and, 160, 163
  - nickel wire, 158, 159, 161
  - platinum, 158, 160, 162
  - Stablohm wire
    - anode wire, 152-154
    - argon-ethane-isoproanol gas, 154-155, 160
    - argon-ethane-methanol gas, 156, 157, 158-159
  - stainless steel, 158, 160, 162
- Wire efficiency, 83
- Wires
  - glow problems, 40-41, 68-69, 164
- Wires (continued)
  - impact parameter errors, 44
  - MAC vertex chamber, 116, 118
  - MARK II drift chamber, 51-53
    - design, 53, 53-55
    - field cage design, 58-60
    - mechanical construction, 60-63
  - MARK-J detector, 73, 88, 90
  - multidrift tubes, 102-105
    - coordinates, 109
    - reduced number, 110
  - PEP vertex chamber, 39-40
  - tilted, 16
  - TEC at HERA, 97, 98

- Xenon, 117, 118

- Z^o, 33, 35
  - B reconstruction, 4, 5
  - B tagging at, 24, 26
  - charm tagging at, 26-27
  - momentum at, 16
  - track pair resolution, 64
  - vector and axial vector couplings, 19, 20, 29-30
- Z coordinate measurement, one TeV system, 12-14
- ZEUS, 95-99
- Z-vertex detectors
  - ALFIE, 279, 281, 282
  - SCIFI, 229
- Z-vertex positions, super-colliders, 268